T0281439

Practicing Safe Sects

Philosophical Studies in Science and Religion

VOLUME 9

The titles published in this series are listed at *brill.com/pssr*

Practicing Safe Sects

Religious Reproduction in Scientific and
Philosophical Perspective

By

F. LeRon Shults

BRILL

LEIDEN | BOSTON

Library of Congress Cataloging-in-Publication Data

Names: Shults, F. LeRon, author.
Title: Practicing safe sects : religious reproduction in scientific and
  philosophical perspective / by F. LeRon Shults.
Description: Boston : Brill, 2018. | Series: Philosophical studies in science
  and religion, ISSN 1877-8542 ; VOLUME 9 | Includes bibliographical
  references and index.
Identifiers: LCCN 2017061433 (print) | LCCN 2018004805 (ebook) | ISBN
  9789004360952 (eBook) | ISBN 9789004360945 (hardback : alk. paper)
Subjects: LCSH: Sects. | Christian sects. | Church growth.
Classification: LCC BP603 (ebook) | LCC BP603 .S54 2018 (print) | DDC
  200--dc23
LC record available at https://lccn.loc.gov/2017061433

Typeface for the Latin, Greek, and Cyrillic scripts: "Brill". See and download: brill.com/brill-typeface.

ISSN 1877-8542
ISBN 978-90-04-46167-3 (paperback)
ISBN 978-90-04-36094-5 (hardback)
ISBN 978-90-04-36095-2 (e-book)

to Michael J. Prince

∵

Contents

Preface

This book provides a fresh synopsis and significant expansion of *theogonic reproduction theory*, which hypothesizes that gods (supernatural agent conceptions) are *born* in human minds and *borne* in human cultures as a result of a complex set of reciprocally reinforcing, phylogenetically inherited, and socially sustained cognitive and coalitional biases. The latter were naturally selected for their survival advantage in early ancestral environments, but today the superstitious beliefs and segregative behaviors they engender are maladaptive in a growing number of contexts. As naturalist explanations of the world and secularist inscriptions of society take root within a population, people start to lose interest in engaging in religious sects.

The theoretical framework outlined below is supported by empirical findings and theoretical developments in a wide variety of disciplines, all of which have converged within what I refer to here as the *bio-cultural* study of religion. I cite hundreds of recent studies that contribute to our understanding of the god-bearing and god-dissolving mechanisms at work in human life. It is difficult to keep up with the literature in this highly generative, globally networked, multi-disciplinary discussion. Dozens of relevant studies were published during the few months between my sending in the manuscript and receiving the first proofs. Rather than incorporating these into the text, I have set up a web page where I list (and sometimes comment on) recent publications whose findings are particularly salient for the basic hypotheses of theogonic reproduction theory: www.leronshults.com/my_weblog/safe-sects.html.

A brief preview of the book appears at the end of Chapter 1 which, along with Chapter 12, provides the most comprehensive synthesis of the literature and detailed defense of the theory. Most of the work on this book occurred during the early phases of the "Modeling Religion in Norway" (MODRN) project, which was supported by a grant from the Research Council of Norway (#250499). The central chapters are adaptations of earlier publications, significantly revised and updated to complement the larger, completely new bookend chapters. I am grateful to the publishers of these earlier essays for permission to incorporate material from them into the current book (details are provided in the footnotes of the relevant chapters below). I am also thankful to the two anonymous reviewers, who provided the most careful reading and constructive critique I have ever received as part of a review process. Thanks also to the many colleagues and friends who have discussed these issues with me over the years. Special thanks goes to Michael J. Prince, to whom this volume is dedicated, for his patient listening and insightful questioning during countless coffee breaks and sushi dinners.

CHAPTER 1

Having "The Talk" about Religious Reproduction

> Atheism is a religion like abstinence is a sex position.
> BILL MAHER

∴

In the academic study of religion the term "sects" is usually reserved for minority religious groups whose beliefs and ritual behaviors are considered (by a majority population) to be abnormal, in contrast to more conventional religious groups such as "churches." Groups that are able to continue reproducing themselves until they are adopted by a dominant culture (as with early 4th century Christianity) or at least adapt to it (as with early 21st century Mormonism), are no longer considered *sects*. My primary interest in this book, however, is not in differentiating between such coalitions by measuring their historical longevity or political centrality. Instead, I focus on the distinctive way in which all "religious" social assemblages are constituted by a partitioning (or *sectioning*) of humanity that is authorized by appeals to supernatural agents who are putatively engaged in rituals performed by the members (or elites) of an in-group.

In this sense, coalitions as diverse as the Peoples Temple cult, the Islamic State, and the Roman Catholic Church, are *religiously* sectarian. I will use the phrase "religious sects" to refer to all such god-bearing groups and argue that this type of social intercourse, which did indeed help (some of) our ancestors thrive in earlier contexts, is becoming increasingly maladaptive in our contemporary global environment. In order to survive as a social species, we humans are likely to continue needing some kind of sects (as well as sex) for quite some time. But can we learn how to practice *safe* sects? That is, can we learn how to live together in social networks without bearing gods – without reproducing the superstitious beliefs and segregative behaviors that are engendered and nurtured by shared ritual engagement with imagined supernatural beings?

Below I will explore some of the reasons why a growing number of people around the globe are no longer *religiously* sectually active. In many contexts, it is becoming easier and easier to make sense of the world and to act sensibly in society without referring to supernatural agents and authorities. Nevertheless, most people on the planet today still like having religious sects. As we will see, some of the evolved cognitive and coalitional biases that lead to the reproduction of religious beliefs and behaviors played an important role in increasing the chances of individual survival and strengthening group cohesion in early ancestral human environments. Supernatural conceptions are regularly born in contemporary human minds – and readily borne in contemporary human cultures – as a result of an aggregate of covertly operating, phylogenetically inherited and socially reinforced mechanisms that kept our progenitors alive long enough to reproduce sexually and transmit these tendencies to us. And so here we are.

Why fight the urge now? The challenges we face today are quite different than those confronting small-scale societies of Paleolithic hunter-gatherers or Neolithic sedentary-agriculturalists. Most of us live in densely populated, pluralistic, large-scale societies, and all of us live in the Anthropocene – a global environment whose ecological instability is due, at least in part, to the astonishing success of (some) human coalitions in competing for resources. I will argue that participating in religious sects is making things worse, and that we need to find and foster new, explicitly non-religious strategies for living together. Discussing religion (like sex, or politics) can make people anxious and even angry. Nevertheless, if we are interested in contributing to a more peaceable and sustainable environment for everyone, we can no longer put off having "the talk" about religious reproduction.

Where Do Gods Come From?

Empirical findings and theoretical developments within the bio-cultural study of religion have converged in support of the claim that "gods" (in the broad sense of the term explained below) are engendered within the mental and social life of human beings as a result of naturally evolved, hyper-sensitive biases that activate *inferences* about hidden human-like forms and *preferences* for distinctive in-group norms, especially when people are confronted with ambiguous or frightening phenomena. Moreover, these cognitive and coalitional mechanisms *reciprocally reinforce* one another within religious sects. In other words, the evolved biases that generate ideas about supernatural agents and nurture them through ritual practices are mutually intensifying.

These are the basic hypotheses of *theogonic reproduction theory*. This theoretical framework has been worked out in some detail elsewhere.[1] In the current volume, I provide additional evidence for these god-bearing (theogonic)

---

1   Shults, *Theology after the Birth of God: Atheist Conceptions in Cognition and Culture* (Palgrave Macmillan, 2014); Shults, "Theology after the Birth of God: A Response to Commentators," *Syndicate: A New Forum for Theology* 2, no. 3 (2015). For a more detailed philosophical

mechanisms and further reflections on the implications of the theory and its component parts. Apologists operating out of a *religious* perspective will typically answer questions about the birth of (their) gods by appealing to mythical narratives in (their in-group's) holy texts, which themselves are alleged to be of supernatural origin. As suggested by the subtitle of the current book, I will try to explain how gods are born(e) in human minds and cultures from a (non-religious) *philosophical* and *scientific* perspective.

The primary goal of my citation strategy in this volume is to provide the reader with examples of scientific literature that illustrate the kind of empirical findings and theoretical developments relevant at each stage of the book's ongoing argument that learning how to practice safe sects is one of the most important challenges facing humanity today. This literature is so vast that I cannot take the space here to outline the details of every intramural and intermural disciplinary debate. The footnotes point the reader interested in these details in the right direction. For the most part, I try to keep the focus in the main text on the more general grounds and warrants that support the broader philosophical and pragmatic claim of the book: for a growing number of individuals, in a growing number of contexts, bearing gods is more trouble than it is worth.

The coordinate grid depicted in *Figure 1* provides a heuristic framework for conceptualizing and discussing the relationships among the mechanisms

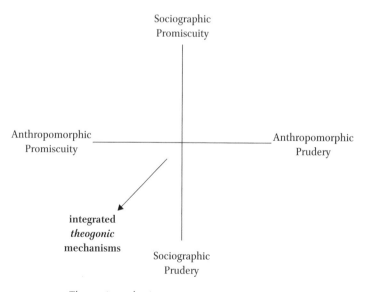

FIGURE 1   *Theogonic mechanisms*

exploration of the mechanisms of religious reproduction, see Shults, *Iconoclastic Theology: Gilles Deleuze and the Secretion of Atheism* (Edinburgh University Press, 2014).

postulated by theogonic reproduction theory. I spell out this framework in more detail in several of the chapters that follow (especially 2, 3, 4, 6, 7, 9, and 12).

Briefly, the horizontal line represents a spectrum on which one can indicate the extent to which a person is open to causal interpretations that appeal to supernatural agents. Most human beings are somewhat naturally drawn into *anthropomorphic promiscuity*, intuitively attributing causality to hidden gods, especially when anxious or excited. The vertical line represents a spectrum on which one can indicate the extent to which a person is likely to privilege social inscriptions that appeal to supernatural authorities privileged by his or her in-group. In this case, most of the human population tends to slip into *sociographic prudery*, intuitively associating with the moral norms of the religious sect with which they identify (in which they were raised or to which they have converted).

The integration of these attributive and associative predispositions is depicted in the lower left quadrant of *Figure 1*. Together these god-bearing mechanisms help to explain why and how gods are born(e) in human imagination. As we will see below, the opposing tendencies represented at the other end of each spectrum have a more or less dissolutive effect on religion.

But what, precisely, is "religion?" Definitions of this concept are highly contested both in the academy and in the public sphere.[2] My interest here is in phenomena commonly identified as "religious" in everyday life and popular discourse, and my argument will be based on philosophical reflections on empirical findings and theoretical developments in a variety of scientific disciplines that explore such phenomena. For reasons that will soon become clear, in this context I will use the term religion to designate *shared imaginative engagement with axiologically relevant supernatural agents*. I parse out this phrase in some detail in several of the following chapters.[3]

---

2   For a discussion of concerns about essentialist and colonialist uses of the term "religion," see Shults, *Theology after the Birth of God*, 8–10. For other recent discussions of the appropriate use of this contentious term in religious studies and related fields, see Warner, "In Defense of Religion: The 2013 H. Paul Douglass Lecture," *Review Of Religious Research* 56, no. 4 (2014); De Muckadell, "On Essentialism and Real Definitions of Religion," *Journal of the American Academy of Religion* 82, no. 2 (2014); Bergunder, "What Is Religion?" *Method & Theory in the Study of Religion* 26, no. 3 (2014); Schilbrack, "A Realist Social Ontology of Religion," *Religion* (July 25, 2016); and Hanegraaff, "Reconstructing 'Religion' from the Bottom Up," *Numen* 63, no. 5–6 (2016).

3   See also *Theology after the Birth of God*, pp. 5–8, 18–44, 195–199. The modifier "axiologically relevant" is meant to differentiate those supernatural agents whose existence or intentions bear on the normative value judgements and ritual behaviors of a particular coalition from those who do not, or who no longer do (e.g., Zeus). See, e.g., Gervais and Henrich, "The Zeus

At this stage, I simply want to point out that this way of utilizing the concept is quite typical in the scientific (and especially the bio-cultural) study of religion. The latter is often used in such contexts as a

> pithy rhetorical prop to cue readers to the kinds of interrelated phenomena that require explanation. The religious package is a statistical pattern governed by specific hypotheses, rather than a predefined concept with necessary or sufficient features. There is therefore no expectation of a single over-arching definition of religion or clear semantic boundaries, because the package of traits that gets labeled "religion," while containing recurrent elements, culturally mutates in a predictable fashion, taking different shapes in different groups and at different historical times.[4]

Instead of treating religion as a top-down category into which various phenomena are forced to fit, this sort of approach begins from the bottom-up, identifying component parts of phenomena generally referred to as "religious" and operationalizing them for particular research purposes.

This process is sometimes called *fractionating*. While there are a diversity of "culturally distributed dogmas and practices that have been collectively labeled 'religion,'" these are "shaped and constrained by a finite but disparate set of evolved cognitive predispositions." In this sense, religion can be "fractionated into distinct components with stable cognitive underpinnings," such as hyperactive agency detection, theory of mind, teleofunctional explanations, the ritual stance, and kinship detection.[5] There are many different ways to fractionate "religion." For example, some scholars emphasize "beliefs about nonhuman agents, religious rituals, community structures, and moral concerns and values"[6] as key dimensions of religion, while others focus on "believing, bonding, behaving and belonging."[7] Another influential approach hypothesizes

---

Problem: Why Representational Content Biases Cannot Explain Faith in Gods," *Journal of Cognition and Culture* 10, no. 3 (2010).

4 Norenzayan et al., "The Cultural Evolution of Prosocial Religions," *Behavioral and Brain Sciences* 39 (2016): 17.

5 McKay and Whitehouse, "Religion and Morality," *Psychological Bulletin* 141, no. 2 (2015): 454;. See also Whitehouse and Lanman, "The Ties That Bind Us," *Current Anthropology* 55, no. 6 (2014).

6 Johnson et al., "Fundamental Social Motives and the Varieties of Religious Experience," *Religion, Brain & Behavior* 5, no. 3 (2015).

7 Saroglou, "Believing, Bonding, Behaving, and Belonging," *Journal of Cross-Cultural Psychology* 42, no. 8 (2011).

three kinds of mechanisms (cognitive, motivational, and cultural learning) that give rise to (or intensify) religious beliefs and behaviors.[8]

My distinction between anthropomorphic promiscuity and sociographic prudery should be understood in this sense as an initial fractionation of "religion." As we will see, each of these mechanisms can (and will) be further fractionated into component mechanisms that contribute to the reproduction of beliefs about gods and ritual behaviors intended to engage them. Breaking a functioning system into its component parts to figure out how they work together is a kind of "reverse engineering." But it is equally important to show how these "building blocks" of a religious system can be put back together.[9] Using language derived from the "new mechanism" in recent philosophy of science, we could say that a robust explanation of religious reproduction will require us to make sense of the relevant mechanisms not only by "looking down" (through reduction), but also by "looking around" (at their organization) and "looking up" (at their situation).[10]

The terminology in *Figure 1* is new, but this kind of fractionation is neither methodologically nor materially novel. In the various fields that contribute to the bio-cultural study of religion, stipulated definitions that emphasize these two components are commonplace. Scholars who approach religion from an evolutionary point of view typically "focus on symbolically and emotionally laden beliefs and practices regarding *superhuman* powers, and on the institutions that maintain and transmit such *beliefs and practices*."[11] Another example: "whatever else might differentiate the religious from the non-religious, claims to the *authority of superhuman agency* that recruit and legitimate otherwise ordinary human behaviors and ideas and that motive their practice and perseverance would seem to characterize all religion."[12] Even more succinctly: the "specific components of religion [are] commitment to *supernatural agents and ritual behaviors*."[13]

---

8   Norenzayan and Gervais, "The Origins of Religious Disbelief," *Trends in Cognitive Sciences* 17, no. 1 (2013): 22.

9   Taves, "Reverse Engineering Complex Cultural Concepts: Identifying Building Blocks of 'Religion,'" *Journal of Cognition and Culture* 15, no. 1–2 (2015).

10   Bechtel, "Looking Down, Around, and Up: Mechanistic Explanation in Psychology," *Philosophical Psychology* 22, no. 5 (2009).

11   Bulbulia et al., "The Cultural Evolution of Religion," in *Cultural Evolution: Society, Technology, Language and Religion*, ed. Richerson and Christiansen (Cambridge: MIT Press, 2013). Emphasis added.

12   Martin, *Deep History, Secular Theory: Historical and Scientific Studies of Religion* (Berlin: De Gruyter, 2014), 178.

13   Purzycki et al., "Extending Evolutionary Accounts of Religion beyond the Mind: Religions as Adaptive Systems," in *Evolution, Religion and Cognitive Science: Critical and Constructive Essays* ed. Watts (Oxford: Oxford University Press, 2014), 84. Emphasis added.

It is important to be clear here at the beginning that anthropomorphic promiscuity and sociographic prudery are *not* meant to refer to "natural kinds." The latter sort of category intends to carve up nature by the joints. Botanists, for example, may differentiate between flowering and non-flowering plants. However, they would have no use *qua* botanists for the term "weeds." The latter is a social construction used by humans for a specific purpose: indicating plants they prefer not to be in their gardens. When I refer to two "kinds" of theogonic mechanism, I am combining a variety of different cognitive and coalitional sub-mechanisms (which may or may not turn out to be "natural" kinds) into broader categories for a *heuristic* purpose: indicating types of god-bearing dispositions that are increasingly problematic in human life.

These evolved tendencies will be evoked and manifested in quite different ways depending on individual and contextual variances. In other words, such mechanisms are not "mechanistic" in a deterministic sense, and their effects are not unidimensional. However, as we work to fractionate them further, and observe the way in which their component parts interact as we put them back together, we may hope to discover recurrent patterns in the emergent effects of participating in religious sects.

What are the Consequences of "Doing It"?

When having the talk about sexual reproduction, it is not enough to explain how "it" works. It is equally important to explain the consequences – and costs – of bearing children. So too with religious reproduction. The material and devotional costs of ritually caring for gods cognitively conceived in religious sects can be enormous. Moreover, "doing it" can have profoundly negative consequences of the sort that most sectually active individuals hardly ever consider. As we will see, imaginatively engaging the supernatural agents of one's in-group can exacerbate problems related to extreme climate change (Chapter 4), excessive consumer capitalism (Chapter 6), and escalating cultural conflict (Chapters 2, 3, 9, and 11), especially under conditions that intensify personal stress and competition for limited resources in threatening environments. At this stage, however, I want to point out three more general consequences of unsafe sects.

First, engaging in religious sects promotes *superstitious interpretations* of nature. But is this really so dangerous? What harm could it do for people to believe in hidden supernatural forces and converse with imaginary friends? After all, exploring counterfactual worlds through fiction can be fun. The problem, of course, is that the failure to distinguish between fact and fiction can lead to serious problems in the real world. When people are confused about the causes of HIV, for example, they may spend unnecessary energy worrying about witches, demons, or divine punishment, or searching for magical cures, instead of focusing on the actual causes of AIDS and ways to prevent it.[14] This applies to mental illness as well as physical illness; help-seeking behavior for the former is also impacted when its aetiology is perceived to be somehow related to supernatural forces (e.g., demons, divine punishment).[15] Humans often process information by appealing to both natural or supernatural causes at the same time. This is why educational programs are necessary "to eradicate erroneous beliefs at the same time that they work to establish correct beliefs."[16]

As we will see in more detail below, religious individuals are statistically more prone to a wide array of cognitive biases, and are more likely to score low on measures of critical reflection (and higher on measures of core ontological confusion) than non-religious individuals.[17] This has led some scholars to call for educational interventions that help correct people's misconceptions about how reality operates with regard to issues like the efficacy of prayer and other "silly beliefs" in phenomena such as lucky charms, in order to "protect them against buying into sham treatments such as homeopathy."[18] People with non-reflective thinking styles are particularly vulnerable to the negative consequences of unsafe sects, since they are far more likely to attribute

---

14   See e.g., Kelly-Hanku et al., "'We Call It a Virus but I Want to Say It's the Devil inside': Redemption, Moral Reform and Relationships with God among People Living with HIV in Papua New Guinea," *Social Science & Medicine* 119 (2014): 106, and Svensson, "God's Rage: Muslim Representations of Hiv/Aids as a Divine Punishment from the Perspective of the Cognitive Science of Religion," *Numen* 61, no. 5–6 (2014).

15   Ramkissoon et al., "Supernatural versus Medical: Responses to Mental Illness from Undergraduate University Students in Trinidad," *International Journal of Social Psychiatry* 63, no. 4 (2017).

16   Legare and Gelman, "Bewitchment, Biology, or Both: The Co-Existence of Natural and Supernatural Explanatory Frameworks across Development," *Cognitive Science* 32, no. 4 (2008): 640; see also Gelman and Legare, "South African Children's Understanding of AIDS and Flu: Investigating Conceptual Understanding of Cause, Treatment and Prevention," *Journal of Cognition and Culture* 9, no. 3–4 (2009).

17   See, e.g., Lindeman and Lipsanen, "Diverse Cognitive Profiles of Religious Believers and Nonbelievers," *The International Journal for the Psychology of Religion* 26, no. 3 (2016). Other examples are provided below and in Chapter 12.

18   Lobato et al., "Examining the Relationship between Conspiracy Theories, Paranormal Beliefs, and Pseudoscience Acceptance among a University Population," *Applied Cognitive Psychology* 28, no. 5 (2014): 624.

supernatural causality to uncanny events. This makes them "especially vulnerable to scammers who attempt to leverage paranormal beliefs into profits."[19]

Another reason that religious beliefs are difficult to challenge is that people often confuse their own thoughts with the thoughts of the supernatural agents ritually engaged by the religious in-group with which they identify. "Correlational, experimental, and neuroimaging methodologies all suggest that religious believers are particularly likely to use their own beliefs as a guide when reasoning about God's beliefs compared to when reasoning about other people's beliefs."[20] This makes it extremely difficult to think critically about such beliefs. This confusion between one's own beliefs and beliefs attributed to one's god is more than merely correlational. Psychological experiments have shown that *manipulation* of people's beliefs "affected their estimates of God's beliefs more than it affected estimates of other people's beliefs, demonstrating that estimates of God's beliefs are *causally* influenced at least in part by one's own beliefs."[21] This causal connection is also supported by the fact that children from religious backgrounds are less capable of distinguishing between reality and fiction than children from non-religious backgrounds.[22]

If the consequences of superstitious attributions of supernatural causality were limited to internal mental states, we might not worry about it too much. However, when people's biases are rewarded over time by participation in a particular religious belief system this can actually alter the way they attend to and process the features of complex visual stimuli.[23] Neuroscientific experiments indicate that the areas of the brain associated with empathy are less activated when perceiving the suffering of out-group members (compared to in-group members).[24] In-group biases can be easily activated when religious

---

19   Bouvet and Bonnefon, "Non-Reflective Thinkers Are Predisposed to Attribute Supernatural Causation to Uncanny Experiences," *Personality and Social Psychology Bulletin* 41, no. 7 (2015): 960.

20   Hodges et al., "Nearer My God to Thee: Self–God Overlap and Believers' Relationships with God," *Self and Identity* 12, no. 3 (2013).

21   Epley et al., "Believers' Estimates of God's Beliefs Are More Egocentric than Estimates of Other People's Beliefs," *Proceedings of the National Academy of Sciences* 106, no. 51 (2009). Emphasis added.

22   Corriveau et al., "Judgments About Fact and Fiction by Children From Religious and Non-religious Backgrounds," *Cognitive Science* 39, no. 2 (2015).

23   Colzato et al., "Losing the Big Picture: How Religion May Control Visual Attention," *PLoS ONE* 3, no. 11 (2008);. Colzato et al., "God: Do I Have Your Attention?" *Cognition* 117, no. 1 (2010).

24   Henry et al., "Death on the Brain: Effects of Mortality Salience on the Neural Correlates of Ingroup and Outgroup Categorization," *Social Cognitive and Affective Neuroscience* 5, no. 1 (2010).

individuals make judgements about punishment and reward. For example, a person's religiosity can affect their reasoning about immanent and ultimate justice; believers' perception of deservingness is skewed by their moral perception and evaluation of a person.[25]

Like all self-validating belief systems, religious ideologies are shaped by belief perseverance, immunizing strategies, motivated reasoning, and other cognitive biases.[26] Unlike "factual" beliefs, religious beliefs are generally unfalsifiable, more susceptible to free elaboration, and more vulnerable to appeals to special authority.[27] Unfortunately, pointing out the implausibility of such beliefs can make things worse. Psychological studies show that believers often automatically render religious beliefs *more* unfalsifiable when confronted by facts that challenge them.[28] The unfalsifiabilty of religious beliefs has concrete implications for intergroup conflict. Supernatural beliefs and narratives "can be unfalsifiably asserted without the inconvenience of evidence... They can be bolstered merely by appeals to traditional authority and social consensus... religious unfalsifiability may therefore be an important affordance that both consolidates zeal and spurs confident and militant action."[29] In other words, religious beliefs, like religious behaviors, can facilitate the process of radicalization into violent extremism.[30]

This brings us to a second major consequence of "doing it." Ritually engaging in religious sects can also amplify violent *segregative inscriptions* of society.

25   Harvey and Callan, "The Role of Religiosity in Ultimate and Immanent Justice Reasoning," *Personality and Individual Differences* 56, no. 1 (2014).

26   Boudry and Braeckman, "How Convenient! The Epistemic Rationale of Self-Validating Belief Systems," *Philosophical Psychology* 25, no. 3 (2012).

27   For a discussion of the similarities and differences between religious and factual belief, see van Leeuwen, "Religious Credence Is Not Factual Belief," *Cognition* 133, no. 3 (2014); Boudry and Coyne, "Fakers, Fanatics, and False Dilemmas: Reply to Van Leeuwen," *Philosophical Psychology* 29, no. 4 (2016).

28   Friesen et al., "The Psychological Advantage of Unfalsifiability: The Appeal of Untestable Religious and Political Ideologies," *Journal Of Personality And Social Psychology* 108, no. 3 (2015). See also Batson, "Rational Processing or Rationalization? The Effect of Disconfirming Information on a Stated Religious Belief," *Journal of Personality and Social Psychology* 32, no. 1 (1975).

29   McGregor et al., "Motivation for Aggressive Religious Radicalization: Goal Regulation Theory and a Personality x Threat x Affordance Hypothesis," *Frontiers In Psychology* 6 (2015): 13.

30   Shults, "Can We Predict and Prevent Religious Radicalization?" in *Processes of Violent Extremism in the 21st Century: International and Interdisciplinary Perspectives*, ed. Gwenyth Øverland et al., (Cambridge Scholars Press, forthcoming).

Differentiations of some sort are a necessary part of any social organization. Unfortunately, the social field has all too often been inscribed in ways that have led to the oppression of some groups within the population. Social constructions of race, class, and gender have been the most common bases for such oppressive segregation. For much of human history, racist, classist, and sexist biases "worked" in the sense that they helped differentiate and organize increasingly large societies. As human populations grew, the internalization of such biases reinforced social distinctions that supported the conditions for the ongoing biological reproduction and resource production the species needed to survive.

Today, however, in many contexts these biases are increasingly being challenged as a growing number of people are resisting modes of organizing society that reinforce unequal distributions of power based on ethnic background, financial status, or sexual identity. What does this have to do with religion? A plethora of scientific studies has established a relationship between religion and each of these biases. "Research has consistently identified *religiosity* as a predictor of various types of *prejudice*, including that which is expressed as racism, homophobia, intolerance toward women, and debasement of political or religious out-groups."[31]

This research suggests that racist, classist, and sexist biases are undergirded or amplified by *theist* biases (anthropomorphic promiscuity and sociographic prudery). The fact that all of these biases are the outputs of cognitive and coalitional tendencies that helped our species survive in the past is no reason to assume they will help us thrive in our current environment. If we really want to weaken the impact of racism, classism, and sexism as forces of oppressive segregation in our societies, it will be necessary to pay attention to the way in which they can be *intensified* by religious beliefs and behaviors. The mutual entanglement of these biases has deep evolutionary roots and a long history of social reinforcement. The theogonic mechanisms that engender theism are so

---

31   Banyasz et al., "Predicting Religious Ethnocentrism: Evidence for a Partial Mediation Model," *Psychology of Religion and Spirituality* 8 no. 1 (2014): 25. Emphasis added. For a summary of recent literature on this theme, see Rowatt et al., "Religion, Prejudice and Intergroup Relations," *Religion, Personality, and Social Behavior* (2013). For a discussion of factors that mediate the relationship between religiosity and prejudice, see Chambers, "Religiosity and Modern Prejudice: Points of Convergence and Points of Departure" (Dissertation, Columbia University, 2016). For an example of the way in which individual variation in religiosity differentially relates to prejudice, see Streib and Klein, "Religious Styles Predict Interreligious Prejudice: A Study of German Adolescents with the Religious Schema Scale," *International Journal for the Psychology of Religion* 24, no. 2 (2014): 151–163.

deeply embedded and woven into these other biases that their role in promoting socially constructed discriminations often goes unnoticed.

For at least the last 40,000 years religious sects have held (some) societies together by increasing stratification based on racist, classist, and sexist biases. Religion provided supernatural sanction for the higher status afforded to one ethnic group over another, to the rich over the poor, to males over females. Religious elites have (more or less unconsciously) exploited such biases "to extract resources from lower-ranking group members."[32] Like many other primate species, early humans often formed themselves into male-dominated social groups. With the advent of religion, alpha males became sponsored by alpha gods.[33] The high status of (male) priests, backed by the high status afforded to the supernatural agents to whom they allegedly had access, was an expression of "social dominance psychology in a context for which it did not evolve: high-density populations made possible by agriculture."[34]

Scientific research has consistently shown that theism is strongly correlated with racism. A meta-analytic review of fifty-five independent studies in the literature on religious racism concluded that "the motives to be religious are also a *motivator of racism*, and these motives appear to be broadly applicable as a framework for understanding religious racism."[35] The amplification effect of religiosity on racism has also been demonstrated cross-culturally through cooperative game experiments.[36] Similar findings hold for the link between religion and classism. Cooperation game experiments indicate that class biases and religious variables such as frequency of thinking about one's gods and ritual performances interact in the promotion of in-group favoritism.[37] Research in India on ingroup evaluations and intergroup attitudes using implicit association tests suggests that religion may also play a role in helping children

32   Soler and Lenfesty, "Coerced Coordination, Not Cooperation," *Behavioral and Brain Sciences* 39 (2014): 39. Social stratification was also supported by forms of religious violence such as ritual human sacrifice; see Watts et al., "Ritual Human Sacrifice Promoted and Sustained the Evolution of Stratified Societies," *Nature* 532, no. 7598 (2016).

33   Garcia, *Alpha God: The Psychology of Religious Violence and Oppression* (New York: Prometheus Books, 2015).

34   Wilkins, "Gods Above: Naturalizing Religion in Terms of Our Shared Ape Social Dominance Behavior," *Sophia* 54, no. 1 (2015): 77.

35   Hall et al., "Why Don't We Practice What We Preach? A Meta-Analytic Review of Religious Racism," *Personality and Social Psychology Review* 14, no. 1 (2010): 135. Emphasis added.

36   Chuah et al., "Religion, Discrimination and Trust across Three Cultures," *European Economic Review* 90 (2016).

37   Purzycki and Kulundary, "Buddhism, Identity, and Class: Fairness and Favoritism in the Tyva Republic," *Religion, Brain & Behavior*, (2017).

accept and affirm their own class (or caste), thereby promulgating unfair social systems.[38]

What about sexism? Regression analysis of cross-national survey variables has shown that, across religious traditions, "intensity of religious belief and the frequency of religious participation" is consistently negatively correlated with an individual's support for gender equality. This suggests that religiosity "contributes to and perpetuates hierarchical gender ideology, norms, and stereotypes."[39] Similar conclusions apply at the country level. Whether or not levels of religiosity in a country cause gender equality or vice versa, the correlation is clear: "the more non-religious people in a country, the more gender equal that country tends to be."[40] It almost goes without saying that prejudice against homosexuals and other non-heterosexual gender identities is strongly correlated with high levels of religiosity.[41] But it should not go without saying; discussing the function of religion in exacerbating this sort of oppression is an important part of having "the talk."

To make things worse, these biases are also implicated within the causal nexus of intra- and intergroup violence. Religious beliefs and norms seem to have co-evolved with mate-guarding and controlling behaviors, along with other sexual selection pressures, in such a way that they "lower the threshold for violence, as well as explicitly promote and conveniently justify violent actions in evolutionarily relevant contexts."[42] Sociological analyses at the national and international levels have consistently found significant correlations between variables such as religious intensity and violent assault.[43] It is clearly not the case that any religious variable will always predict violence for every individual in all contexts. Unfortunately, however, the conditions under which violence

---

38   Dunham et al., "Religion Insulates Ingroup Evaluations: The Development of Intergroup Attitudes in India," *Developmental Science* 17, no. 2 (2014). For an assessment of the therapeutic implications of the relation between religion and social class bias, see Ali and Gaasedelen, "Religion, Social Class, and Counseling," in *The Oxford Handbook of Social Class in Counseling* (Oxford University Press, 2013).

39   Seguino, "Help or Hindrance? Religion's Impact on Gender Inequality in Attitudes and Outcomes," *World Development* 39, no. 8 (2011).

40   Schnabel, "Religion and Gender Equality Worldwide: A Country-Level Analysis," *Social Indicators Research* 129, no. 2 (2016): 893.

41   See, e.g., Herek and McLemore, "Sexual Prejudice," *Annual Review of Psychology* 64 (2013).

42   Sela et al., "When Religion Makes It Worse: Religiously Motivated Violence as a Sexual Selection Weapon," in *The Attraction of Religion: A New Evolutionary Psychology of Religion*, ed. Slone and van Slyke (London: Bloomsbury Academic, 2015), 112.

43   See, e.g., Corcoran, et al., "A Double-Edged Sword: The Countervailing Effects of Religion on Cross-National Violent Crime," *Social Science Quarterly*, (2017).

and other societal dysfunctions are correlated with religiosity are those that characterize the environments in which the majority of the human population now live.[44]

As we will see in more detail below, many of the components of anthropomorphic promiscuity and sociographic prudery are more than *correlationally* linked; there is significant evidence for *causal* relationships between cognitive and coalitional god-bearing biases. The third, and all too often unnoticed, consequence of religious sects is that some of the theogonic mechanisms that engender the misattribution of events to supernatural agents and some of the theogonic mechanisms that intensify conflict among groups with different supernatural authorities are *reciprocally reinforcing*. In other words, the superstitious interpretations of the world and the segregative inscriptions of society born(e) within religious social assemblages can become entwined in a spiral of mutual amplification. Before exploring all three of these consequences in more detail, let me back up and reiterate the importance of disciplinary and methodological pluralism when approaching phenomena as complex as those that lead us into the temptations of religious sects.

Explaining the Nature *and* Nurture of Supernatural Conceptions

Learning how to practice *safe* sects – how to live together without reproducing sectarian gods – will require ongoing discussion about *both* where ideas of supernatural agents come from *and* why people spend so much time and

---

44   See, e.g., Paul, "Cross-National Correlations of Quantifiable Societal Health with Popular Religiosity and Secularism in the Prosperous Democracies," *Journal of Religion & Society* 7 (2005); Jensen, "Religious Cosmologies and Homicide Rates among Nations: A Closer Look," *Journal of Religion & Society* 8 (2006); Svensson, "Fighting with Faith – Religion and Conflict Resolution a in Civil Wars," *Journal Of Conflict Resolution* 51, no. 6 (2007): 930–949; Paul, "The Chronic Dependence of Popular Religiosity upon Dysfunctional Psychosociological Conditions," *Evolutionary Psychology* 7, no. 3 (2009); Delamontagne, "High Religiosity and Societal Dysfunction in the United States during the First Decade of the Twenty-First Century," *Evolutionary Psychology* 8, no. 4 (2010); Campbell and Vollhardt, "Fighting the Good Fight: The Relationship between Belief in Evil and Support for Violent Policies," *Personality and Social Psychology Bulletin* 40, no. 1 (2014); Schubert and Lambsdorff, "Negative Reciprocity in an Environment of Violent Conflict," *Journal of Conflict Resolution* 58, no. 4 (2014); Vüllers, et al., "Measuring the Ambivalence of Religion: Introducing the Religion and Conflict in Developing Countries (RCDC) Dataset," *International Interactions* 41, no. 5 (2015); Chon, "Religiosity and Regional Variation of Lethal Violence Integrated Model," *Homicide Studies* 20, no. 2 (2016); Sadique and Stanislas, eds., *Religion, Faith and Crime: Theories, Identities and Issues* (Palgrave, 2016).

energy cultivating and caring for them. Traditionally, there has been a tension in the scientific study of religion between scholars who focus on the biological or psychological forces that engender mental constructions, on the one hand, and scholars who focus on the social constructions produced by cultural or contexual forces, on the other. Within and across the disciplines that converge in the bio-cultural study of religion, however, a growing number of scholars are formulating scientific hypotheses that emphasize the interactions among cognitive *and* coalitional mechanisms.

> Religious ideas can successfully colonise human minds thanks to their ability to parasitize on biologically evolved human *cognitive* structures ... due to their counterintuitive properties, this colonization can only succeed if those ideas are *culturally* transmitted through a special language.[45]

> (Locally specific socioecological) coordination problems forge the relationship between religious *cognition* and *ritual* ... religious concepts will converge around those problems and this heightens the retention and stability of religious concepts.[46]

> It is the convergence of *cognitive* and *motivational* vectors that determine the overwhelming presence and resilience of supernatural narratives around the world.[47]

> Innate *cognitive* content biases explain how people mentally represent gods, and *cultural* evolutionary models explain why people come to believe and commit to the particular supernatural beliefs that they do.[48]

---

45 Salazar, "Religious Symbolism and the Human Mind: Rethinking Durkheim's Elementary Forms of Religious Life," *Method and Theory in the Study of Religion* 27 (2015): 82. Emphasis added.

46 Purzycki and Sosis, "The Extended Religious Phenotype and the Adaptive Coupling of Ritual and Belief," *Israel Journal of Ecology & Evolution* 59, no. 2 (2013). Emphasis added. See also, Purzycki et al., "Religion," in *Emerging Trends in the Social and Behavioral Sciences*, ed. Scott and Kosslyn (New York: John Wiley & Sons, 2015), 1. "The cognitive science of religion documents the *mental* organization and structure of religious thought, while the behavioral science of religion focuses on ritual behavior as the building block of *sociality*." Emphasis added.

47 Norenzayan et al., "Memory and Mystery: The Cultural Selection of Minimally Counterintuitive Narratives," *Cognitive Science: A Multidisciplinary Journal of Artificial Intelligence, Linguistics, Neuroscience, Philosophy, Psychology* 30, no. 3 (2006): 551. Emphasis added.

48 Gervais et al., "The Cultural Transmission of Faith: Why Innate Intuitions Are Necessary, but Insufficient, to Explain Religious Belief," *Religion* 41, no. 3 (2011): 389. Emphasis added.

This is the kind of explanatory and methodological pluralism that will be required if we hope to make sense of both the *origin* and the *maintenance* of religious beliefs and behaviors.[49] Some sub-disciplines are good at explaining one but not the other, and vice versa. This is why we need multi-disciplinary *bio-cultural* approaches to the study of religion.[50]

The nature of supernatural conceptions is constrained by the cognitive mechanisms of the minds that engender them. The nurture of supernatural conceptions is constrained by the coalitional mechanisms within the groups that ritually engage them. Instead of separating these constitutive and regulative mechanisms, scholars of religion are increasingly using phrases like "social minds" and "mental cultures,"[51] or "normative cognition,"[52] in order to emphasize the inextricable link between the innate dispositions and the processes of enculturation that give rise to religiosity. Cognition *is* "embodied and embrained, situated, extended, distributed, materialized and deeply cultural,"[53] and the evolution of religion can only be understood in light of the interrelationships among multi-level (epigenetic, cognitive-developmental, sociohistorical) "landscapes."[54]

Although the same sorts of theogonic mechanisms are at work in all religious sects, they can be manifested in a wide variety of ways. This diversity of expression is due, in part, to variance at the *individual* and the *contextual* level.

See also Granqvist and Nkara, "Nature Meets Nurture in Religious and Spiritual Development," *British Journal of Developmental Psychology* 35, no. 1 (2017).

49   Bourrat, "Origins and Evolution of Religion from a Darwinian Point of View: Synthesis of Different Theories," in *Handbook of Evolutionary Thinking in the Sciences* (Springer, 2015), 761.

50   Wildman, *Science and Religious Anthropology: A Spiritually Evocative Naturalist Interpretation of Human Life* (Ashgate, 2009); Wildman, *Religious Philosophy as Multidisciplinary Comparative Inquiry: Envisioning a Future for the Philosophy of Religion* (SUNY Press, 2011); Carroll et al., "Biocultural Theory: The Current State of Knowledge," *Evolutionary Behavioral Sciences*, 2015.

51   McCorkle, Jr. and Xygalatas, "Social Minds, Mental Cultures – Weaving Together Cognition and Culture in the Study of Religion," in *Mental Cultures: Classical Social Theory and the Cognitive Science of Religion*, ed. Xygalatas and McCorkle, Jr. (Durham, UK: Acumen, 2013), 1–10.

52   Jensen, "Normative Cognition in Culture and Religion," *Journal for the Cognitive Science of Religion* 1, no. 1 (2013).

53   Geertz, "Long-Lost Brothers: On the Co-Histories and Interactions Between the Comparative Science of Religion and the Anthropology of Religion," *Numen* 61, no. 2–3 (2014): 263.

54   Whitehouse, "Rethinking Proximate Causation and Development in Religious Evolution," in *Cultural Evolution: Society, Technology, Language and Religion*, ed. Richerson and Christiansen (Cambridge Mass.: MIT Press, 2013).

In other words, both personality factors and environmental factors can affect the extent to which – and the way in which – shared imaginative engagement with supernatural agents dominates the lives of human beings. As we will see below, some people find it easier to believe in gods than others. And some ecological contexts are more likely to evoke ritual participation with gods than others. Analysis of the relationships among genetic and dispositional factors (such as quest orientation, humility, and coping style), on the one hand, and environmental and situational factors (such as ecological duress, level of existential insecurity, and cultural plurality), on the other, in mediating and moderating "religiosity" is one of the most active areas of research in the scientific study of religion.[55] Throughout the following chapters we will have the opportunity to observe the role that these sorts of individual and contextual variations play in encouraging (or discouraging) religious reproduction.

One way to approach the interactions among the various cognitive and ecological factors at work within religious sects is to think of the latter as "complex adaptive systems."[56] In fact, in the sense in which I am using the term, a "religious" social assemblage can be considered a textbook case of this sort

---

55   See, e.g., van Tongeren et al., "Toward an Understanding of Religious Tolerance: Quest Religiousness and Positive Attitudes Toward Religiously Dissimilar Others," *The International Journal for the Psychology of Religion* 26, no. 3 (2016); Gebauer et al., "Big Two Personality and Religiosity Across Cultures: Communals as Religious Conformists and Agentics as Religious Contrarians," *Social Psychological And Personality Science* 4, no. 1 (2013) ; Gebauer et al., "Cross-Cultural Variations in Big Five Relationships With Religiosity: A Sociocultural Motives Perspective," *Journal of Personality and Social Psychology* 107, no. 6 (2014); Rosenkranz and Charlton, "Individual Differences in Existential Orientation: Empathizing and Systemizing Explain the Sex Difference in Religious Orientation and Science Acceptance," *Archive for the Psychology of Religion* 35, no. 1 (2013); Lane and Harris, "Confronting, Representing, and Believing Counterintuitive Concepts: Navigating the Natural and the Supernatural," *Perspectives on Psychological Science* 9, no. 2 (2014); Silvia et al., "Blessed Are the Meek? Honesty–humility, Agreeableness, and the HEXACO Structure of Religious Beliefs, Motives, and Values," *Personality and Individual Differences* 66 (2014): 19–23; Schmitt and Fuller, "On the Varieties of Sexual Experience: Cross-Cultural Links Between Religiosity and Human Mating Strategies," *Psychology of Religion and Spirituality* 7, no. 4 (2015): 314–326; Maltseva, "Prosocial Morality in Individual and Collective Cognition," *Journal of Cognition and Culture* 16, no. 1–2 (2016); Kandler and Riemann, "Genetic and Environmental Sources of Individual Religiousness: The Roles of Individual Personality Traits and Perceived Environmental Religiousness," *Behavior Genetics* 43, no. 4 (2013).

56   See, e.g. Sosis and Kiper, "Religion Is More Than Belief: What Evolutionary Theories of Religion Tell Us About Religious Commitment," in *Challenges to Religion and Morality: Disagreements and Evolution*, 2014, 262; Purzycki and McNamara, "An Ecological Theory

of system: relatively robust, complex, emergent networks composed of highly interdependent heterogeneous agents.[57] Explaining such systems requires attending to the interactions among agents and the emergent properties of the system as a whole. Approaching the study of religion in this way also lends itself to computer simulation methodologies, to which I will return at the end of the book. However, it is important to emphasize that some complex systems (including "religions") might be *adaptive* even if they are not *adaptations*.[58] One does not have to accept the controversial notion of "group selection"[59] in order to see how a religious sect might adapt (in a general sense) to its contemporary environment.

There is a lively debate among scholars in these fields about whether religion itself is an adaptation or merely a by-product of other adaptations.[60] This depends, of course, on what we mean by religion "itself." A growing

---

of Gods' Minds," in *Advances in Religion, Cognitive Science, and Experimental Philosophy*, ed. De Cruz and Nichols (London: Bloomsbury Academic, 2016).

57   Miller and Page, *Complex Adaptive Systems: An Introduction to Computational Models of Social Life* (Princeton: Princeton University Press, 2007).

58   "[A]daptations arrive through natural selection in order to solve ecological problems organisms face for survival and propagation .... An adaptive trait confers survival or reproductive benefits. But, an adaptive trait may or may not be an adaptation; adaptive traits are useful but not necessarily a result of evolution." Smith, *Thinking about Religion: Extending the Cognitive Science of Religion* (New York: Palgrave Macmillan, 2014), 70.

59   For recent challenges to the idea of group selection by evolutionary psychologists, see Krasnow et al., "Group Cooperation without Group Selection: Modest Punishment Can Recruit Much Cooperation," *PLoS ONE* 10, no. 4 (2015), Kundt, *Contemporary Evolutionary Theories of Culture and the Study of Religion* (Bloomsbury Academic, 2015), and Krasnow and Delton, "Are Humans Too Generous and Too Punitive? Using Psychological Principles to Further Debates about Human Social Evolution," *Frontiers in Psychology* 7 (2016). For defenses of the group selection hypothesis, see Wilson et al., "The Nature of Religious Diversity: A Cultural Ecosystem Approach," *Religion, Brain & Behavior*, (2016), and Richerson et al., "Cultural Group Selection Plays an Essential Role in Explaining Human Cooperation: A Sketch of the Evidence," *Behavioral and Brain Sciences*, (2014). It is important to remember that "group" selection is not the same as "social" selection, which can simply mean "genetic selection that is accomplished by the social, as opposed to the natural, environment." Boehm, "The Moral Consequences of Social Selection," *Behaviour* 151, no. 2–3 (2014): 168. See also Davis, "Group Selection in the Evolution of Religion: Genetic Evolution or Cultural Evolution?" *Journal of Cognition and Culture* 15 (2015): 247, Morgan, "Testing the Cognitive and Cultural Niche Theories of Human Evolution," *Current Anthropology* 57, no. 3 (2016), and Claidiere et al., "How Darwinian Is Cultural Evolution?" *Philosophical Transactions of the Royal Society B: Biological Sciences* 369, no. 1642 (2014).

60   See, e.g., Pyysiäinen and Hauser, "The Origins of Religion: Evolved Adaptation or by-Product?" *Trends in Cognitive Sciences* 14, no. 3 (2010), and Singh and Chatterjee, "The

number of scholars are attempting to move past this impasse by focusing on the components of religious systems: distinguishing between those components that appear to be by-products and those that fit the more technical definition of "adaptation," or even proposing "exaptation" models to help explain the evolution of religion.[61] As we go along, it should become clear how fractionating "religion" in the way I have been describing helps us escape this false dichotomy by recognizing that some of its fractionated components may be by-products while others are adaptations (or exaptations).

It should also become increasingly clear why I find it so important to describe religious sects as complex adaptive *theogonic* systems – as social networks whose cohesion is dependent in large part on shared imaginative engagement with *supernatural agents*. This attempt to include a somewhat neo-Tylorian emphasis on ideas about "gods" (or "spirits," in the broadest sense) in discussions about "religion" is often resisted by religious apologists, as well as by many scholars in the humanities and in religious studies departments, who prefer a more limited neo-Durkheimian emphasis on its cultural dimensions. The good news is that we do not have to decide between these approaches. We can emphasize both the cognitive *and* the cultural dynamics that contribute to shared imaginative engagement with axiologically relevant supernatural agents within religious systems.

Because I am exploring *scientific* perspectives on theogonic reproduction, I will continue to use the term "religion" in the way it commonly functions in the literature of the relevant disciplines that contribute to the bio-cultural study of religion. This decision is further warranted by recent research involving factor analyses of psychological measures of religiosity. It turns out that "supernatural-related belief/practice" is "the only *unique* diagnostic feature of religiosity ... and empirically distinct from sociability, virtue, hope, etc."[62]

Evolution of Religious Belief in Humans: A Brief Review with a Focus on Cognition," *Journal of Genetics* 96, no. 3 (2017): 517–524.

61   See, e.g., Girotto et al., "Supernatural Beliefs: Adaptations for Social Life or by-Products of Cognitive Adaptations?" *Behavior* 151 (2014); Ma-Kellams, "When Perceiving the Supernatural Changes the Natural: Religion and Agency Detection," *Journal of Cognition and Culture* 15, no. 3–4 (2015); Davis, "The Goldberg Exaptation Model: Integrating Adaptation and By-Product Theories of Religion," *Review of Philosophy and Psychology* 8, no. 3 (2017).

62   Schuurmans-Stekhoven, "Are We, like Sheep, Going Astray: Is Costly Signaling (or Any Other Mechanism) Necessary to Explain the Belief-as-Benefit Effect?" *Religion, Brain & Behavior* (2016): 36. See also Lindeman, Blomqvist, and Takada, "Distinguishing Spirituality From Other Constructs: Not A Matter of Well-Being but of Belief in Supernatural Spirits," *The Journal of Nervous and Mental Disease* 200, no. 2 (2012); Schuurmans-Stekhoven, "Measuring Spirituality as Personal Belief in Supernatural Forces: Is the

Statistical analysis of the results of psychological experiments designed to measure the implicit beliefs of believers and skeptics suggests that "supernatural content" is "the *only thing* that distinguishes religiosity from non-religiosity."[63] Perhaps more surprisingly, this is not only true of "religiosity," but of "spirituality" as well. Although it is also correlated with other variables (such as values, and paranormal beliefs), regression anlaysis suggests that spirituality – like religiosity – is primarily predicted by "belief in supernatural spirits."[64] Factor analyses of survey data also confirm that beliefs and practices related to *supernatural* forces form a relatively independent cluster of variables.[65]

I am also interested in providing a *philosophical* perspective on religious reproduction, and so will allow myself to indulge in retroductive as well as abductive argumentation (see Chapter 5 for a discussion of these terms). Explaining *can* sometimes mean *explaining away*. Explaining the biological mechanisms that lead to sexual reproduction, for example, render the "stork hypothesis" implausible. Even if one cannot logically *prove* that babies are not sometimes brought by storks, the latter notion is explained away once people understand the processes of copulation, gestation, and parturition. What caused that creaking noise in the attic? Understanding the effects of wind on old wood can explain away the "ghost hypothesis." What caused that tsunami? Understanding the effects of seismic activity on ocean waves can help explain away hypotheses about angry sea gods. The first step, however, is to explain where these sorts of supernatural conceptions come from in the first place.

The Mechanisms of Anthropomorphic Promiscuity

There are many empirically well-validated theories within the bio-cultural study of religion about the mechanisms that engender belief in gods. Some

Character Strength Inventory-Spirituality Subscale a Brief, Reliable and Valid Measure?" *Implicit Religion* 17, no. 2 (2014).

63   Lindeman et al., "Skepticism: Genuine Unbelief or Implicit Beliefs in the Supernatural?" *Consciousness and Cognition* 42 (2016): 225. Emphasis added.

64   Lindeman et al., "Distinguishing Spirituality From Other Constructs: Not A Matter of Well-Being but of Belief in Supernatural Spirits," *The Journal of Nervous and Mental Disease* 200, no. 2 (2012): 172.

65   Schofield et al., "Mental Representations of the Supernatural: A Cluster Analysis of Religiosity, Spirituality and Paranormal Belief," *Personality and Individual Differences* 101 (2016). Lemos, et al., "Exploratory and Confirmatory Analyses of Religiosity: A Four-Factor Conceptual Model," (2017): arXiv: 1704.06112.

of the most commonly discussed involve hypotheses about teleological reasoning, intuitive mind/body dualism, memory systems related to minimally counterintuitive ideas, the theory of mind mechanism, and the so-called hyper-sensitive agency detection device. I have discussed these mechanisms in earlier presentations of theogonic reproduction theory.[66] In this section, I survey some of the more recent theoretical developments and empirical findings that shed light on these and other cognitive biases that are operative in the religious reproduction of supernatural conceptions. This contributes to the ongoing scientific process of fractionating the phenomena associated with "religion."

As in all generative research programs, there is ongoing debate about these features of the human mind and their relative importance in the production of religious beliefs.[67] My concern here, however, is not with defending a narrow position on any one of these mechanisms, but with the plausibility of the wider claim that a suite of evolved cognitive tendencies fosters the attribution of intentionality in general (and personification in particular) to non-living entities, forces, or patterns, thereby contributing to the emergence of god-beliefs. Despite disagreement on the details, there is a broad consensus that a set of interrelated cognitive biases engenders "belief in supernatural agents such as gods and spirits, and related phenomena ... Equipped with these cognitive biases, human minds gravitate towards religious and religious-like beliefs and intuitions."[68]

---

66   Some of these are also described in the chapters that follow, and in more detail in Shults, *Theology after the Birth of God.*

67   Lisdorf, "What's HIDD'n in the HADD?" *Journal of Cognition and Culture* 7, no. 3–4 (2007); Hornbeck and Barrett, "Refining and Testing 'Counterintuitiveness' in Virtual Reality: Cross-Cultural Evidence for Recall of Counterintuitive Representations," *The International Journal for the Psychology of Religion* 23, no. 1 (2013): 15; Purzycki and Willard, "MCI Theory: A Critical Discussion," *Religion, Brain & Behavior* (2015); Gervais, "Perceiving Minds and Gods," *Perspectives on Psychological Science* 8, no. 4 (2013); White, "The Cognitive Foundations of Reincarnation," *Method & Theory in the Study of Religion* (2016); Bastian et al., "Moral Vitalism: Seeing Good and Evil as Real, Agentic Forces" 41, no. 8 (2015): 1069. Maij et al., "The Boundary Conditions of the Hypersensitive Agency Detection Device: An Empirical Investigation of Agency Detection in Threatening Situations," *Religion, Brain & Behavior* (2017), 1–29; van Leeuwen and van Elk, "Seeking the Supernatural: The Interactive Religious Experience Model," *Religion, Brain & Behavior*, in press. See the critical commentaries on the target article by Sterelny, "Religion Re-Explained," *Religion, Brain & Behavior* (2017) for representative examples of key positions in some of these debates.

68   Willard and Norenzayan, "Cognitive Biases Explain Religious Belief, Paranormal Belief, and Belief in Life's Purpose," *Cognition* 129, no. 2 (2013).

The mechanisms that lead to the false detection of the gods of one's own in-group are the result of natural selection (or by-products, or exaptations of earlier adaptations). Like many other heritable traits, these tendencies have been passed on through hundreds of generations. What makes religious cognitive biases distinctive is that they are based on "a *strategic distortion* of reality as *god infested*."[69] For reasons soon to be spelled out, mistakenly believing that nature is filled with disembodied (or ambiguously-embodied, or at least ontologically confused, in the sense explained below) intentional forces provided a survival advantage to our ancestors. This is why some of the most powerful "cognitive attractors" today are supernatural concepts; most contemporary humans are naturally attracted to these "ideal" forms of representation because they are based on phylogenetically inherited cognitive architectures.[70]

Seeing faces in the clouds is not necessarily a religious experience; seeing the clouds as god infested *is*. It is important to emphasize that *religious* cognition, in the sense we are using the term here, has to do with *supernatural* causality.

> *Anthropomorphic promiscuity*: the tendency to appeal to the causal efficacy of supernatural agents when trying to make sense of the world.

But how would cognitive biases that engender supernatural beliefs have given our ancestors a strategic advantage? Many of the mechanisms that contribute to the mistaken detection of disembodied intentional forces behind natural patterns or events are the result of *error* management strategies.[71]

It is important to emphasize that it was not *making* errors per se that would have helped our progenitors survive, but their capacity to *manage* errors. Humans make mistakes all the time, but some errors in judgment can be fatal. Survival requires some relatively quick way of deciding whether it is worthwhile

---

69   Bulbulia, "Religious Costs as Adaptations That Signal Altruistic Intention," *Evolution and Cognition* 10, no. 1 (2004): 38. Emphases added.

70   Sperber, *Explaining Culture: A Naturalistic Approach* (Oxford, UK ; Cambridge, Mass.: Blackwell Publishers, 1996); Sperber and Wilson, *Relevance: Communication and Cognition*, 2 edition (Cambridge, MA: Wiley-Blackwell, 1996). This can be illustrated in the cognitive attractors of ghosts, vampires and zombies, which activate different mental systems: Bahna, "Explaining Vampirism: Two Divergent Attractors of Dead Human Concepts," *Journal of Cognition and Culture* 15 (2015).

71   For a concise introduction to error management theory, see Haselton et al., "The Evolution of Cognitive Bias," *The Handbook of Evolutionary Psychology* (2005), and Haselton et al., "The Paranoid Optimist: An Integrative Evolutionary Model of Cognitive Biases," *Personality and Social Psychology Review* 10, no. 1 (2006).

spending energy to find out whether one has made a mistake. We do not have time to consider every possible error, so we need a way of sorting out those perceptive judgments that require further critical reflection from those that we ought to just accept and move on. One way to save time is to believe the people around us; credulity toward caretakers, then, would have provided a survival advantage. Early humans who were consistently credulous in response to the false promises of enemies, on the other hand, would be less likely to survive long enough to pass on their genes.

What is the *optimal* level of credulity? That depends both on the environment and on the capacities of the sentient organism trying to adapt to it. When dealing with adaptive tasks in a complex and ambiguous environment, the equilibrium for optimal credulity involves an asymmetric ratio between erroneous credulity (believing information that is false, or Type I errors) and eroneous incredulity (failing to believe information that is true, or Type II errors). The basic claim of error management theory is that making Type I errors is the price our ancestors paid for avoiding costlier Type II errors, especially when confronted by natural or social hazards. Natural selection "crafted learner's minds so as to be more credulous toward information concerning hazards ... together these biases constitute attractors that should shape cultural evolution via the aggregated effects of learner's differential retention and transmission of information."[72]

The logic of error management is not maximized for accuracy, but "systematically biased to commit the *least costly* error."[73] Detecting harmful agents where there are none is, by definition, an error. However, the capacity to quickly calculate asymmetric error costs under uncertain or dangerous conditions would have helped our ancestors adapt in hazardous environments. In other words, "people believe in supernatural agents which do not exist for the same reason that birds sometimes mistake harmless birds passing overhead for raptors."[74] Better safe than sorry. Better to believe a punitive god is watching you and behave, than not to believe, misbehave, and be eternally (or even

72   Fessler et al., "Negatively-Biased Credulity and the Cultural Evolution of beliefs.(Report)," *PLoS ONE* 9, no. 4 (2014).

73   Haselton et al., "Adaptive Rationality: An Evolutionary Perspective on Cognitive Bias," *Social Cognition* 27, no. 5 (2009); emphasis added. Nola, "Do Naturalistic Explanations of Religious Beliefs Debunk Religion?" in *A New Science of Religion*, ed. Dawes and Maclaurin (London: Routledge, 2013).

74   Wilkins et al., "Evolutionary Debunking Arguments in Three Domains," *A New Science of Religion* 23 (2012): 143. For a discussion of supernatural predation, see Kazanas and Altarriba, "Did Our Ancestors Fear the Unknown? The Role of Predation in the Survival Advantage," *Evolutionary Behavioral Sciences*, (2016).

just temporally) punished. Or, so the argument goes. While this logic may have worked for our ancient ancestors, the problem today is that error management biases that "evolved to deal with environments of our evolutionary past are likely to cause *damaging behavior in the modern environment*."[75] Several of the following chapters are devoted to assessing the damages.

The first major hypothesis of theogonic reproduction theory is that the mechanisms that contribute to anthropomorphic promiscuity are the result of cognitive biases that engender mistaken attributions of intentionality. If this hypothesis were true, we would expect to find particular patterns in – and to be able to make specific predictions about – the available empirical data. We would expect to find evidence that the emergence of many of these god-bearing mechanisms was *phylogenetically* ancient in the human species, that their cross-cultural manifestation today is *ontogenetically* early in human development, and that their distribution in human populations is *individually* variant.

And this is exactly what we do find. The predisposition to attribute characteristics of animacy or intentionality to nonliving things or random patterns is an evolved bias that is evident not only in all primates, but in many other animals as well.[76] When it comes to the various anthropomorphizing tendencies that shape religious beliefs, a reconstruction of ancestral character traits based on a time-calibrated phylogenetic supertree (using ethnographic and other datasets) suggests that "the oldest trait of religion, present in the most recent common ancestor of present-day hunter-gatherers, was animism... Belief in an afterlife emerged [later], followed by shamism and ancestor worship."[77] Evidence for the early influence of anthropomorphism in the evolution of religion can also be found in archaeological reconstructions of upper Paleolithic human societies which, in conjunction with ethnographic research on contemporary hunter-gatherer cultures, indicate that dreams, hallucinations, and other altered states of consciousness were an important source of supernatural agent concepts in ancestral environments, just as they are today.[78]

---

75   Johnson et al., "The Evolution of Error: Error Management, Cognitive Constraints, and Adaptive Decision-Making Biases," *Trends in Ecology and Evolution* 28, no. 8 (2013): 480. Emphasis added.

76   Guthrie, "Animal Animism: Evolutionary Roots of Religious Cognition," in *Current Approaches in the Cognitive Science of Religion* (London: Continuum, 2002), 46.

77   Peoples et al., "Hunter-Gatherers and the Origins of Religion," *Human Nature*, 2016; Peoples and Marlowe, "Subsistence and the Evolution of Religion," *Human Nature* 23, no. 3 (2012).

78   McNamara and Bulkeley, "Dreams as a Source of Supernatural Agent Concepts," *Frontiers In Psychology* 6 (2015). See also Bulkeley, *Big Dreams: The Science of Dreaming and the Origins of Religion* (New York: Oxford University Press, 2016). For an archaeological and

If the biases that contribute to anthropomorphic promiscuity in the human species are ancient, phylogenetically inherited tendencies, then we would expect them to be manifested cross-culturally and relatively early in the ontogentic development of individual human beings. Psychological experiments consistently show that this is the case. Infants scan for faces in the first days of life, and very small children already think of God in egocentric ways (for example, as a relatively ignorant man in the sky).[79] In fact, the hyper-sensitive anthropomorphic tendencies that can lead to the sort of Type I supernaturalist errors described above may already be developing in the womb.[80] Property attribution experiments on children and adults indicate that the former are even more likely than the latter to attribute human-like qualities to religious beings.[81] The bias toward teleological explanations, especially under processing constraints, also appears to be cross-culturally robust.[82] Regardless of context, young children somewhat automatically personify non-living natural entities. However, they eventually come to describe imagined non-natural entities as "intentional" in ways that increasingly reflect the distinctive anthropomorphic characteristics of the supernatural agents postulated within their religious family of origin.[83]

---

anthropological discussion of the role of altered states of consciousness in the origin of religion see Lewis-Williams, *Conceiving God: The Cognitive Origin and Evolution of Religion* (Thames & Hudson, 2010). Dreams and hallucinations can involve supportive and protective gods or dangerous and demonic supernatural agents such as succubi and incubi; both types can play a role in regulating biological sex within religious sects. See Braxton, "Policing Sex: Explaining Demons in the Cognitive Economies of Religion," *Journal of Cognition and Culture* 8, no. 1–2 (2008).

79   Wigger et al., "What Do Invisible Friends Know? Imaginary Companions, God, and Theory of Mind," *International Journal for the Psychology of Religion* 23, no. 1 (2013); Heiphetz et al., "How Children and Adults Represent God's Mind," *Cognitive Science* 40, no. 1 (2016); Kiessling and Perner, "God–Mother–Baby: What Children Think They Know," *Child Development* 85, no. 4 (2014).

80   Reid et al., "The Human Fetus Preferentially Engages with Face-like Visual Stimuli," *Current Biology* 27, no. 12 (2017).

81   Shtulman, "Variation in the Anthropomorphization of Supernatural Beings and Its Implications for Cognitive Theories of Religion," *Journal of Experimental Psychology: Learning, Memory, and Cognition* 34, no. 5 (2008).

82   See, e.g., Rottman et al., "Cultural Influences on the Teleological Stance: Evidence from China," *Religion, Brain & Behavior* 7, no. 1 (2017), and Schachner et al., "Is the Bias for Function-Based Explanations Culturally Universal? Children from China Endorse Teleological Explanations of Natural Phenomena," *Journal of Experimental Child Psychology* 157 (2017).

83   See, e.g., Barrett and Keil, "Conceptualizing a Nonnatural Entity: Anthropomorphism in God Concepts," *Cognitive Psychology* 31, no. 3 (1997); Barrett, "Cognitive Constraints on

This does not mean that children (or adults) in different religious in-groups have the "same" belief, but that beliefs in certain types of supernatural agents function as cognitive attractors and can be measured across cultures with general invariance.[84] As we will see in Chapter 12, this does not mean that children are naturally religious, only that they are naturally susceptible to inculcation into belief in the supernatural agents imaginatively engaged by their parents and other authorities. The early ontogenetic emergence of anthropomorphic promiscuity is also supported by cross-cultural research that shows children recall counterintuitive repesentations of the religious sort more preferentially than adults, suggesting there may be a "window of opportunity for religiosity."[85] The fact that so many young human beings go through this window can be partially explained by the function of a neonatal survival instinct that predisposes an infant to look for a being that matches its innate neural model of a caregiver, whose existence is presupposed. In other words, the illusion of the presence of gods may be a by-product of supernormal stimuli that "fill an emotional and cognitive vaccum left over from human infancy."[86]

The material content of this illusion, and the level of confidence with which it is believed, will be influenced by religious context, and especially by levels of parental anthropomorphism.[87] This brings us to the third general prediction related to the mechanisms behind anthropomorphic promiscuity; like all phylogenetically inherited biases, the traits that engender erroneous credulity toward claims about supernatural agents should be distributed in human populations in such a way that their manifestation in individuals *varies*. From an evolutionary point of view, we would expect that individuals in a population who more strongly believe in gods would also more strongly exhibit the relevant biases. Cross-cultural psychological experiments have confirmed

Hindu Concepts of the Divine," *Journal for the Scientific Study of Religion* 37, no. 4, (1998); Knight et al., "Children's Attributions of Beliefs to Humans and God: Cross-Cultural Evidence," *Cognitive Science* 28, no. 1 (2004); Johnson et al., "Fuzzy People: The Roles of Kinship, Essence, and Sociability in the Attribution of Personhood to Nonliving, Nonhuman Agents," *Psychology of Religion and Spirituality* 7, no. 4 (2015).

84   Bluemke et al., "Measuring Cross-Cultural Supernatural Beliefs with Self- and Peer-Reports," *PLoS ONE* 11, no. 10 (2016).

85   Gregory and Greenway, "Is There a Window of Opportunity for Religiosity? Children and Adolescents Preferentially Recall Religious-Type Cultural Representations, but Older Adults Do Not," *Religion, Brain & Behavior* 7, no. 2 (2017).

86   Wathey, *The Illusion of God's Presence: The Biological Origins of Spiritual Longing* (Amherst, NY: Prometheus Books, 2016), 68.

87   Richert et al., "The Role of Religious Context in Children's Differentiation between God's Mind and Human Minds," *British Journal of Developmental Psychology* 35, no. 1 (2017).

this prediction with reference to a wide variety of cognitive mechanisms. For example, several studies have suggested that mentalizing (or hyperactive theory of mind) plays an important role in supernatural belief.[88]

Other studies suggest that high mentalization per se is not the key variable; religious belief is predicted more strongly by a person's tendency toward *ontological confusion* – the propensity to make category mistakes in which "the distinctive properties of the superordinate categories of mental and physical, animate and inanimate, and living and lifeless are inappropriately mixed."[89] Confusing the attributes of mental, physical, and biological entities appears to be a core element of superstitious beliefs in ghosts, gods and other paranormal phenomena.[90] Hierarchical regression analyses of the results of psychological experiments suggest that supernatural believers have a poorer understanding of the physical world. This association seems to be related primarily to a kind of confusion between mental and physical properties of the sort that is typical among "ancient people and small children."[91]

At this stage, some might object that I am "infantilizing" religious people. The point of this research, however, is not that supernaturalists are less mature than naturalists. And despite the connotations associated with the metaphor of having "the talk" with believers about religious reproduction, the call for such conversations does not imply their puerilization. In fact, quite the opposite: it implies taking them seriously as individuals who are capable of challenging their biases once they understand the deleterious effects they are having on their own lives and those around them. Pointing out theist biases is no more infantilizing than pointing out racist, classist, and sexist biases.

Why are supernatural believers more likely to commit (and accept) ontological confusions? Experimental research suggests that it is because they tend to have weaker inhibitory control (compared with skeptics). In fact, "the weaker one's inhibitory control is, the more ontological confusions one accepts, and when the reasoning system is subjected to cognitive load, this relationship is

---

88 E.g., Willard and Norenzayan, "Cognitive Biases Explain Religious Belief, Paranormal Belief, and Belief in Life's Purpose." *Cognition* 129/2 (2013), and Norenzayan et al., "Mentalizing Deficits Constrain Belief in a Personal God" 7, no. 5 (2012) *PLoS ONE*.

89 Lindeman et al., "Ontological Confusions but Not Mentalizing Abilities Predict Religious Belief, Paranormal Belief, and Belief in Supernatural Purpose," *Cognition* 134 (2015): 65. See also Barber, "Believing in a Purpose of Events: Cross-Cultural Evidence of Confusions in Core Knowledge," *Applied Cognitive Psychology* 28 (2014).

90 Lindeman and Aarnio, "Superstitious, Magical, and Paranormal Beliefs: An Integrative Model," *Journal of Research in Personality* 41, no. 4 (2007).

91 Lindeman and Svedholm-Häkkinen, "Does Poor Understanding of Physical World Predict Religious and Paranormal Beliefs?" *Applied Cognitive Psychology* 30, no. 5 (2016): 740.

strongly accentuated."[92] Other experimental studies show that when individuals are primed to feel a lack of control, they are more likely to believe in gods and the efficacy of supernatural rituals.[93] Here too we find evidence not only of correlation but of a *causal* link between weak inhibition (relative incapacity to correct errors) and belief in gods (or God). This is the sort of result that we would expect to find if our hypothesis about the role of error management in religion were true.

We would also expect to find that religious individuals make more erroneous attributions of intentionality when confronted by random movements (compared to less- or non-religious individuals). This aspect of the anthropomorphic promiscuity hypothesis is confirmed by studies that use experimental manipulations to increase thoughts of randomness, which in turn appear to cause an increase in belief about supernatural sources of control. This suggests that "the anxiety engendered by a lack of control plays a crucial role in the generation of compensatory belief in a controlling God."[94] Another series of studies found that religious participants, as well as participants who read about a controlling God or were reminded of their own strong beliefs in a controlling God, reported higher goal commitment, but *only* when their sense of self-efficacy had been lowered through an experimental manipulation.[95]

The hypothesis that anthropomorphic promiscuity is partially the result of inherited error management biases would also lead us to expect that religious individuals in a population would be more likely to make mistakes about purposiveness in nature or life events. A variety of studies using diverse methodologies, including fMRI experiments, Stroop tasks, sorting tests, and laboratory reasoning experiments, have shown that believers do consistently make more of this sort of mistake than non-believers. These findings indicate that the errors associated with supernatural beliefs are not simply a weakness

92   Svedholm and Lindeman, "The Separate Roles of the Reflective Mind and Involuntary Inhibitory Control in Gatekeeping Paranormal Beliefs and the Underlying Intuitive Confusions," *British Journal of Psychology* 104, no. 3 (2013): 317.

93   Boucher and Millard, "Belief in Foreign Supernatural Agents as an Alternate Source of Control When Personal Control Is Threatened," *The International Journal for the Psychology of Religion* 26, no. 3 (2016); Legare and Souza, "Searching for Control: Priming Randomness Increases the Evaluation of Ritual Efficacy," *Cognitive Science*, (2013).

94   Kay et al., "Religious Belief as Compensatory Control," *Personality and Social Psychology Review* 14, no. 1 (2010): 39. See also Kay, Moscovitch, and Laurin, "Randomness, Attributions of Arousal, and Belief in God," *Psychological Science* 21, no. 2 (2010): 217.

95   Khenfer et al., "When God's (Not) Needed: Spotlight on How Belief in Divine Control Influences Goal Commitment," *Journal of Experimental Social Psychology* 70 (2017).

in probabilistic reasoning but arise "from a specific bias associated with perception of randomness (misrepresentation of chance)."[96]

Statistical analysis of survey data indicates that individuals with a stronger cognitive bias toward seeing design in the world are more attracted to theistic and paranormal beliefs, which in turn encourage and strengthen an uncritical attitude toward teleological intuitions.[97] Non-theists, on the other hand, are more likely to be able to effortfully override biases that foster the mistaken detection of purposiveness.[98] In other words, religious individuals typically make more errors about purposiveness and the coherence of meaningful beliefs than non-religious individuals, but are less stressed about those errors, in part because their supernatural content is unfalsifiable – which makes it easier to keep believing.[99]

It is not simply that theists are more receptive to false reasoning about intentionality. Experimental studies have shown that religious people are also less able to *detect* conflicts in their own reasoning, especially conflicts dealing with material and allegedly immaterial causes in the world, suggesting that a weak conflict detection mechanism contributes to supernatural belief.[100] This is confirmed by neuroscientific research in which the priming of religious ideas is found to lower anxious *reactions* to self-generated, generic errors – but only in people who already believe in axiologically relevant supernatural agents. Religious fundamentalism predicts even lower sensitivity to anxiogenic cues

---

96   Dagnall et al., "Misperception of Chance, Conjunction, Belief in the Paranormal and Reality Testing: A Reappraisal," *Applied Cognitive Psychology* 28, no. 5 (2014): 716. See also, e.g., Riekki et al., "Supernatural Believers Attribute More Intentions to Random Movement than Skeptics: An fMRI Study," *Social Neuroscience* 9, no. 4 (2014), Lindeman et al., "Is Weaker Inhibition Associated with Supernatural Beliefs?" *Journal of Cognition and Culture* 11, no. 1–2 (2011); Miyazaki, "Being Watched by Anthropomorphized Objects Affects Charitable Donations in Religious People" *Japanese Psychological Research* 59, no. 3 (2017): 221–229.

97   Banerjee and Bloom, "Why Did This Happen to Me? Religious Believers' and Non-Believers' Teleological Reasoning about Life Events," *Cognition* 133, no. 1 (2014): 298.

98   Heiphetz et al., "In the Name of God: How Children and Adults Judge Agents Who Act for Religious versus Secular Reasons," *Cognition* 144 (2015): 134.

99   Inzlicht et al., "The Need to Believe: A Neuroscience Account of Religion as a Motivated Process," *Religion, Brain & Behavior* 1, no. 3 (2011); Good et al., "God Will Forgive: Reflecting on Gods Love Decreases Neurophysiological Responses to Errors," *Social Cognitive and Affective Neuroscience* 10, no. 3 (2013).

100   Pennycook et al., "Cognitive Style and Religiosity: The Role of Conflict Detection," *Memory and Cognition* 42, no. 1 (2014); Pennycook et al., "On the Reception and Detection of Pseudo-Profound Bullshit," *Judgment and Decision Making* 10, no. 6 (2015): 549.

that one holds conflicting ideas.[101] All of this is complicated by the fact that
even when people are able to *detect* their errors about superstitious, paranormal, and religious beliefs, cognitive dissonance (and other) biases can lead
them to choose not to *correct* them anyway.[102]

Another individual variable strongly correlated to religiosity is high schizotypy. This term is used to describe a wide range of traits, but is most often
linked to magical thinking, unusual cognitive-perceptual experiences, hallucinations (such as hearing voices), and paranoid ideation – all of which can
be associated with religious and spiritual experiences. At the high end of the
scale, schizotypy can be manifested in schizophrenic hyperreligiosity, which is
a common feature of obsessive-compulsive disorder, temporal-lobe epilepsy
and similar disorders related to dopaminergic systems in humans.[103] Neuroimaging experiments indicate that a heightened susceptibility to religiosity can
be conferred by brain abnormalities associated with non-clinical psychosis.[104]
Throughout human history, shamanic leaders and priestly elites exhibiting
schizotypic traits have played an important role in the emergence and cohesion of religious groups, including monotheistic traditions (e.g., Moses, Paul,
Mohammed).[105]

Like other traits commonly associated with religiosity, susceptibility to hallucinations is also distributed normally in human populations.[106] One study
showed, for example, that reported occurrences of hallucinations "increased
significantly from normal controls through evangelical Christians to psychotic

---

101   Kossowska et al., "Anxiolytic Function of Fundamentalist Beliefs: Neurocognitive Evi-
        dence," *Personality and Individual Differences* 101 (2016).

102   Risen, "Believing What We Do Not Believe: Acquiescence to Superstitious Beliefs and
        Other Powerful Intuitions," *Psychological Review* 123, no. 2 (2015). For a discussion of "The
        Implicit Associations Between Religious and Nonreligious Supernatural Constructs,"
        see Weeks and Gilmore, *The International Journal for the Psychology of Religion* 27, no. 2
        (2017).

103   Previc, "The Role of the Extrapersonal Brain Systems in Religious Activity," *Conscious-
        ness and Cognition* 15, no. 3 (2006). See also Mauzay et al., "Devils, Witches, and Psychics:
        The Role of Thought-Action Fusion in the Relationships between Obsessive-Compulsive
        Features, Religiosity, and Paranormal Beliefs," *Journal of Obsessive-Compulsive and Related
        Disorders* 11 (2016).

104   Pelletier-Baldelli et al., "Orbitofrontal Cortex Volume and Intrinsic Religiosity in Non-
        Clinical Psychosis," *Psychiatry Research: Neuroimaging* 222 (2014).

105   Murray et al., "The Role of Psychotic Disorders in Religious History Considered," *The Jour-
        nal of Neuropsychiatry and Clinical Neurosciences* 24, no. 4 (2012).

106   Willard and Norenzayan, "'Spiritual but Not Religious': Cognition, Schizotypy, and Con-
        version in Alternative Beliefs," *Cognition* 165 (2017); Bronkhorst, "Can Religion Be Ex-
        plained? The Role of Absorption in Various Religious Phenomena," *Method and Theory in
        the Study of Religion* 29 (2017).

individuals."[107] Even in its non-psychotic manifestations, higher levels of schizoptypy consistently correlate strongly with higher scores on conspiracy mentality, susceptibility to hallucinations, and hyperactive agency detection.[108] A recent multiple regression analysis of survey results indicated that *all types of prayer* are uniquely predicted by the magical thinking component of the schizotypal personality scale.[109] Although they can lead to full-blown psychotic symptoms at the high end of the scale (e.g., confidence in being *the* "chosen" one), schizotypal traits like hyperactive agency detection and paranoid ideation in the mid- to high-range of the normal distribution can lead to comforting experiences of having a special connection to a supernatural entity who provides significance to one's life.[110]

The mechanisms that contribute to anthropomorphic promiscuity – only some of which have been fractionated here – are phylogenetically inherited tendencies that emerge early in ontogenetic development but whose intensity is variably distributed and expressed across individuals in human populations. Clarifying the causes and effects of these theistic *credulity* biases is an important part of having "the talk" about religious reproduction. However, understanding where supernatural conceptions come from is not enough. We must also explain the theistic *conformity* biases that enable people to continue nurturing such ideas as they engage in religious sects.

---

107 Davies et al., "Affective Reactions to Auditory Hallucinations in Psychotic, Evangelical and Control Groups," *British Journal of Clinical Psychology* 40, no. 4 (2001): 366.

108 van Der Tempel and Alcock, "Relationships between Conspiracy Mentality, Hyperactive Agency Detection, and Schizotypy: Supernatural Forces at Work?" *Personality and Individual Differences* 82 (2015); see also Gearing et al., "Association of Religion with Delusions and Hallucinations in the Context of Schizophrenia: Implications for Engagement and Adherence," *Schizophrenia Research* 126, no. 1 (2011), Maltby et al., "Religious Orientation and Schizotypal Traits," *Personality and Individual Differences* 28, no. 1 (2000), Maltby and Day, "Religious Experience, Religious Orientation and Schizotypy," *Mental Health, Religion & Culture* 5, no. 2 (2002), Barnes and Gibson, "Supernatural Agency: Individual Difference Predictors and Situational Correlates," *International Journal for the Psychology of Religion* 23, no. 1 (2013), Crespi and Summers, "Inclusive Fitness Theory for the Evolution of Religion," *Animal Behaviour* 92 (2014), and Wlodarski and Pearce, "The God Allusion: Individual Variation in Agency Detection, Mentalizing and Schizotypy and Their Association with Religious Beliefs and Behaviors," *Human Nature* 27, no. 2 (2016).

109 Breslin and Lewis, "Schizotypy and Religiosity: The Magic of Prayer," *Archive for the Psychology of Religion* 37, no. 1 (2015).

110 Unterrainer and Lewis, "The Janus Face of Schizotypy: Enhanced Spiritual Connection or Existential Despair?" *Psychiatry Research* 220, no. 1–2 (2014): 236. See also Saavedra, "Function and Meaning in Religious Delusions: A Theoretical Discussion from a Case Study," *Mental Health, Religion & Culture* 17, no. 1 (2014).

The Mechanisms of Sociographic Prudery

Empirical research from a wide variety of disciplines that contribute to the bio-cultural study of religion also continues to shed light on the mechanisms that nurture the tendency to presuppose the normativity of the ritual and moral practices observed by one's own religious in-group. Some of the most commonly discussed theories that bear on these mechanisms include hypotheses about "supernatural monitoring" and "supernatural punishment," which focus on the role that belief in watchful, punitive gods plays in keeping everyone in line and promoting in-group cooperation and coordination.[111] Another influential set of hypotheses in the field attempt to illuminate prudish sociography by applying "costly signaling" theory to religion; here the focus is on the ways in which believers signal their commitment to their religious coalition with costly behaviors that provide no obvious survival advantage for the individual, but reinforce the cohesion of the coalition.[112] I have outlined and

---

111   See, e.g., Saleam and Moustafa, "The Influence of Divine Rewards and Punishments on Religious Prosociality," *Frontiers in Psychology*, 7 (2016), Purzycki et al., "Moralistic Gods, Supernatural Punishment and the Expansion of Human Sociality," *Nature*, 530 (2016), Johnson, *God Is Watching You: How the Fear of God Makes Us Human* (New York: Oxford University Press, 2015). Shariff and Norenzayan, "Mean Gods Make Good People: Different Views of God Predict Cheating Behavior," *International Journal for the Psychology of Religion* 21, no. 2 (2011); Atkinson and Bourrat, "Beliefs about God, the Afterlife and Morality Support the Role of Supernatural Policing in Human Cooperation," *Evolution and Human Behavior* 32, no. 1 (2011); Gervais and Norenzayan, "Like a Camera in the Sky? Thinking about God Increases Public Self-Awareness and Socially Desirable Responding," *Journal of Experimental Social Psychology* 48, no. 1 (2012); Bourrat, et al., "Supernatural Punishment and Individual Social Compliance across Cultures," *Religion, Brain & Behavior* 1, no. 2 (2011); Johnson, "The Wrath of the Academics: Criticisms, Applications, and Extensions of the Supernatural Punishment Hypothesis," *Religion, Brain & Behavior* 7 (2017); Laurin, "Religion and Its Cultural Evolutionary by-Products," in *The Science of Lay Theories* (Springer, 2017); Xygalatas et al., "Big Gods in Small Places: The Random Allocation Game in Mauritius," *Religion, Brain & Behavior* 7 (2017). Atkinson, "Religion and Expanding the Cooperative Sphere in Kastom and Christian Villages on Tanna, Vanuatu," *Religion, Brain & Behavior* 7 (2017); Nordin, "Indirect Reciprocity and Reputation Management in Religious Morality Relating to Concepts of Supernatural Agents," *Journal For The Cognitive Science Of Religion* 3, no. 2 (2016); Gray and Watts, "Cultural Macroevolution Matters," *Proceedings of the National Academy of Sciences* 114, no. 30 (2017): 7846–7852.

112   See, e.g., Power, "Discerning Devotion: Testing the Signaling Theory of Religion," *Evolution and Human Behavior*, 38 (2017); Bulbulia and Sosis, "Signalling Theory and the Evolution of Religious Cooperation," *Religion* 41, no. 3 (2011); Schell et al., "Religious-Commitment Signaling and Impression Management amongst Pentecostals: Relationships to Salivary Cortisol and Alpha-Amylase," *Journal of Cognition and Culture* 15, no. 3–4

offered philosophical reflections on these theories in the chapters below and elsewhere.[113]

The purpose of this section is to introduce some additional scientific evidence for and theoretical analysis of these and other mechanisms, thereby contributing to the further clarification and fractionation of sociographic prudery. Here too the generativity of the field is evident in the lively debates over the weighting of causal factors and other details within and across the sometimes competitive and often complementary research programs that shape the dialogue. My goal is not to defend a narrow reading of any one of these hypotheses but to point out the plausiblility of the general consensus that naturally evolved biases are at work in producing (and reproducing) the tendency of human individuals to defend the norms and engage in the supernatural ritual practices of their in-group, which under certain conditions also activates anxiety about and even antagonism toward out-group members.

Like religious credulity biases, religious conformity biases are the result of natural selection (or by-products, or exaptations of earlier adaptations). The basic argument here is that such tendencies provided a strategic advantage among early hominids by, for example, enhancing cooperation and commitment within in-groups, which in turn made them more competitive in relation to out-groups. The individuals in such cohesive groups were more likely to survive and pass on their genetically based dispositions to later generations. Exploring and parsing the evidence for this type of claim is part of the ongoing task of fractionating, or reverse engineering, the causal dynamics of religious sects.

Preferring the norms of one's in-group and feeling prejudice toward members of an out-group are not necessarily religious experiences. Discriminating between people on the basis of their participation in supernatural rituals and enforcing segregations based on the failure of some to conform to laws or taboos allegedly revealed through interactions with axiologically relevant disembodied intentional forces *are*. What sets apart "religious" sociography from

---

(2015); Dengah, "Being Part of the Nação: Examining Costly Religious Rituals in a Brazilian Neo-Pentecostal Church," *Ethos* 45, no. 1 (2017). Sexual selection theory can also shed light on some of the sociographically prudish characteristics of religious individuals and groups; see Shaver, "Why and How Do Religious Individuals, and Some Religious Groups, Achieve Higher Relative Fertility?" *Religion, Brain & Behavior* 7, no. 4 (2017), and van Slyke, "Can Sexual Selection Theory Explain the Evolution of Individual and Group-Level Religious Beliefs and Behaviors?" *Religion, Brain & Behavior* 7, no. 4 (2017).

113   See especially Chapters 2 and 3 of *Theology after the Birth of God*, and Chapters 2, 3, 6, 7 and 9 below.

other segregative attitudes and behaviors is that the former is related to super-natural forces putatively engaged by a person's in-group.

> *Sociographic prudery*: the tendency to appeal to the moral normativity of supernatural authorities when trying to act sensibly in society.

Why might this sort of disposition have provided a survival advantage to human beings in an early ancestral environment? Broadly speaking, the mechanisms of the kind we are interested in here can be understood in relation to what we might call risk management theory.

In other words, sociographically prudent tendencies were most likely naturally selected because they worked as good *risk* management strategies. It would have been the capacity to *manage* risks – and not the *taking* of risks per se – that contributed to survival. Life is a risky business, and the ability to decide quickly whether a particular action was safe or beneficial would have aided survival. Our hunter-gatherer ancestors encountered natural hazards, predators, and other subsistence challenges that many readers of this book will never have to face. Nevertheless, our phylogenetic inheritance still includes cognitive and coalitional mechanisms that foster relatively quick judgments about what sorts of risk to take, some of which are correlated with a variety of religious beliefs and behaviors.[114]

As we will see in the chapters below, there is growing evidence that shared imaginative engagement with ambiguously accessible, potentially punitive supernatural agents has played – and continues to play – an important role in fostering behaviors that reinforce in-group cohesion. Such behaviors are initially puzzling because evolutionary theory leads us to expect organisms to act judiciously in their attempts to capture and conserve energy long enough to reproduce. Why then has the unforgiving economy of natural selection rewarded the inefficient and wasteful activities found so prominently in religious sects? This puzzle is resolved when one recognizes that such behaviors are the result of "assurance mechanisms" that contribute to the process of cooperative niche construction in an evolutionary landscape.[115] By conforming to the norms of

114 Nielsen et al., "Risk Aversion and Religious Behaviour: Analysis Using a Sample of Danish Twins," *Economics and Human Biology* 26 (2017). See also Kahan, "Cultural Cognition as a Conception of the Cultural Theory of Risk," in *Handbook of Risk Theory*, ed. Roeser et al. (New York: Springer, 2011).

115 Bulbulia, "Spreading Order: Religion, Cooperative Niche Construction, and Risky Coordination Problems," *Biology & Philosophy* 27, no. 1 (2012): 23. See also Kurzban et al., "The Evolution of Altruism in Humans," *Annual Review of Psychology* 66, no. 1 (2015).

the people around them humans assure one another that they can be relied upon in risky situations.

But what is the *optimal* level of conformity within a group? This depends, of course, on the ecological context in which and to which human organisms are attempting to adapt. Affiliative decisions that lean toward normative conformity in all cases can increase risk for the individual, especially if the social environment turns out to be filled with cheaters, free-loaders, or even hidden enemies. The risk is even higher for infants and small children who do not yet have enough experience to help them make wise decisions about when to conform and with whom to affiliate. This is why human societies have "design features" like the behavioral attachment system that leads children to reach out toward their caregivers when stressed (and leads most caregivers to respond). Elsewhere I have discussed research on attachment theory and religion that shows how this evolved system plays a role in shaping individual's relationships to their imagined divine caregivers.[116] The main point here is that attachment-related mechanisms reinforce the tendency of human beings to seek out and conform to conservative, group-based, religious ideals as a compensatory defense strategy in anxious circumstances.[117]

Survival in ancestral environments would also have required the capacity to make quick decisions about affiliating with potentially contagious individuals who may increase susceptibility to disease. Recent research on the behavioral immune system has revealed a cluster of psychological mechanisms that enhance the capacity to avoid disease. Structural equation modeling based on experimental studies suggests that religious conservativism can be understood as an "evolutionarily evoked disease-avoidance strategy."[118] Evolved tendencies to feel disgust toward certain types of persons – including those with divergent religious beliefs and rituals – can function well as risk aversion strategies. However, they can also engender and exacerbate intergroup conflict and violence within and across religions.[119]

---

116   See *Theology after the Birth of God*, pp. 67–69.

117   McGregor et al., "Reactive Approach Motivation (RAM) for Religion," *Journal of Personality and Social Psychology* 99, no. 1 (2010); McGregor et al., "Approaching Relief: Compensatory Ideals Relieve Threat-Induced Anxiety by Promoting Approach-Motivated States," *Social Cognition* 30, no. 6 (2012).

118   Terrizzi et al., "Religious Conservatism: An Evolutionarily Evoked Disease-Avoidance Strategy," *Religion, Brain & Behavior* 2, no. 2 (2012); Terrizzi et al., "Does the Behavioral Immune System Prepare Females to Be Religiously Conservative and Collectivistic?" *Personality and Social Psychology Bulletin* 40, no. 2 (2014).

119   Fincher and Thornhill, "Parasite-Stress Promotes In-Group Assortative Sociality: The Cases of Strong Family Ties and Heightened Religiosity," *Behavioral and Brain Sciences* 35, no. 2 (2012); Choma et al., "Avoiding Cultural Contamination: Intergroup Disgust

Even in contemporary, pluralistic, scientifically literate contexts many adults continue to follow religious risk management strategies. Why would someone today affiliate – and maintain affiliation – with a religious in-group that requires strict conformity with idiosyncratic supernatural beliefs and causally opaque ritual practices that are emotionally and cognitively expensive and yet provide no obvious survival benefit? A layperson in a religious coalition has to calculate the risk that a ritual officer will fail to provide the supernatural products or services promised. This calculation is rarely explicit; it occurs somewhat automatically through implicit mechanisms that evolved to maintain group cohesion. Analysis of social network externalities implies that interaction with "like-minded and committed adherents reinforces and bolsters the average adherents subjective estimate" of the probability that a faith intermediary will not engage in opportunistic behavior.[120] This perceived reduction of affiliation risk helps to explain why individuals accept the sacrifice and stigma associated with strict religious affiliation and conform to the demands of "faith intermediaries" such as priests.

Participation in religious *rituals* also helps individuals manage affiliational and other risks by fostering in-group conformity. Here too fractionation is important; several mechanisms are at work in the production of "social cohesion" via "ritual." Underlying these commonly used terms "are a diverse set of phenomena including causally opaque conventions, synchrony, dysphoric and euphoric arousal, identity fusion, and group identification."[121] Empirical evidence from ethnographic and experimental research suggests that rituals intensify prosocial behaviors (like cooperation) by orchestrating bodily motions in ways that produce overlapping task-representation, increase perceptions of oneness with fellow ritualists, and amplify commitment to sacred values.[122]

Sensitivity and Religious Identification as Predictors of Interfaith Threat, Faith-Based Policies, and Islamophobia," *Personality and Individual Differences* 95 (2016); Ritter and Preston, "Gross Gods and Icky Atheism: Disgust Responses to Rejected Religious Beliefs," *Journal of Experimental Social Psychology* 47, no. 6 (2011). See also Cohen et al., "Religion, Synchrony, and Cooperation," *Religion, Brain & Behavior* 4, no. 1 (2014).

120   Raynold, "Sacrifice and Stigma: Managing Religious Risk," *Journal for the Scientific Study of Religion* 53, no. 4 (2014): 835.

121   Whitehouse and Lanman, "The Ties That Bind Us," 5.

122   Atran and Henrich, "The Evolution of Religion: How Cognitive By-Products, Adaptive Learning Heuristics, Ritual Displays, and Group Competition Generate Deep Commitments to Prosocial Religion," *Biological Theory* 5 (2010); Fischer et al., "How Do Rituals Affect Cooperation?" *Human Nature* 24, no. 2 (2013); Paul Reddish et al., "Let's Dance Together: Synchrony, Shared Intentionality and Cooperation," *PLoS ONE* 8, no. 8 (2013).

Humans tend to adopt the "ritual stance" when they perceive an action as goal-demoted, opaque and yet deliberate (as opposed to accidental). Once this stance is adopted, conventional and affiliative motivational reasoning shapes normative inferences that inform subsequent social behavior.[123] As components of sociographic prudery, the mechanisms that engender the ritual stance are exceptionally effective. Even the repetition in a laboratory of relatively elaborate novel rituals devoid of historical or religious meaning seems to promote intergroup bias.[124] In the case of high-ordeal rituals, such as those involving body piercing and walking on swords, it seems that simply observing such behaviors promotes conformity to social norms and in-group altruism.[125]

Religiosity is a consistent predictor of in-group favoritism in a wide variety of contexts.[126] Why? Shared imaginative engagement with axiologically relevant supernatural agents (whether animal-spirits, ancestor-ghosts or high gods) is produced by – and contributes to the further reproduction of – evolved cognitive and coalitional biases that helped increase group conformity and collaboration in ancestral environments. Ethnographic analysis of religious totemism suggests that "ancestor manipulation" may have functioned as a mechanism for promoting in-group altruism. Clan totemism can be seen as a "long-term descendant leaving strategy that increased the ability of ancestors to leave descendents because it promoted the identification of codescendants... and encouraged altruism toward these codescendants in many subsequent generations."[127] In several of the following chapters we will further examine the roles that religious affiliation and ritual participation play in promoting coalition-favoring attitudes and behaviors within groups.

Like the error management strategies discussed above, the risk management strategies associated with religion can have negative consequences. As we saw

123   Kapitány and Nielsen, "The Ritual Stance and the Precaution System: The Role of Goal-Demotion and Opacity in Ritual and Everyday Actions," *Religion, Brain & Behavior* (2016), Kapitany and Nielsen, "Adopting the Ritual Stance: The Role of Opacity and Context in Ritual and Everyday Actions," *Cognition* 145 (2015). These articles provide a survey of the recent literature supporting this claim.

124   Hobson et al., "When Novel Rituals Lead to Intergroup Bias: Evidence From Economic Games and Neurophysiology," *Psychological Science* 28, no. 6 (2017).

125   Xygalatas et al., "Extreme Rituals Promote Prosociality," *Psychological Science* 24, no. 8 (2013); Mitkidis et al., "The Effects of Extreme Rituals on Moral Behavior: The Performers-Observers Gap Hypothesis," *Journal of Economic Psychology* 59 (2017).

126   Dunkel and Dutton, "Religiosity as a Predictor of in-Group Favoritism within and between Religious Groups," *Personality and Individual Differences* 98 (2016).

127   Palmer et al., "Totemism and Long-Term Evolutionary Success," *Psychology of Religion and Spirituality* 7, no. 4 (2015): 286.

above, one of the downsides of religious ritual is that it can hinder self-control. Unfortunately, it can also increase antisociality.[128] The flip side of parochial altruism is out-group antagonism. While much of the early research in the bio-cultural study of religion focused on its "prosocial" aspects, scholars are increasingly turning their attention to its anti-social or segregative aspects. Religious prosociality is always "assortative."[129] The symbolic and ideological barriers erected by ritual interaction can improve the morale within a religious sect by minimizing internal conflicts, but they also enforce group boundaries in a way that directs conflict outward toward nonmembers.[130] In other words, the cultivation of conformity to the supernaturally sanctioned norms of a religious in-group also cultivates the sorting of human beings into out-groups. Moreover, just as heightened religious belief can decrease anxiety about making errors, so ritual participation can decrease neural responses to performance failure.[131]

So, the second major hypothesis of theogonic reproduction theory is that some of the mechanisms that engender sociographic prudery are evolved cognitive and coalitional biases that were naturally selected in early ancestral environments because they contributed to the management of affiliative risks. If this hypothesis were true, what sort of predictions would follow? We would expect to find evidence in the data that these god-bearing biases have a *phylogenetically* ancient heritage in the human species, are *ontogenetically* manifested relatively early in human development, and are *contextually* variant in their expression across cultures.

The data do not disappoint. With respect to phylogenetic antiquity, we can begin by noting that the tendency to respond to risk by engaging in spontaneous ritualized behaviors is not unique to humans. It is quite common for all sorts of organisms to return to familiar low-entropy states by engaging in

---

128   Hobson and Inzlicht, "Recognizing Religion's Dark Side: Religious Ritual Increases Antisociality and Hinders Self-Control" 39 (2016).

129   Martin and Wiebe, "Pro-and Assortative-Sociality in the Formation and Maintenance of Religious Groups," *Journal for the Cognitive Science of Religion* 2, no. 1 (2014).

130   Abbink et al., "Parochial Altruism in Inter-Group Conflicts," *Economics Letters* 117, no. 1 (2012); Puurtinen, et al., "The Joint Emergence of Group Competition and within-Group Cooperation," *Evolution and Human Behavior* 36, no. 3 (2015); Draper, "Effervescence and Solidarity in Religious Organizations," *Journal for the Scientific Study of Religion* 53, no. 2 (2014): 233.

131   Hobson et al., "Rituals Decrease the Neural Response to Performance Failure," *PeerJ* 5 (2017).

repetitive action as a way of regaining a sense of control.[132] When humans are confronted by natural disasters, or predation and contagion hazards, normally autonomous behaviors suddenly require cognitive control, which swamps working memory systems. A variety of phylogenetically ancient mechanisms (e.g., threat detection, hazard caution, and security motivation systems) have evolved as ways of reacting to the stress involved in risk management.[133] Of particular importance is the "cognitive coalitional system" which evolved in order to "garner support from conspecifics, organize and maintain alliances, and increase an alliance's chance of success against rival coalitions."[134] Although such systems evolved long before *Homo sapiens*, their expression in human life is uniquely entangled within shared ritual engagement with culturally postulated supernatural agents.[135] The covert operation of these systems helps to explain why church attendance rapidly rises (and then slowly falls) in a human population that is threatened by a major natural hazard like an earthquake.[136]

Second, our evolutionary "reverse-engineering" argument would lead us to expect that some of the inherited biases that promote acceptance of in-group norms and participation in ritualized behaviors would be manifested relatively early in human ontogenetic development. In fact, most children do become sociographically prudish within the first few years of life. Even when they are given explicit perceptual cues about the causal ineffectiveness of a ritual behavior, children's "propensity for overimitation" leads them to more readily adopt what appear to be conventional behaviors.[137] This early-emerging sensitivity

---

132 Lang et al., "Effects of Anxiety on Spontaneous Ritualized Behavior," *Current Biology* 25, no. 14 (2015).

133 Eilam et al., "Threat Detection: Behavioral Practices in Animals and Humans," *Neuroscience & Biobehavioral Reviews* 35, no. 4 (2011); Hinds et al., "The Psychology of Potential Threat: Properties of the Security Motivation System.," *Biological Psychology* 85, no. 2 (2010); Woody and Szechtman, "Adaptation to Potential Threat: The Evolution, Neurobiology, and Psychopathology of the Security Motivation System," *Neuroscience and Biobehavioral Reviews* 35, no. 4 (2011).

134 Boyer et al., "Safety, Threat, and Stress in Intergroup Relations" 10, no. 4 (2015).

135 Liénard and Lawson, "Evoked Culture, Ritualization and Religious Rituals," *Religion* 38, no. March (2008); Fux, "Cultural Transmission of Precautionary Ideas: The Weighted Role of Implicit Motivation," *Journal of Cognition and Culture* 16, no. 5 (2016); Vail, III, et al., "A Terror Management Analysis of the Psychological Functions of Religion," *Personality and Social Psychology Review* 14, no. 1 (2010): 84–94.

136 Sibley and Bulbulia, "Faith after an Earthquake: A Longitudinal Study of Religion and Perceived Health before and after the 2011 Christchurch New Zealand Earthquake" *PLoS ONE* 7, no. 12 (2012).

137 Nielsen et al., "The Perpetuation of Ritualistic Actions as Revealed by Young Children's Transmission of Normative Behavior," *Evolution and Human Behavior* 36, no. 3 (2015).

to ritual is linked to the formation of group boundaries. Participation in novel rituals, even when they have no obvious religious content, increases children's sense of affiliation with their in-group.[138] Already at the age of five, groups of children are able to deal with the free-rider problem and achieve higher-order cooperation by preferring and judging enforcers of conventional norms more positively, and sharing more resources with them than non-enforcers.[139] Like religious adults, religious children are more likely to have an implicit negative bias against religious out-group members; when group differences are large, the bias becomes explicit.[140]

This brings us to our third prediction: the expression of the inherited theistic biases that nurture the tendency to conform to the norms of a religious in-group will *vary across contexts*. This variance can even be found between religious denominations within the same general context.[141] The key point here is that contextual conditions that affect the social and environmental risk faced by a human population can intensify or dampen sociographically prudish tendencies. In fact, there seems to be a strong correlation between the prevalence of religiosity and the lack of "existential security" in a population.[142] As with all of the factors we have been discussing so far, the strength and the mode of the association between insecurity and religiosity varies across contexts as well as individuals.

However, research in this field quite consistently shows that high religiosity is typically correlated with a wide variety of kinds of insecurity (existential and economic, past and present, individual and contextual).[143] For example,

---

138   Wen et al., "Ritual Increases Children's Affiliation with in-Group Members," *Evolution and Human Behavior* 37, no. 1 (2016).

139   Vaish et al., "Preschoolers Value Those Who Sanction Non-Cooperators," *Cognition* 153 (2016).

140   Heiphetz et al., "Patterns of Implicit and Explicit Attitudes in Children and Adults: Tests in the Domain of Religion," 142, no. 3 (2013).

141   van Elk et al., "Why Are Protestants More Prosocial Than Catholics? A Comparative Study Among Orthodox Dutch Believers," *International Journal for the Psychology of Religion* 27, no. 1 (2017).

142   Norris and Inglehart, "Are High Levels of Existential Security Conducive to Secularization? A Response to Our Critics," in Brunn (ed), *The Changing World Religion Map* (Springer, 2015). See also Norris and Inglehart, *Sacred and Secular: Religion and Politics Worldwide* (Cambridge University Press, 2011); Inglehart and Welzel, *Modernization, Cultural Change, and Democracy: The Human Development Sequence* (Cambridge University Press, 2005).

143   Immerzeel and van Tubergen, "Religion as Reassurance? Testing the Insecurity Theory in 26 European Countries," *European Sociological Review* 29, no. 2 (2013); Healy and Breen, "Religiosity in Times of Insecurity: An Analysis of Irish, Spanish and Portuguese European Social Survey Data, 2002–12," *Irish Journal of Sociology* 22, no. 2 (2014).

a recent factor analysis of global data showed that harsh and demanding environments are correlated with high levels of religiosity, and the dominance of religiously sanctioned mate-guarding strategies.[144] Another cross-national bioclimatic data analysis found that beliefs in (punitive) moralizing high gods were "more prevalent among societies that inhabit poorer environments and are more prone to ecological duress."[145] Similar findings emerge in small-scale, local ethnographic analyses. Psychological experiments in Fiji, for example, which utilized an economic game designed to measure in-group favoritism, found that when material insecurity is high, belief in "a punitive supernatural agent focused on the ingroup may *amplify* local favoritism, to the *detriment of outsiders*."[146]

It makes sense that scarcity of resources would predict the increase in a population's religiosity, because the latter can enhance a group's cohesion and competitiveness with other groups. Insofar as religious sects operate on the basis of imagined (and allegedly scarce) resources such as the good-will of a god, however, they (re)produce the sort of anxiety that strengthens antagonism toward out-group members.[147] This also helps to explain why (generally speaking) religious individuals tend to be less risk-tolerant than atheists, even in relatively healthy socioeconomic environments.[148] An analysis of all five waves of the World Values Survey suggests that this holds in most contexts: after controlling for standard sociodemographic variables (including country), greater religiosity was found to be almost uniformly and very significantly associated with less favorable views of innovation.[149]

---

144   Pazhoohi et al., "Religious Veiling as a Mate-Guarding Strategy: Effects of Environmental Pressures on Cultural Practices," *Evolutionary Psychological Science* 3, no. 2 (2017).

145   Botero et al., "The Ecology of Religious Beliefs," *Proceedings of the National Academy of Sciences* 111, no. 47 (2014).

146   McNamara et al., "Supernatural Punishment, in-Group Biases, and Material Insecurity: Experiments and Ethnography from Yasawa, Fiji," *Religion, Brain & Behavior* (2014): 17. Emphasis added.

147   Avalos, "Religion and Scarcity: A New Theory for the Role of Religion in Violence," in *The Oxford Handbook of Religion and Violence*, ed. Juergensmeyer, et al., (Oxford: Oxford University Press, 2013).

148   Bartke and Schwarze, "Risk-Averse by Nation or by Religion? Some Insights on the Determinants of Individual Risk Attitudes," *SOEPpaper*, no. 131 (2008). See also Coccia, "Socio-Cultural Origins of the Patterns of Technological Innovation: What Is the Likely Interaction among Religious Culture, Religious Plurality and Innovation? Towards a Theory of Socio-Cultural Drivers of the Patterns of Technological Innovation," *Technology in Society* 36 (2014): 13.

149   Bénabou et al., "Religion and Innovation," *The American Economic Review* 105, no. 5 (2015).

One of the major factors shaping our contemporary social environment is globalization. At least two mechanisms are at work in the impact of globalization on sociographic prudery – and they work at cross purposes. On the one hand, cross-national analysis suggests that globalization provides new opportunities for social contact (even if only through the Internet), which can facilitate new favorable attitudes toward members of other religious groups. On the other hand, globalization "increases the salience of religious identity, which may pit social groups against each other to generate unfavorable views of the religious other."[150] As I argue in later chapters, dealing with this dilemma is one of the greatest challenges facing our species.

The way in which contextual variance influences the expression of sociographically prudish theistic biases is also evident at the national level. Analysis of the relationship between quantifiable social health indicators and popular religiosity across nations indicates that highly secular democracies have the lowest rates of societal dysfunction. The data suggest both that dysfunctional socioeconomic conditions favor mass religiosity and that conservative religious ideology contributes to societal dysfunction (measured by factors such homicide and poverty rates).[151] Among the dysfunctional factors upon which high levels of religious socialization seem to depend is a high level of income inequality. The latter provides people with an incentive to "invest in relational social capital that might be provided by religious organizations." Moreover, when resources are unequally distributed across a society, the "signaling of trustworthiness or religious conformity might prove more important" than in contexts with more fair distributions.[152]

Another contextual factor that affects the variation in levels of religious affiliation and conformity across societies is the extent to which individuals in the population are confronted by "credibility enhancing displays" (CREDS) such as costly participation in ritual behaviors. It appears that watching respected others within one's in-group regularly engage in this sort of CRED is one of the most powerful predictors of commitment to religious belief and practice later in life.[153] A recent study comparing the Czech Republic and Slovakia found that

---

150   Ciftci et al., "Globalization, Contact, and Religious Identity: A Cross-National Analysis of Interreligious Favorability," *Social Science Quarterly* 97, no. 2 (2016).

151   Paul, "The Chronic Dependence of Popular Religiosity upon Dysfunctional Psychosociological Conditions," *Evolutionary Psychology* 7, no. 3 (2009): 398.

152   Müller, et al., "Which Societies Provide a Strong Religious Socialization Context? Explanations Beyond the Effects of National Religiosity," *Journal for the Scientific Study of Religion* 53, no. 4 (2014): 739.

153   Lanman and Buhrmester, "Religious Actions Speak Louder than Words: Exposure to Credibility-Enhancing Displays Predicts Theism," *Religion, Brain & Behavior* 7, no. 1 (2015);

70% of the difference in belief in God and 80% of the difference in religious practice between these countries was mediated by exposure to CREDs and church attendance in childhood.[154] In other words, the extent to which one's environment offers intensive and extensive displays of sociographically prudish behaviors plays a key role in shaping the cutural variance of religious variables across populations.

High CRED contexts may enhance affiliation and commitment within supernatural groups but, as we have seen, religious variables are globally associated with a host of other problematic variables such as existential insecurity, societal dysfunction, and ecological duress. Unfortunately, they are also associated with intergroup conflict and violence. Data analysis provides strong and consistent evidence that religion really can make things worse. Conflicts in which at least one side explicitly makes claims rooted in its religious tradition are "significantly less likely than others to be terminated through negotiated settlement."[155] When the stakes of a conflict, the social production of "hypercommitted selves," the mobilization of justifications and rewards, and the perceptions on either side of the "right" social order are distinctively religious, this really can "authorize, legitimate, enable, and even require violent action in the face of urgent threats, profanations of sacred symbols, and extreme otherhood."[156]

Evidence that religious bonds can be even more important than ethnic bonds for shaping in-group and out-group bias[157] might help to explain why religious identity seems to play such a powerful role in escalating intergroup conflict. Regression analysis of a dataset covering 130 developing countries over a twenty year period showed that the overlap of religious identity and other (e.g., ethnic) identities made it six times more likely that armed intergroup conflict would occur.[158] This does not mean that religious affiliation

---

Gervais and Najle, "Learned Faith: The Influences of Evolved Cultural Learning Mechanisms on Belief in Gods," *Psychology of Religion and Spirituality* 7, no. 4 (2015): Henrich, "The Evolution of Costly Displays, Cooperation and Religion: Credibility Enhancing Displays and Their Implications for Cultural Evolution," *Evolution and Human Behavior* 30, no. 4 (2009).

154   Willard and Cingl, "Testing Theories of Secularization and Religious Belief in the Czech Republic and Slovakia," *Evolution and Human Behavior* 38, no. 5 (2017).

155   Svensson, "Fighting with Faith," *Journal of Conflict Resolution* 51, no. 6 (2007): 930.

156   Brubaker, "Religious Dimensions of Political Conflict and Violence," *Sociological Theory* 33, no. 1 (2015): 12.

157   Willard, "Religion and Prosocial Behavior among the Indo-Fijians," *Religion, Brain & Behavior* 7 (2017).

158   Basedau et al., "Bad Religion? Religion, Collective Action, and the Onset of Armed Conflict in Developing Countries" *Journal of Conflict Resolution* 60, no. 2 (2016): 226.

necessarily causes violence in every context, or even that religious violence
always looks the same in every context. Expressions of religious extremism
and violence in intergroup conflict can be conceptualized as examples of
"niche construction," the result of "longtime interactions between natural and
cultural environments and human animals."[159]

It is important to emphasize once again that the problem is not simply with
fundamentalists in stressful situations, although that combination is especially problematic. An experimental study using a prisoner's dilemma game
found that fundamentalism in particular was a stronger predictor of out-
group prejudice, but religiosity *in general* also amplified group biases.[160] Other
studies have shown that religiosity is predictive not only of actual prosocial
behavior toward in-groups but also of aggressive *antisocial behavior* toward
out-groups. For example, laboratory experiments have found that generally
religious (and not merely fundamentalist) participants are more likely to engage in overt and direct aggression toward value-threatening out-group targets
(in comparison to neutral targets). This unfortunate in-group/out-group paradox "can be identified at the *heart of religion* rather than only in its margins
(fundamentalism)."[161]

It is also important to differentiate between extremist religious ideology and
extremist religious behaviors. A host of dispositional and situational factors
play a role in pushing (or pulling) a person with strong religious beliefs across
the threshold of commiting violent actions.[162] Research in Ghana found that
the intensity and frequency of incidences of radicalization and violence were
influenced by a variety of factors such as the presence of a demographic youth
bulge, external financial support, and preaching methodologies. However, the
most important driver in that context was "first and foremost a struggle among
the dominant religious groups for doctrinal pre-eminence."[163]

The drivers of religious intergroup conflict are indeed extremely complex.
Elsewhere I have identified some of the most relevant micro-, meso- and

---

159 Saroglou, "Intergroup Conflict, Religious Fundamentalism, and Culture," *Journal of Cross-Cultural Psychology* 47, no. 1 (2016): 33.

160 Chuah et al., "Religion, Ethnicity and Cooperation: An Experimental Study," *Journal of Economic Psychology* 45 (2014): 42.

161 Blogowska et al., "Religious Prosociality and Aggression: It's Real," *Journal for the Scientific Study of Religion* 52, no. 3 (2013): 534. Emphasis added.

162 Borum, "Radicalization into Violent Extremism I: A Review of Social Science Theories," *Journal of Strategic Security* 4, no. 4 (2011); McCauley and Moskalenko, "Understanding Political Radicalization: The Two-Pyramids Model," *American Psychologist* 72, no. 3 (2017).

163 Aning and Abdallah, "Islamic Radicalisation and Violence in Ghana," *Conflict, Security & Development* 13, no. 2 (2013) 163.

macro-level factors at work in this phenomenon and described ways in which computer modeling and simulation techniques can make this complexity more tractable.[164] We will explore the potential of some of these methodologies in more detail at the end of Chapter 12. At this stage, however, the important point is that religiosity does in fact seem *endogenous* to the level of violence in many contexts. For example, research on the Palestinian conflict suggests that politically motivated violence leads both Jews and Muslims in the region to identify more strongly as religious, *and* that this increased religiosity itself heightens the hostility between the groups.[165] Another recent study suggests that the relationship between war and religiousness is *bidirectional*: "war strengthens individual's religiousness through worries about war, while fundamentalist religious beliefs result in violent conflicts and war."[166]

This is only one example of the way in which the mechanisms of anthropomorphic promiscuity and sociographic prudery can amplify one another within the complex adaptive systems of religious reproduction.

The Reciprocal Reinforcement of Theogonic Mechanisms

Another secret to the success of god-bearing social assemblages is the boosting effect that their component parts can have on one another. This dynamic interaction among variables within complex religious systems can enhance the stability and transmissibility of in-group beliefs and behaviors, which helps to explain why so many human minds across human cultures keep on engaging in religious sects. In later chapters I provide more detailed arguments for my claim that fueling anthropomorphic promiscuity and sociographic prudery is no longer adaptive in our current, globally interconnected, ecologically fragile environment. In this section I introduce some more evidence for the third (and perhaps most important) major hypothesis of theogonic reproduction theory: under a wide variety of conditions the covert operations of theistic credulity and conformity biases are *reciprocally reinforcing*.

If this scientific hypothesis were correct, what would we expect to find in the data? Claims about the mutual fortification of some of the fractionated component mechanisms that contribute to anthropomorphic promiscuity

---

164   Shults, "Can We Predict and Prevent Religious Radicalization?"
165   Zussman, "The Effect of Political Violence on Religiosity," *Journal of Economic Behavior and Organization* 104 (2014): 76.
166   Du and Chi, "War, Worries, and Religiousness," *Social Psychological and Personality Science* 7, no. 5 (2016): 449.

and sociographic prudery would be corroborated by the discovery of statistical patterns of *correlation* between various aspects of these theogonic biases, by empirical evidence of the mutual *amplification* of these mechanisms under certain conditions, and by the detection of both individual and contextual *variation* in their reciprocal consolidation.[167] A brief review of some of the relevant research literature provides ample evidence for corroborating this third hypothesis.

First, survey analyses consistently find strong statistical correlations between dimensions of religious belief and religious parochialism. For example, one recent study found that variables related to belief in God were most strongly associated with variables like "ingroup parochialism" and fewer interactions with individuals from other religious groups.[168] Multivariate analysis of another survey data set found that the stability of belief in a supernatural creator was predicted by respondents' embeddedness within a "social network of co-religionists."[169] The strength of the correlation between believing in supernatural agents and feeling (or acting) prejudicially toward out-groups can vary based on individual personality differences such as right wing authoritarianism (RWA) or social dominance orientation (SDO), as well as on socio-economic contextual differences such as the proportion of the adherents of a religious majority in a population or country GDP.[170]

---

167   In the previous two sections, I emphasized the *individual* variance of anthropomorphic promiscuity distributed across human populations and the *contextual* variance of sociographic prudery under different environmental conditions. As should become increasingly clear in this section and in Chapter 12, however, both sorts of theogonic mechanism are susceptible to both sorts of variance.

168   Galen et al., "Nonreligious Group Factors Versus Religious Belief in the Prediction of Prosociality," *Social Indicators Research* 122, no. 2 (2015): 411.

169   Hill, "Rejecting Evolution: The Role of Religion, Education, and Social Networks," *Journal for the Scientific Study of Religion* 53, no. 3 (2014): 575. Belief in *evil* supernatual agents is also strongly correlated to parochial religious prosociality; see Martinez, "Is Evil Good for Religion? The Link between Supernatural Evil and Religious Commitment," *Review of Religious Research* 55, no. 2 (2013).

170   Mavor et al., "Religion, Prejudice, and Authoritarianism: Is RWA a Boon or Bane to the Psychology of Religion?" *Journal for the Scientific Study of Religion* 50, no. 1 (2011); Meeusen et al., "Generalized and Specific Components of Prejudice: The Decomposition of Intergroup Context Effects: Intergroup Context, Generalized and Specific Components of Prejudice," *European Journal of Social Psychology* 47, no. 4 (2017); Kanas et al., "Religious Identification and Interreligious Contact in Indonesia and the Philippines: Testing the Mediating Roles of Perceived Group Threat and Social Dominance Orientation and the Moderating Role of Context," *European Journal of Social Psychology* 46, no. 6 (2016); Bohman and Hjerm, "How the Religious Context Affects the Relationship between Religiosity and Attitudes towards Immigration," *Ethnic and Racial Studies* 37, no. 6 (2014); Doebler, "Relationships Between Religion and Intolerance Towards Muslims and Immigrants

There is no doubt that both dispositional and situational variances must always be taken into account. However, a larger survey analysis using cross-national data sets, more representative samples, and advanced statistical techniques demonstrated that the correlation between religiosity in general and prejudice at the *population* level is consistently high; when statistically controlling for factors like RWA and SDO, religiosity clearly predicts levels of prejudice toward specific target groups such as homosexuals.[171] For decades social psychologists have noted the apparently paradoxical way in which religion so often "makes" prejudice but also sometimes seems to "unmake" it.[172] So much depends on precisely how "religion" is being defined and which variables are being measured.

Not surprisingly, survey analyses typically find that fundamentalist belief in God is particularly highly correlated with intolerance toward value-violating out-groups, especially racial and religious others, as well as to an "irrational need for comfort and having derogatory thoughts about other people."[173] One study found, as expected, that higher commitment to religious belief amplified the effect of the correlation between perceived dissimilarity in belief and "greater religious intergroup bias," increasing the extent to which the former predicts the latter.[174] Another cross-cultural survey analysis with even larger samples found that extreme fundamentalism was highly correlated with active

in Europe: A Multilevel Analysis," *Review of Religious Research* 56, no. 1 (2014); Doebler, "Love Thy Neighbor? Relationships between Religion and Racial Intolerance in Europe" *Politics and Religion* 8, no. 4 (2015).

171   Newheiser et al., "Social-Psychological Aspects of Religion and Prejudice: Evidence from Survey and Experimental Research," in Clarke, et al., *Religion, Intolerance, and Conflict: A Scientific and Conceptual Investigation* (Oxford University Press, 2013). For a discussion of the ways in which RWA and SDO can lead to different sorts of religious prejudice based on distinct underlying psychological mechanisms, see Ng and Gervais, "Religion and Prejudice," in *The Cambridge Handbook of the Psychology of Prejudice*, ed. Sibley and Barlow (Cambridge: Cambridge University Press, 2017).

172   For an introduction to earlier literature on this theme, see Rock, "Introduction: Religion, Prejudice and Conflict in the Modern World," *Patterns of Prejudice* 38, no. 2 (2004), and Hunsberger and Jackson, "Religion, Meaning, and Prejudice," *Journal of Social Issues* 61, no. 4 (2005); see also Shaver et al., "Religion and the Unmaking of Prejudice toward Muslims: Evidence from a Large National Sample" *PLoS ONE* 11, no. 3 (2016).

173   Mora et al., "Religious Fundamentalism and Religious Orientation Among the Greek Orthodox," *Journal of Religion and Health* 53, no. 5 (2014): 1510. See also Kossowska and Sekerdej, "Searching for Certainty: Religious Beliefs and Intolerance toward Value-Violating Groups," *Personality and Individual Differences* 83 (2015).

174   Maxwell-Smith et al., "Individual Differences in Commitment to Value-Based Beliefs and the Amplification of Perceived Belief Dissimilarity Effects," *Journal of Personality* 83, no. 2 (2014): 137.

participation in religious services and the extent to which participant's indicated their willingness to obey their own religion's sacred texts (which include passages that encourage killing out-group members).[175]

Complex analyses of variance in another survey involving American university students found that those who scored more highly in fundamentalism and dogmatism were more likely to attribute causality to supernatural agents when faced with ambiguous vignettes. However, the study also found that participants with greater *intrinsic* religious orientation and those who viewed God as more *loving* were also more likely to explain outcomes by appealing to supernatural causation, especially when the outcomes of the vignettes were considered positive.[176] A structural equation modeling analysis of survey data from 194 groups around the world found that religious infusion was a significant and independent predictor of increased prejudice and intergroup conflict. That is to say, the extent to which ideas about supernatural agents pervade private and public life within a population was related to the extent to which those groups were "especially prejudiced against those groups that held incompatible values," and "likely to discriminate against such groups."[177]

In other words, theistic credulity biases and theistic conformity biases are quite often highly correlated within human populations. But can we say anything about the *causal* links between these variables? Do we have evidence that the mechanisms of anthropomorphic promiscuity and sociographic prudery can actually *amplify* one another? Earlier in this chapter I claimed that the potential *consequences* of engaging in religious sects can include an inceased susceptibility to superstitious beliefs and an increased proclivity toward segregative behaviors, each of which in turn can causally increment the other. We have already noted several studies that utilize structural equation modeling, which is at least suggestive of causality.

However, one of the best ways to render claims about causality more plausible is through scientific measurement methodologies that use some kind of *manipulation* in the context of controlled experimentation. Altering some component of anthropomorphic promiscuity through "priming" or other

175   Ellis, "Religious Variations in Fundamentalism in Malaysia and the United States: Possible Relevance to Religiously Motivated Violence," *Personality and Individual Differences* 107 (2017); for a discussion of the role of sacred texts in promoting violence, see Nelson-Pallmeyer, *Is Religion Killing Us? Violence in the Bible And the Quran* (New York: Continuum, 2005).

176   Vonk and Pitzen, "Religiosity and the Formulation of Causal Attributions," *Thinking & Reasoning* 22, no. 2 (2016): 138.

177   Neuberg et al., "Religion and Intergroup Conflict," *Psychological Science* 25, no. 1 (2014): 198.

experimental techniques utilized in neuroscientific, psychological, ethno-graphic, and economic game theoretic research in order to determine what effect this has on some component of sociographic prudery (or vice versa) can shed light on the extent to which – and the ways in which – these theogonic mechanisms are reciprocally reinforcing.

A wide variety of priming studies have shown that implictly or explicitly triggering religious ideas or beliefs about supernatural agents can increase in-group religious prosociality and out-group prejudice. And vice versa: triggering anxiety about out-groups can increase belief in gods or other misattributions of intentionality. For example, in one study participants subliminally primed with Christian concepts (like gospel, Bible, and Jesus) displayed both more co-vert and overt racial prejudice toward African-Americans than those that re-ceived neutral primes.[178] In another series of studies, researchers reported that activating Christian concepts in laboratory experiments (as well as exposure to real-life Christian contexts such as a cathedral) causally increases participants intolerance of ambiguity and certainty about their own judgments.[179]

The in-group protecting and prejudice-enhancing effects of religious primes seem to be operative cross-culturally. In a Singaporean study, both Christians and Buddhists became increasingly prejudicial toward out-groups when sub-liminally primed with in-group religious terms.[180] An experimental study in Mauritius, which utilized contextual, ecologically relevant primes (religious locations with images of supernatural agents vs. non-religious locations) found that participants in the religious priming condition were more generous toward anonymous members of their own community in a post-experimental charity task.[181] A multi-national contextual priming study in western Europe discovered that participants in a religious context reported significantly more negative attitudes toward out-groups as well as higher levels of religiosity.[182]

178 Johnson et al.,"Priming Christian Religious Concepts Increases Racial Prejudice," *Social Psychological and Personality Science* 1, no. 2 (2010).

179 Sagioglou and Forstmann, "Activating Christian Religious Concepts Increases Intolerance of Ambiguity and Judgment Certainty," *Journal of Experimental Social Psychology* 49, no. 5 (2013): 933.

180 Ramsay et al., "Rethinking Value Violation: Priming Religion Increases Prejudice in Singaporean Christians and Buddhists," *International Journal for the Psychology of Religion* 24, no. 1 (2014).

181 Xygalatas et al., "Location, Location, Location: Effects of Cross-Religious Primes on Proso-cial Behavior" *The International Journal for the Psychology of Religion* 26, no. 4 (2016).

182 Labouff et al., "Differences in Attitudes Toward Outgroups in Religious and Nonreligious Contexts in a Multinational Sample: A Situational Context Priming Study," *International Journal for the Psychology of Religion* 22, no. 1 (2012).

The causal relation between religious ideation and in-group prosocial behavior (or out-group prejudice) has also been demonstrated in priming experiments using images, symbols, and music rather than words or places.[183] Other studies have explored the effectiveness of holy texts as religious primes. For example, one such study hypothesized that reading biblical passages in which God sanctions violence would lead participants to behave more aggressively under experimental conditions. The authors concluded that reading violence-justifying texts did indeed cause increased aggression, and more so with religious believers who attributed the words to a supernatural source.[184] A meta-analysis of 93 priming studies showed that religious priming has robust effects on a variety of measures, especially in-group prosociality. From the fact that non-religious participants were not reliably affected by such primes, the authors concluded that the effect of the latter is dependent on "the cognitive activation of culturally transmitted religious beliefs."[185]

Other sorts of psychological experiments that manipulate religious beliefs or behaviors also shed light on the reciprocal causality at work among the mechanisms of anthropomorphic promiscuity and sociographic prudery. For example, several recent economic game experiments have revealed more of the "dark side" of ritual. In one study participants who engaged in mock rituals increased their out-group discrimination (compared to those in the no-ritual condition). The authors concluded that "while rituals galvanize groups by *binding* us to the ingroup, they do so at a cost of *blinding* us towards those who don't belong."[186] In another series of five laboratory experiments,

183   Cavrak and Kleider-Offutt, "Pictures Are Worth a Thousand Words and a Moral Decision or Two: Religious Symbols Prime Moral Judgments," *The International Journal for the Psychology of Religion* 25, no. 3 (2015); Lang et al., "Music As a Sacred Cue? Effects of Religious Music on Moral Behavior," *Frontiers in Psychology* 7 (2016): 814.

184   Bushman et al., "When God Sanctions Killing: Effect of Scriptural Violence on Aggression," *Psychological Science* 18, no. 3 (2007).

185   Shariff et al., "Religious Priming: A Meta-Analysis with a Focus on Prosociality" *Personality and Social Psychology Review* 20, no. 1 (2016). For a critical discussion of the importance of experimental replication in such studies, see van Elk et al., "Meta-Analyses Are No Substitute for Registered Replications: A Skeptical Perspective on Religious Priming," *Frontiers In Psychology* 6 (2015).

186   Hobson et al., "Mock Ritual Leads to Intergroup Biases in Behavior and Neurophysiology," in *Annual Meeting of the Association for Psychological Science, New York, NY*, (2015) 1. Emphasis added. See also Hobson and Inzlicht, "Recognizing Religion's Dark Side: Religious Ritual Increases Antisociality and Hinders Self-Control" *Behavioral and Brain Sciences* 39 (2016): 30.

participants with stronger religious worldviews showed higher levels of religious prejudice, derogation of other religious viewpoints, and even support for aggression toward members of religious out-groups. More importantly, mediational analyses revealed that those religious participants "expressed heightened prejudice *because* of the worldview threat posed by religious out-group members."[187]

The activation of religiously relevant behaviors can also causally increase religiously relevant beliefs. In a betting game experiment, for example, researchers showed that when a confederate performed a credibility-enhancing display by betting money on the truth of a story with minimally counterintuitive content (which are characteristic of religious narratives), participants were nearly seven times more likely to bet on that same story.[188] This indicates that CREDs play a causal role in promoting belief in supernatural agents and concepts. Another set of experiments demonstrated that participants who received induced-compliance dissonance manipulations were not only more likely to punish norm violators, but also to espouse a stronger belief in God.[189]

Uncertainty seems to play a key role in the triggering of some of these mechanisms. For example, in an economic game experiment carried out in Fiji, researchers found that uncertainty about material resources moderated beliefs in supernatural agents (as well as beliefs that secular agents like police would be more punitive). As expected, local favoritism increased along with uncertainty. Based on the effect that belief in punitive local spirits played in ratcheting up this effect, the researchers concluded that seeing one's resources as "too uncertain might lead to supernatural agent beliefs that further bolster preference for less-risky investment in locals."[190] Other studies have shown that self-report in religiosity goes up in response to manipulated experiences of uncertainty. It seems that many human beings have a tendency to protect themselves from feelings about uncertainty by

187   Goplen and Plant, "A Religious Worldview Protecting One's Meaning System Through Religious Prejudice," *Personality and Social Psychology Bulletin* 41, no. 11 (2015). Emphasis in original.

188   Willard et al., "Memory and Belief in the Transmission of Counterintuitive Content," *Human Nature* 27, no. 3 (2016).

189   Daniel Randles et al., "Is Dissonance Reduction a Special Case of Fluid Compensation? Evidence That Dissonant Cognitions Cause Compensatory Affirmation and Abstraction," *Journal of Personality and Social Psychology* 108, no. 5 (2015).

190   McNamara and Henrich, "Jesus vs. the Ancestors: How Specific Religious Beliefs Shape Prosociality on Yasawa Island, Fiji," *Religion, Brain & Behavior*, (2017): 16.

embracing unfalsifiable beliefs in supernatural beings that are watching over their group.[191]

As we might expect, priming experiments also show that religious biases related to *error* management can intensify biases related to *risk* management and vice versa. We have already seen several examples of research studies demonstrating that individuals who affiliate with religious groups (a strategy for managing risk) are statistically more likely to make errors on a variety of tests. One study found that the nature of such errors can also be distinctive. When subliminally primed with religious terms (e.g., messiah, Christ, prayer), highly religious individuals reported more false perceptions of a religious type. The authors argued that individual variance in the idiosyncratic content of hallucinations can be the result of a mechanism by which "context affects the content of false perceptions through the activation of stored beliefs and values."[192] Another study found that individuals with more religious supernatural beliefs are more susceptible to religious cues when they are trying to manage their affiliation risks.[193]

In our discussion of some of the components of sociographic prudery above, we noted that religious biases seem to have evolved alongside a complex set of risk management mechanisms. Under a variety of conditions, religious individuals tend to be more risk averse. However, there are some conditions under which religious variables may increase people's willingness to take risks. For example, the priming of religious imagery, when coupled with a confederate's encouragement to engage in risky behavior (in a laboratory risk task), actually increased participants' risk-taking behaviors in one study.[194] Another study found that a willingness to engage in certain kinds of risks (in nonmoral domains) was enhanced among participants whose

191   Wichman, "Uncertainty and Religious Reactivity: Uncertainty Compensation, Repair, and Inoculation," *European Journal of Social Psychology* 40, no. 1 (2010); Hogg, et al., "Religion in the Face of Uncertainty: An Uncertainty-Identity Theory Account of Religiousness," *Personality and Social Psychology Review* 14, no. 1 (2010). See also Fergus and Rowatt, "Uncertainty, God, and Scrupulosity: Uncertainty Salience and Priming God Concepts Interact to Cause Greater Fears of Sin," *Journal of Behavior Therapy and Experimental Psychiatry* 46 (2015).

192   Reed and Clarke, "Effect of Religious Context on the Content of Visual Hallucinations in Individuals High in Religiosity." *Psychiatry Research* 215, no. 3 (2014): 594.

193   Periss and Bjorklund, "Playing for God's Team: The Influence of Belief in the Supernatural on Perceptions of Religious, Spiritual, and Natural Cues," *Journal of Cognition and Culture* 16, no. 3–4 (2016).

194   Shenberger et al., "The Effect of Religious Imagery in a Risk-Taking Paradigm," *Peace and Conflict: Journal of Peace Psychology* 20, no. 2 (2014).

scrambled-sentence tasks included words conceptually related to God (e.g., spirit, divine).[195] There is some debate among scholars over the extent to which priming different sorts of religious concepts (e.g., God, religious leaders, religious institutions, religious ideas) have divergent effects.[196] If being primed to think of "God," as opposed to "religion" in general, had the effect of stretching parochial prosocial attitudes or behaviors toward out-groups, this could be explained in terms of impression management vis-à-vis an imagined divine attachment figure.

Another way of expressing this third major hypothesis of theogonic reproduction theory is by claiming that *compliance* with supernaturally authorized norms and *reliance* on supernatural explanations are reciprocally reinforcing. Participants in one experiment became more compliant with confederate leaders (especially charismatic leaders) after they were primed with religion, which made them more susceptible to changing their attitude toward funding green energy initiatives.[197] In another study, feelings of personal authorship (or agency) in relation to a lexical task were decreased for individuals who received subliminal "God" primes (as opposed to "self" and "computer" primes) – but *only* for those who already believed in God.[198] The sense of authorship for non-believers did not differ between primes. In other words, for religious individuals, thinking about God may lower their sense of self-agency. Another experiment revealed that simply priming typically religious contractive postures such as praying and kneeling seems to increase expressed agreement with conventional religious beliefs.[199]

195   Kupor et al., "Anticipating Divine Protection? Reminders of God Can Increase Nonmoral Risk Taking," *Psychological Science* 26, no. 4 (2015): 381–382.

196   Preston and Ritter, "Different Effects of Religion and God on Prosociality With the In-group and Outgroup," *Personality and Social Psychology Bulletin* 39, no. 11 (2013), found that "religion" primes increased prosociality toward in-groups while "God" primes could enhance out-group prosociality, which they interpreted in light of moral impression management theory. However, more recent studies have not been able to replicate these effects; see Ramsay et al., "A Puzzle Unsolved: Failure to Observe Different Effects of God and Religion Primes on Intergroup Attitudes," *PLoS ONE* 11, no. 1 (2016), and Batara et al., "Effects of Religious Priming Concepts on Prosocial Behavior Towards Ingroup and Outgroup," *Europe's Journal of Psychology* 12, no. 4 (2016).

197   Smith and Zárate, "The Effects of Religious Priming and Persuasion Style on Decision-Making in a Resource Allocation Task," *Peace and Conflict: Journal of Peace Psychology* 21, no. 4 (2015).

198   Dijksterhuis et al., "Effects of Subliminal Priming of Self and God on Self-Attribution of Authorship for Events," *Journal of Experimental Social Psychology* 44 (2008).

199   Fuller and De Montgomery, "Body Posture and Religious Attitudes," *Archive for the Psychology of Religion-Archiv Fur Religionspsychologie* 37, no. 3 (2015).

This research suggests that people who typically rely on appeals to a supernatural agent to make sense of their world are more likely to comply with supernatural authorities when primed to think about God. Neuroscientific studies indicate that the causality goes the other direction as well. For example, fMRI studies have shown that religious believers, unlike their secular peers, deactivate their frontal network (including the medial and dorsolateral prefrontal cortex, areas known to be operative in critical, executive thinking) when listening to speakers with allegedly miraculous abilities. In other words, they "hand over" their executive functioning to a person they perceive to be a charismatic religious leader.[200] Understanding the complex bio-cultural systems that foster this sort of compliance requires attention to the "reciprocal causality" or "bi-directional interaction" between neurophysiological mechanisms involved in rhythmic entrainment processes (susceptibility to which varies among individuals) and situational properties associated with ritual interaction.[201]

The natural selection of the mechanisms that contributed to the "large-scale hypnotic effects" of religious rituals may have been linked to the benefits of "large-scale cooperative interactions," but charismatic religious leaders can easily take advantage of such hypnotic cultures.[202] Moreover, research on autobiographical and episodic memory and narrative processing suggests that the emotional nature of ritual experiences "typically leads to memories that are high in vividness and confidence but *low in accuracy*... [and] reconstructed in a way that allows the bearer to maintain a consistent identity and life story, and to fit his/her *current* goals and knowledge."[203] This sort of proclivity toward confirmation bias and mistaken memory among those who participate in religious sects would make sense if reduced prediction error monitoring in relation to both interoceptive and exteroceptive events was one of the neurological bases of religious beliefs and experiences.[204]

200   Schjoedt et al., "The Power of Charisma – Perceived Charisma Inhibits the Frontal Executive Network of Believers in Intercessory Prayer," *Social Cognitive and Affective Neuroscience* 6, no. 1 (2011): 119; see also Schjoedt et al., "Cognitive Resource Depletion in Religious Interactions," *Religion, Brain & Behavior* 3, no. 1 (2013).

201   Heinskou and Liebst, "On the Elementary Neural Forms of Micro-Interactional Rituals: Integrating Autonomic Nervous System Functioning Into Interaction Ritual Theory," *Sociological Forum* 31, no. 2 (2016).

202   Bulbulia and Schjoedt, "Religious Culture and Cooperative Prediction under Risk: Perspectives from Social Neuroscience," *Religion, Economy, and Cooperation* 49 (2010): 56.

203   van Mulukom, "Remembering Religious Rituals: Autobiographical Memories of High-Arousal Religious Rituals Considered from a Narrative Processing Perspective," *Religion, Brain & Behavior*, (2017), 9. Emphasis added.

204   van Elk and Aleman, "Brain Mechanisms in Religion and Spirituality: An Integrative Predictive Processing Framework," *Neuroscience and Biobehavioral Reviews* 73 (2017): 359–378.

Scientists have long known about the neural *correlates* associated with various religious and spiritual experiences.[205] But can mechanisms involved in religious rituals, such as emotionally arousing synchronic movement and the downregulation of the capacity for critical reflection, actually *cause* people to become more vulnerable to opportunists and less accurate in their recollection of what actually happened during such events? So it seems. Causality can be plausibly established using experimental intervention techniques. Mystical experiences can indeed be elicited in the laboratory, with participants reporting that such experiences are highly authentic and have lasting effects on memory and attribution.[206]

Even more importantly, recent neuroscientific manipulation experiments show that altering specific brain regions can increase (or decrease) belief in supernatural agents. Transcranial magnetic stimulation of the inferior parietal lobe and dorsolateral prefrontal cortex, for example, has been demonstrated to *increase* implicit religious and spiritual beliefs (measured through implicit association tests).[207] Another experiment that combined theta burst stimulations and an implicit association test designed to catch implicit, self-referential representations of religiosity and spirituality found that the latter can be "*decreased* by enhancing the excitablity of right IPL [inferior parietal lobe] and increased after disruption of IPL activity."[208] Yet another study using a neuromodulation experimental design established causality by showing that downregulating the posterior medial frontal cortex, which plays a role in ramping up ideological responses to threat, leads to a significant decrease in *both* belief in God, angels and heaven, *and* derogation toward out-groups, following a reminder of death.[209]

Why would death reminders be relevant here? The relationship between religiosity and mortality salience is one the most extensively researched topics in the psychology of religion. Here too priming studies and other experimental manipulation techniques have played an important role. It

205   See, e.g., Cristofori et al., "Neural Correlates of Mystical Experience," *Neuropsychologia* 80 (2016).

206   See Andersen et al., "Mystical Experience in the Lab," *Method & Theory In The Study Of Religion* 26, no. 3 (2014); and Arzy et al., "Induction of an Illusory Shadow Person," *Nature* 443, no. 7109 (2006).

207   Crescentini et al., "Virtual Lesions of the Inferior Parietal Cortex Induce Fast Changes of Implicit Religiousness/Spirituality," *Cortex* 54 (2014).

208   Crescentini et al., "Excitatory Stimulation of the Right Inferior Parietal Cortex Lessens Implicit Religiousness/Spirituality," *Neuropsychologia* 70 (2015): 77. Emphasis added.

209   Holbrook et al., "Neuromodulation of Group Prejudice and Religious Belief," *Social Cognitive and Affective Neuroscience* 11, no. 3 (2016).

turns out that when people are triggered (even subliminally) to think about their own death, they become more likely to believe in supernatural agency.[210] Unfortunately, the terror management system activated by anxiogenic death primes can also make people more antagonistic and aggressive toward value-threatening religious others.[211] Not surprisingly, this anxiogenic activation causes stronger belief in the god(s) of the participants' *own* religious in-group and stronger disbelief in the god(s) of out-groups. In other words, thinking about death can cause Christians to believe more strongly in God (or Jesus) and to deny Allah more strongly – but it has the inverse effect on Muslims.[212]

This amplification of pro-ingroup and anti-outgroup biases seems to be mediated by a variety of personality variables such as extrinsic religiosity, and the effect of death awareness on religious belief seems to be mediated by specific beliefs such as acceptance of the idea of an afterlife and apocalyptic prophecies.[213] Although there are important differences in the way in which mortality salience affects implicit (and explicit) religious beliefs and religious prejudice, and in the way in which supernatural primes affect theists and skeptics, there is little doubt that getting people to think about death and the possibility of an afterlife (a common theme in religious contexts) can causally boost their theistic credulity and conformity biases.[214] In other words,

---

210   Norenzayan and Hansen, "Belief in Supernatural Agents in the Face of Death," *Personality & Social Psychology Bulletin* 32, no. 2 (2006); Norenzayan et al., "An Angry Volcano? Reminders of Death and Anthropomorphizing Nature," *Social Cognition* 26, no. 2 (2008).

211   McGregor et al., "Terror Management and Aggression: Evidence That Mortality Salience Motivates Aggression Against Worldview-Threatening Others," *Journal of Personality and Social Psychology* 74, no. 3 (1998).

212   Vail, III, et al., "Exploring the Existential Function of Religion and Supernatural Agent Beliefs among Christians, Muslims, Atheists, and Agnostics," *Personality & Social Psychology Bulletin* 38, no. 10 (2012).

213   See, e.g., van Tongeren et al., "Ebola as an Existential Threat? Experimentally-Primed Ebola Reminders Intensify National-Security Concerns among Extrinsically Religious individuals." *Journal of Psychology and Theology* 44, no. 2 (2016): Lifshin et al., "It's the End of the World and I Feel Fine Soul Belief and Perceptions of End-of-the-World Scenarios" *Personality and Social Psychology Bulletin* 42, no. 1 (2016), and Routledge et al., "Death and End Times: The Effects of Religious Fundamentalism and Mortality Salience on Apocalyptic Beliefs," *Religion, Brain & Behavior* 6, no. 2 (2016).

214   See, e.g., Jackson et al., "Testing the Causal Relationship between Implicit Religiosity and Death Anxiety," *Religion, Brain & Behavior* 6, no. 2 (2016); Halberstadt and Jong, "Scaring the Bejesus into People: The Role of Religious Belief in Managing Implicit and Explicit Anxiety" in Forgas and Harmon-Jones, *Motivation and its Regulation: The Control Within* (New York: Psychology Press, 2014); Norenzayan et al., "Mortality Salience and Religion:

gods are more likely to be born in human minds and borne in human cultures when people become anxious about their own safety or the safety of their kith and kin.

One does not have to prime thoughts about death in order to increase people's tendency to detect gods and protect in-group norms. As we saw above, simply priming randomness or a sense of uncertainty can cause changes in religious belief. The results of experimental manipulations that prime perceived threats to *meaning* suggest that individuals high in religiosity tend to respond by increasing their propensity for detecting supernatural forces.[215] Another similar experiment found that causally manipulating feelings of meaninglessness *increased* belief in miraculous stories involving (for example) guardian angels or audible divine guidance.[216] Religion seems to help individuals with security-focused orientations lower their existential anxiety when primed with meaning threats, but unfortunately this occurs at the expense of openness and tolerance toward ideological others.[217]

Belief in supernatural agents can also be increased by priming the *need to belong* or *feelings of exclusion*. Cognitive scientists have found that manipulatively causing an individual to experience a perceived loss of social connection can elicit complex representations of imagined beings, suggesting that this sort of compensation is one of the ways in which "people may benefit from their attachment to different types of fantasy companions."[218] Psychological studies have also found that superstitious and conspiratorial beliefs can be causally increased in a laboratory simply by priming a sense of social exclusion.[219] Another psychological study that involved priming "close others" demonstrated that "belief in God can be temporarily altered by a persuasive message within a laboratory session."[220]

Divergent Effects on the Defense of Cultural Worldviews for the Religious and the Non-religious," *European Journal of Social Psychology* 39, no. 1 (2009).

215   Routledge et al., "An Existential Function of Evil: The Effects of Religiosity and Compromised Meaning on Belief in Magical Evil Forces," *Motivation and Emotion* 40, no. 5 (2016).

216   Routledge et al., "Miraculous Meaning: Threatened Meaning Increases Belief in Miracles," *Journal of Religion and Health* 56, no. 3 (2017).

217   van Tongeren et al., "Security Versus Growth: Existential Tradeoffs of Various Religious Perspectives" *Psychology of Religion and Spirituality* 8, no. 1 (2016).

218   Niemyjska and Drat-Ruszczak, "When There Is Nobody, Angels Begin to Fly: Supernatural Imagery Elicited by a Loss of Social Connection," *Social Cognition* 31, no. 1 (2013): 68.

219   Graeupner and Coman, "The Dark Side of Meaning-Making: How Social Exclusion Leads to Superstitious Thinking," *Journal of Experimental Social Psychology* 69 (2017).

220   Gebauer and Maio, "The Need to Belong Can Motivate Belief in God," *Journal of Personality* 80, no. 2 (2012): 491.

So, apparently, can commitment to affiliate with a religious in-group. The results of a series of five experiments consistently indicated that priming social exclusion heightens reported levels of religious affiliation and intention to engage in religious behaviors.[221] Another set of experiments has shown that manipulating participants' sense of *loneliness* also leads to higher belief in ghosts, angels, the Devil, God, etc. Unfortunately, concluded the authors, this research also suggests that a higher sense of social connection can lead people to be "more likely to *dehumanize* those to whom they are not socially connected."[222] This would help to explain the powerful prejudice that many religious people have against atheists. Another experimental study involving threat manipulations found that individuals with high levels of orthodox religious belief increased their expressions of prejudice toward atheists when exposed to threatening worldviews. This indicates that religious prejudice can be a coping or self-regulating strategy for dealing with threat.[223]

In other words, stressors of many sorts activate and amplify both anthropomorphic promiscuity and sociographic prudery, which in turn can amplify one another. However, even stress is not a necessary condition for the amplification of theistic biases. The key is axiological relevance. That is to say, the evolved tendencies toward theistic credulity and conformity are intensified by reminders of shared values and norms that bear on the human drive to survive and thrive in community. For example, perceptual awareness seems to be constrained by a "moral pop-out effect," which leads people "to see evidence of their moral values and beliefs [such as religious iconography] in grilled cheese sandwiches or other perceptually ambiguous stimuli."[224] Studies also show that religious believers will tend to make more causal attributions about God in relation to vignettes involving health-related or socially charged situations than less relevant quotidian events.[225]

221   Aydin et al., "Turning to God in the Face of Ostracism: Effects of Social Exclusion on Religiousness," *Personality and Social Psychology Bulletin* 36, no. 6 (2010).

222   Epley et al., "Creating Social Connection through Inferential Reproduction: Loneliness and Perceived Agency in Gadgets, Gods, and Greyhounds," *Psychological Science* 19, no. 2 (2008): 119. Emphasis in original.

223   Kossowska et al., "From Threat to Relief: Expressing Prejudice toward Atheists as a Self-Regulatory Strategy Protecting the Religious Orthodox from Threat," *Frontiers in Psychology* 8 (2017).

224   Gantman and van Bavel, "The Moral Pop-out Effect: Enhanced Perceptual Awareness of Morally Relevant Stimuli," *Cognition* 132, no. 1 (2014).

225   Cragun and Sumerau, "God May Save Your Life, but You Have to Find Your Own Keys: Religious Attributions, Secular Attributions, and Religious Priming," *Archive for the Psychology of Religion-Archiv Fur Religionspsychologie* 37, no. 3 (2015).

Experiences of awe have also been shown to have "a causal effect on supernatural beliefs," increasing the latter via a mediated effect driven by the influence of awe on uncertainty intolerance.[226] Many religious rituals are (or intend to be) awe-inspiring, and so it is not hard to understand why regularly participating in them with co-religionists would naturally amplify people's intolerance of uncertainty, as well as their belief in the gods allegedly engaged by their in-group. Unfortunately, being primed to feel awe decreases theists' openness to scientific (naturalist) explanations of the world (more so than atheists).[227] Given the importance of science for addressing issues like climate change, unfair social stratifications and economic distributions, and cultural conflicts fueled by religious intolerance, policies for promoting safe sects can no longer ignore the amplifying effect that ongoing participation in religious rituals has on the cognitive and coalitional biases that exacerbate these global challenges.

The claim that theogonic mechanisms are reciprocally reinforcing also receives support from ethnographic reports on a variety of types of rituals in quite different contexts. For example, researchers in the Tyva Republic have argued that there is a "coupling of gods and rituals" in the small-scale societies in the region, a process whereby ideas about gods' minds and local axiological practices (related to respecting the territories of other groups) evolve together in response to ever-shifting ecological and social problems.[228] Studies on ancestral beliefs and practices in rural Madagascar showed that children were more likely to express belief in an afterlife (as opposed to the empirically observed termination of life at the point of death) when primed with a religious narrative about rituals involving ancestor ghosts.[229] A study of Brazilian supernatural rituals called "simpatias" hinted at causality in the other direction: imagining the presence of supernatural agents (a component of anthropomorphic promiscuity) increased participant's evaluation of the efficacy of a culturally normative ritual (a component of sociographic prudery).[230]

It is important to remember that psychological and contextual variances shape the way in which – and the extent to which – "religion" amplifies

226   Valdesolo and Graham, "Awe, Uncertainty, and Agency Detection," *Psychological Science* 25, no. 1 (2014): 173.

227   Valdesolo et al., "Awe and Scientific Explanation," *Emotion* 16, no. 7 (2016).

228   Purzycki, "The Evolution of Gods' Minds in the Tyva Republic," *Current Anthropology* 57, no. S13 (2016).

229   Astuti and Harris, "Understanding Mortality and the Life of the Ancestors in Rural Madagascar," *Cognitive Science* 32, no. 4 (2008): 737.

230   Legare and Souza, "Evaluating Ritual Efficacy: Evidence from the Supernatural," *Cognition* 124, no. 1 (2012).

superstitious beliefs and segregative behaviors. Among the most significant individual variables for predicting intergroup religious violence are sacred values and identity fusion. A sacred value can be operationally defined as "anything that people refuse to treat as fungible with material or economic goods, for example, when people refuse to compromise over an issue regardless of the costs or benefits." Because sacred values have "privileged links to emotions, such as anger and disgust at their violation, leading to moral outrage and increased support for violence," people who are pressured to defend such a value "will resist trading it off for any number of material benefits, or even for peace."[231]

Psychological experiments and ethnographic research guided by identity fusion theory explore ways in which dispositional and situational factors work together to influence extreme behaviors. When personal and social identities are blurred, an individual can come to regard his or her group as functionally equivalent to his or her sense of self (identity fusion). Less fused people may have strong beliefs about what "ought" to be done for their group, but highly fused people are far more willing to act on these beliefs even, or especially, when that involves dying or killing for the group.[232] All of this has rather obvious implications for attempts at peacemaking. When policy-makers or conflict mediators ignore the function of sacred values and identity fusion in intensifying parochial (in-group) parochialism, they pursue strategies that actually make highly fused devoted actors *less likely* to compromise.[233]

Scholars of religious violence debate whether religious *behaviors* or religious *beliefs* play a stronger causal role in the escalation of intergroup conflict. On the one hand, some argue that in-group ritual participation (or religious affiliation) have the most pround effect. They point out that statistical analysis of cross-cultural datasets indicates that "coalitional religiosity" and "devotional

231 Sheikh et al., "Sacred Values in the Israeli–Palestinian Conflict: Resistance to Social Influence, Temporal Discounting, and Exit Strategies," *Annals of the New York Academy of Sciences* 12991, no. 1 (2013): 12, 21.

232 Swann et al., "Dying and Killing for One's Group: Identity Fusion Moderates Responses to Intergroup Versions of the Trolley Problem," *Psychological Science* 21, no. 8 (2010); Swann et al., "What Makes a Group Worth Dying for? Identity Fusion Fosters Perception of Familial Ties, Promoting Self-Sacrifice," *Journal of Personality and Social Psychology* 106, no. 6 (2014); Kiper and Sosis, "Shaking the Tyrant's Bloody Robe," *Politics and the Life Sciences* 35, no. 01 (2016).

233 Ginges and Atran, "What Motivates Participation in Violent Political Action.(Report)," *Annals of the New York Academy of Sciences* 1167 (2009); Ginges et al., "Psychology Out of the Laboratory," *American Psychologist* 66, no. 6 (2011); Atran, "The Devoted Actor: Unconditional Commitment and Intractable Conflict across Cultures," *Current Anthropology* 57 (2016).

religiosity" have opposing relationships to religious intolerance. *Coalitional religiosity*, which is characterized by accepting one's own in-group's beliefs and moral vision as true and considering the views of others as deviant, rather unsurprisingly predicts intolerance. *Devotional* religiosity, however, which is characterized by an "intrinsic" or supernaturally grounded faith, predicts tolerance.[234] A later study of intergroup hostility across religious groups produced similar results. When controlling for "coalitional rigidity" (rigid adherence to in-group norms), intrinsic religiosity (inwardly held religious devotion) was not positively correlated to intergroup hostility (morally impugning or wishing harm upon out-group members).[235]

Does this mean that sociographic prudery is a stronger motivator for religious violence than anthropomorphic promiscuity? This claim seems to be warranted by evidence showing that "the more people participate in religious ritual the more likely they are to report a preference to be a sacred value," an effect that is amplified by perceptions of high threat to the in-group to which a person belongs.[236] There is little doubt that regular attendance at church, temple, or mosque increases the salience of group identity which, under certain conditions, can function as a trigger for defensive or offensive reactions to religious out-groups. Moreover, a series of cross-cultural surveys and priming experiments covering several major religions found that "attendance at religious services" was positively correlated with support for suicide attacks, while "regular prayer" was not. The authors interpreted these findings as providing support for the hypothesis that the association between religion and suicide attacks was a function of "coalitional-commitment" rather than "religious-belief."[237]

On the other hand, some scholars argue against a too quick dismissal of the hypothesis that religious belief, or at least some specific religious beliefs in some contexts, can have a causal impact on intergroup violence. For example,

234   Hansen and Norenzayan, "Between Yang and Yin and Heaven and Hell: Untangling the Complex Relationship between Religion and Intolerance.," in *Where God and Science Meet*, vol. III ed. McNamara and Wildman (New York: Praeger, 2006), 188.

235   Hansen and Ryder, "In Search of 'Religion Proper,'" *Journal of Cross-Cultural Psychology* 47, no. 6 (2016).

236   Sheikh et al., "Religion, Group Threat and Sacred Values," *Judgment and Decision Making* 7, no. 2 (2012): 110. See also Sheikh et al., "The Devoted Actor as Parochial Altruist: Sectarian Morality, Identity Fusion, and Support for Costly Sacrifices," *Cliodynamics* 5, no. 1 (2014), and Hirsch-Hoefler et al., "Radicalizing Religion? Religious Identity and Settlers' Behavior," *Studies in Conflict & Terrorism* 39, no. 6 (2016).

237   Ginges et al., "Religion and Support for Suicide Attacks," *Psychological Science* 20, no. 2 (2009): 230.

the study just cited has been criticized as conceptually flawed and misleading. Especially in some of the Muslim contexts in which the research occurred, the variable "frequency of prayer" does not shed much light on the actual beliefs or commitments of the participants. The rejection of the religious-belief hypothesis seems premature and unwarranted.[238] However, it is important to acknowledge that other studies using experimental priming techniques and survey data analysis in other contexts, such as Lebanon and rural Jamaica, provide additional support for the coalitional-commitment hypothesis. There does seem to be a distinction: "while devotion to religious principles can increase in-group cooperation, the social aspects of religion can generate hostile attitudes towards out-groups."[239]

As we have seen, however, a wide variety of other empirical studies from diverse disciplines have shown that activating some of the mechanisms of anthropomorphic promiscuity *trigger* the intensification of sociographic prudery. All of these mechanisms are entangled within the complex reproductive systems of religious sects. It may turn out that theistic *conformity* biases have a more powerful role in promoting religious violence than theistic *credulity* biases, but it is hard to imagine how these factors could be completely divorced from one another in the real world. Some defenders of the coalitional-commitment hypothesis go out of their way to reject the notion "that there is something special about religious faith... that *invariably* favors promotion of violent intergroup conflict."[240]

However, I do not know of any scholars in the scientific study of religion who would entertain such a notion. If the causality were *invariable*, having "the talk" would be much simpler. Unfortunately, the complexity of individual and contextual variations makes it all too easy for apologists to avoid taking responsibility for the consequences of their own religious reproduction. Even if it turns out that prayer, devotion, or other components of anthropomorphic

---

238   Liddle et al., "Understanding Suicide Terrorism: Premature Dismissal of the Religious-Belief Hypothesis," *Evolutionary Psychology* 8, no. 3 (2010). For a commentary on the response, see Ginges et al., "Religious Belief, Coalitional Commitment, and Support for Suicide Attacks: Response to Liddle, Machluf, and Shackelford," *Evolutionary Psychology* 8, no. 3 (2010).

239   Lynch et al., "Religious Devotion and Extrinsic Religiosity Affect In-Group Altruism and Out-Group Hostility Oppositely in Rural Jamaica," *Evolutionary Psychological Science* 3, no. 4 (2017) 1; see also Hoffman and Nugent, "Communal Religious Practice and Support for Armed Parties," *Journal of Conflict Resolution* 61, no. 4 (2017).

240   Ginges et al., "Thinking from God's Perspective Decreases Biased Valuation of the Life of a Nonbeliever," *Proceedings of the National Academy of Sciences of the United States of America* 113, no. 2 (2016): 318. Emphasis added.

promiscuity do not *directly* cause intergroup conflict, there is little doubt that believing in hidden, person-like, coalition-favoring, ontologically confused agents can elicit behaviors designed to protect the supernaturally authorized norms and boundaries of one's religious in-group.

The argument of this section, and indeed the entire book, is not that anthropomorphic promiscuity (per se) invariably leads to violent intergroup conflict, nor that sociographic prudery (per se) invariably leads to dangerous schizotypal hallucinations. In the ongoing process of exploring the dynamics of theogonic reproduction within religious sects, it will continue to be important to pay close attention to the interdependent relations among individual and contextual factors whose variance influences the quality and the intensity of the reciprocal reinforcement of god-bearing mechanisms. For those of us interested in mitigating the negative consequences of religious biases, it will also be important to pay close attention to the causal dynamics at work among god-dissolving mechanisms.

The Anaphrodisiacal Effects of Science, Philosophy and Theology

For all of the reasons just outlined, and others to be described in the chapters that follow, most human beings throughout history have been attracted to the idea of engaging in religious sects. Even today the reciprocal reinforcement of the evolved cognitive and coalitional biases we have been discussing puts most people in the mood to scan for and have some kind of imaginative ritual intercourse with the supernatural agents of their in-group. The subtitle of this book alludes to two particularly powerful anaphrodisiacs: science and philosophy. Examining religious reproduction from a scientific and philosophical perspective challenges the theistic biases that engender and ritually nurture supernatural conceptions.

As I explain in more detail at the ends of Chapters 3 and 4, the middles of Chapters 6, 7 and 9, and the whole of Chapter 12, science and (non-religious) philosophy are evidence of the human capacity to contest the god-bearing mechanisms that have for so long constrained human thought and canalized human societies. In other words, they promote sociographic promiscuity and anthropomorphic prudery, each of which has a god-dissolving or "theolytic" effect (*Figure 2*). Chapters 2, 5, 8, and 11 provide illustrations of some of the reciprocally reinforcing dynamics at work in the integration of theolytic mechanisms.

The god-dissolving forces of anthropomorphic prudery and sociographic promiscuity contribute to the production of "atheism," by which I simply mean

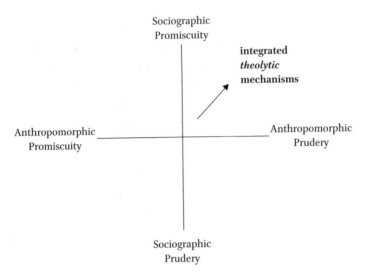

FIGURE 2   *Theolytic mechanisms*

the attempt to make sense of the world and to act sensibly in society without
appealing to supernatural agents or authorities. To put it even more positively,
an atheist is someone who embraces *naturalism* in their construction of causal
explanations in the academic sphere and *secularism* in their proposals for normative inscriptions in the public sphere. As we will see in Chapter 12, this is
not the only way of being (or becoming) atheistic, but this is how I will most
often use the term.

Most varieties of naturalism share a resistance to the inclusion of disembodied intentional forces in interpretations of the evolving cosmos. No doubt
some individual scientists continue to harbor superstitious religious beliefs,
but *qua* scientists the vast majority are *methodologically* naturalistic in the
sense that they exclude god-concepts from their scholarly hypotheses. Most
varieties of secularism share a resistance to the inclusion of supernaturally
authorized sectarian policies in prescriptions for organizing pluralistic societies. No doubt some political philosophers in complex, democratic contexts
maintain membership in religious in-groups, but *qua* philosophers a growing
number are *methodologically* secularist in the sense that they exclude god-
sanctioned commands from their political proposals.

Atheism, in the general sense I am using the term, is the affirmation of
metaphysical naturalism and *metaphysical* secularism: the most plausible
hypotheses and the most feasible strategies are those that incorporate only
axiological dynamics whose actual existence (or existential actualizability)
are inter-subjectively and trans-communally contestable.[241] The distinction

---

241   I spell this out in more detail in Shults, *Theology after the Birth of God*, Chapters 6 & 7.

between the theolytic forces of anthropomorphic prudery and sociographic promiscuity can be understood as an initial fractionation of atheism, the reverse-engineering of which must include further fractionation in order to identify, understand, and explain the recombination of various component parts of complex adaptive god-dissolving systems. I begin to take up this task in earnest below.

As we will see in the following chapters, segregative inscriptions of the social field based on superstitious interpretations of punitive (or otherwise axiologically relevant) gods are becoming more and more problematic in our pluralistic, globalizing context. A growing number of people, and especially young people, are finding it increasingly easy to evaluate explanatory hypotheses and normative proposals without the need for supernatural agents as causal powers or moral regulators. In other words, in many parts of the world we find a growing tendency toward the integration of anthropomorphically prudish and sociographically promiscuous tendencies. We will explore some of the mechanisms behind these cognitive, social, and demographic shifts in more detail in the final chapter. Each of the ten central chapters of this book are adapted versions of previous publications, all of which were intended as "contraceptive" essays with the more or less explicit goal of unveiling god-bearing mechanisms and promoting safe sects.[242]

The first five of these central chapters focus primarily on scientific or philosophical issues. Chapter 2 provides a different sort of overview of recent developments in the bio-cultural study of religion by describing the way in which four scholars (from different disciplines) weave together scientific insights into the god-bearing mechanisms we have been discussing. The third and fourth chapters illustrate the tension between those mechanisms and the theolytic tendencies at work in science in light of recent archaeological research on the Neolithic transition and contemporary climate change (respectively). Chapters 5 and 6 offer more explicitly philosophical perspectives, exploring ways in which uncovering theistic credulity and conformity biases can help us move beyond impasses in theoretical arguments about the existence of God and public debates about which social-machines we ought to build (respectively).

---

242   Several of these earlier articles and essays utilized the conceptual framework depicted in Figure 1 and Figure 2. However, in some cases, the original versions defined anthropomorophic promiscuity and sociographic prudery more generically, as aggregates of cognitive and coalitional mechanisms that did not explicitly refer to supernatural agents or authorities, but which contributed to their detection and protection. All of these have been re-worked to incorporate the new, more precise definitions outlined earlier in this chapter.

Chapters 7 through 11 continue to examine religious reproduction from scientific and philosophical perspectives, but with a more explicit focus on the (an)aphrodisiacal effect of theology. There is no doubt that this discipline has been used by the priestly and intellectual elites of the major monotheistic religions to keep people bearing the gods of their own in-groups. For most of human history, shared imaginative engagement with supernatural agents was mediated by shamans or other relatively informal ritual officers within small-scale societies. "[N]o matter how false or irrational," religious beliefs of the sort that are entangled within shamanic ritual interactions "can be adaptive if they help produce a moral order and orient human communities toward common purposes that help assure survival and reproduction." Large-scale societies required a different kind of mediator for social integration, "exemplifed in the magico-religious functionary, the priest, who appear[s] universally in societies with a primary reliance on agriculture and hierarchical political systems."[243]

Despite all this, the rise of the priestly class also had a (somewhat) god-dissolving effect. When intellectuals in complex, literate religious coalitions began to think analytically about the doctrines of their in-groups, and to express concern about how their beliefs and rituals had an oppressive effect on some members of society, their efforts sometimes loosened the ties that bound people to local supernatural agent conceptions of the sort that evolved to hold small-scale societies together. As we will see, these relatively "iconoclastic" theological efforts have almost always collapsed under the weight of "sacerdotal" pressures to affirm the ontologically confused beliefs and ritual behaviors authorized by a dominant large-scale religious system of the sort that emerged in the wake of the axial age.

Nevertheless, insofar as "theology" raises people's consciousness about the idiosyncracy of their own beliefs and norms, it can have an enervating effect on theistic credulity and congruity biases, which helps to explain why religious laypeople are often so suspicious of theologians. Chapters 7–9 explore some of the ways in which conservative, moderate, and liberal theological reflection can have a (limited) theolytic effect, more or less inadvertently promoting naturalism and secularism. In this way, and usually in spite of themselves, theologians secrete atheism. The tenth and eleventh chapters pay special attention to the psychological and political implications of the integration

---

243   Winkelman, "Shamanism as a Biogenetic Structural Paradigm for Humans' Evolved Social Psychology," *Psychology of Religion and Spirituality* 7, no. 4 (2015): 275. For a cultural evolutionary explanation of the emergence and role of shamanism in human groups, see Manvir Singh, "The Cultural Evolution of Shamanism," *Behavioral and Brain Sciences*, in press.

of god-dissolving mechanisms, whose operation in our contemporary global environment seems to be expanding.

The final chapter explores these mechanisms in more detail, demonstrating the way in which, and the conditions under which, they too can be reciprocally reinforcing. For a long time, religion has been "good for us" in the sense that it has helped members of our species bind themselves into groups that cooperated as they (re)produced supernatural conceptions. Today, however, we are confronted with ecological challenges that are quite different from those faced by our upper Paleolithic ancestors. I conclude with some reasons to be hopeful about our capacity to adapt and to learn how to practice safe sects. First, however, let's fill out our scientific and philosophical exploration of the cognitive and coalitional mechanisms by which the gods are born(e).

CHAPTER 2

Bearing Gods in Mind and Culture

The journal *Religion, Brain & Behavior* was founded in order to foster multidisciplinary research on religion that reaches across fields as diverse as neuroscience, evolutionary biology, moral psychology, cultural anthropology, cognitive archaeology and political science. In what follows I will continue referring to the conceptual arena in which these (and other) disciplines meet and interact to clarify our understanding of religious belief, behavior, and experience as the "bio-cultural study of religion."[1] This nomenclature is not intended to demarcate a new singular academic field or discipline.

Given the astonishing fecundity of the open integration and overlapping application of evolutionary, cognitive, and social scientific theories and research methods to religious phenomena, trying to set such boundaries would seem counterproductive. If we think of this as a "field," the metaphor should be construed not in geographical but in physical terms: a dynamic force field of interconnected and open explanatory events. If we think of this as a "discipline," the focus should not be on deciding its departmental location but on disciplining ourselves to remain interconnected and open during every event of explanation.

This chapter illustrates the weaving together of the two broad conceptual threads introduced above, which are increasingly being integrated into theoretical patterns within the disciplined fields of the bio-cultural study of religion, in four books published in 2010: David Lewis-Williams' *Conceiving God*, Pascal Boyer's *The Fracture of an Illusion*, Scott Atran's *Talking to the Enemy*, and Matt Rossano's *Supernatural Selection*. After comparing and contrasting the approaches of these authors, I conclude by briefly calling attention to the contemporary psychological, political, and philosophical relevance of these developments. We will return to these issues in other chapters throughout the book.

A convergence of insights from a variety of sciences is leading a growing number of scholars to argue that religion (whatever else it may include) typically involves shared imaginative engagement with supernatural agents who

---

1 My use of this phrase is inspired by the name of the Institute that helped found the journal: The Institute for the Bio-Cultural Study of Religion (www.mindandculture.org/focus-areas/ religion). This chapter is an adapted version of "Bearing Gods in Cognition and Culture," originally published in *Religion, Brain & Behavior* 1, no. 2 (2011).

are born through naturally evolved human cognitive functioning and borne by culturally transmitted human coalitional functioning. By "supernatural agents" I mean disembodied wielders of causal power, i.e., discarnate intentional entities that are not susceptible to normal scientific empirical observation or measurement.[2] Although the term "god" commonly refers to beings like Zeus, Yahweh, or Buddhist devas, in this literature one increasingly finds it used to indicate supernatural agents in general, including ghosts, angels, saints, jinn, etc.[3] I follow this usage in this context for the sake of simplicity.

As we saw in Chapter 1, there is a burgeoning literature within the bio-cultural study of religion that supports the claim that gods are born(e) in minds and cultures as a result of evolved cognitive and coalitional biases. This chapter provides a different sort of introduction to this literature: a review of the four influential books identified above, all published in 2010. One reason for choosing monographs instead of articles is that the former provide authors with more space (and license) to explore the wider implications of their empirical efforts. Before briefly expositing, comparing, and contrasting these four representative theoretical contributions, I first outline in more detail the two threads that we will find woven into their work in various ways.

Anthropomorphic Promiscuity and Sociographic Prudery

The books we discuss below are examples of the way in which the bio-cultural study of religion is providing answers to the questions "Where do supernatural agents come from?" and "Why do they stay around?" Religions have their own answers to these questions, sometimes even developing complex theogonies, i.e., narratives about the birth of the gods and their connections to particular human families or other coalitions.

Although scientists have long been interested in naturalistic descriptions of this phenomena, in the last two decades empirical research across disciplines has led to integrative and compelling explanations of what I call the theogonic (god-bearing) mechanisms of anthropomorphic promiscuity and sociographic prudery. As we have seen, these include traits that foster the hyper-sensitive detection of agential forms in nature and the hyper-sensitive protection of

---

2 This use of the phrase "supernatural agent" is derived and adapted from Wildman, *Science and Religious Anthropology*. (Ashgate, 2009).

3 See, e.g., Barrett, *Why Would Anyone Believe in God?* (Walnut Creek, CA: AltaMira Press, 2004); Tremlin, *Minds and Gods* (Oxford University Press, USA, 2010).

coalitional forms in culture.[4] Although they differ in their emphases and formulations of these naturally evolved human tendencies, in one way or another
each of the authors reviewed below has something to say about both sorts of
theogonic mechanism and the complex interactions among them.

Most human beings are naturally promiscuous in their seeking out of human
forms in the natural environment. The natural selection of cognitive processes
that were overly sensitive to detecting agency contributed to our ancestors'
survival. Imagine an early hominid perceiving some ambiguous movement in
the forest. Interpretations of such movements as caused by the presence of a
potential enemy (or a potential mate) will usually be wrong; however, in those
cases where a person is in fact present, failing to guess "relevant agent" can
be fatal (or counter-productive in other ways). Individuals with a perceptual
strategy that led them to guess "person" may have more often been wrong than
individuals who automatically guessed "wind" until more compelling evidence
emerged for an intentional cause. The latter, however, would have been less
likely to survive in the early ancestral environment than the former. And so it
makes sense that this sort of trait, which fostered anthropomorphic promiscuity, would have been naturally selected.

This helps to explain why even today we are ever on the lookout for human (and other) agents, seeing faces in the clouds and mistaking boulders for
bears.[5] Scientists use a variety of terms to refer to this overeager interpretation of ambiguous natural phenomena in terms of agency, intentionality and
purposiveness.[6] All four of the 2010 monographs reviewed below argue that
this hyper-sensitivity played a role in the detection of imagined supernatural
agents, thereby contributing to the origin and evolution of religion.

Gods may be born in human minds through anthropomorphic promiscuity, but it takes a more or less faithful village to raise (maintain and sustain)
them. As we have seen, supernatural agents are borne in human communities
(in part) because their imagined presence helps protects in-group cohesion.
Over-detecting human minds emerge and are implicated within fields of social relations, which are always and already inscribed with proscriptions and
prescriptions. Those groups that survived the last ice age were those whose

---

4   See also Shults, "Science and Religious Supremacy: Toward a Naturalist Theology of Reli-
    gions," in *Science and the World's Religions, Volume III: Religions and Controversies*, ed. Wild-
    man and McNamara (New York: Praeger, 2012).

5   Guthrie, *Faces in the Clouds: A New Theory of Religion* (Oxford University Press, 1993).

6   Pyysiäinen, *Supernatural Agents: Why We Believe in Souls, Gods and Buddhas* (Oxford Uni-
    versity Press, 2009); Kelemen, "Why Are Rocks Pointy? Children's Preference for Teleological
    Explanations of the Natural World," *Developmental Psychology* 35, no. 6 (1999).

inscription of the socius was reinforced by a prudish over-protection of the coalition's norms by its members, including a willingness to punish defectors and use violence against out-group competitors.

These and other components of sociographic prudery were reinforced over time by the transmission of beliefs in gods who were interested in the coalition, watching its members and capable of bringing health or misfortune. Of course gods do not look the same in every culture. No supernatural agent conceptions are immaculate. Ideas of discarnate intentional entities gestate within a particular social matrix whose historical development influences their ontogenesis; cultural birth marks on the human mind. Whether or not ontogeny recapitulates phylogeny, it seems clear that theogony capitulates to ethnogeny. The fact that all known human cultures, past and present, have been characterized by widespread imaginative engagement with a diversity of supernatural agents is partially explained by the fact that all of their members were *Homo sapiens,* whose shared phylogenetic inheritance includes evolved mechanisms that naturally reproduce gods who hold us together – psychologically and politically.

Each of the authors discussed below is sensitive to the fact that their empirical findings and theoretical reflections intensify the apparent opposition between science and religion, although they deal with that tension in different ways. Highlighting the two conceptual threads identified above makes it easier to understand this tension. Scientists (*qua* scientists) tend to be anthropomorphically prudent, resisting explanations that appeal to supernatural agency, and sociographically promiscuous, preferring modes of inquiry that do not depend on appeals to religious authorities. In the conclusion of this chapter, I will return to the wider significance of this tension for contemporary human life, but first let us illustrate the unveiling of theogonic mechanisms in four major works published in 2010.

Conceiving God

I begin with David Lewis-Williams' *Conceiving God: The Cognitive Origin and Evolution of Religion* (2010). Although he does not play with the metaphor of the "birth of God" as I have, his title aptly points to the conceptual pattern briefly introduced above. Building on several of his earlier works,[7] this volume provides a comprehensive presentation of his argument that religion emerged

---

7   Lewis-Williams, *The Mind in the Cave: Consciousness and the Origins of Art* (London: Thames & Hudson, 2002); Lewis-Williams and Pearce, *Inside the Neolithic Mind: Consciousness, Cosmos and the Realm of the Gods,* 1 edition (London: Thames & Hudson, 2009).

as a result of early humans' mistakenly taking the entopic phenomenal and hallucinatory experiences of some altered states of consciousness as really indicating a supernatural realm, and granting social privilege to "shamans" who were adept at (or susceptible to) having such experiences during which it was believed that they engaged with (or even became) powerful spiritual beings within that realm.

Lewis-Williams is aware of the general hypersensitivity of human beings to detecting agents everywhere, but his focus is on the way in which the detection of particular kinds of agents within intensified inner-directed altered states contributed to belief in the real existence of such agents and their impact on the natural and social worlds of (waking) human life during outer-directed, problem-oriented conscious thought. Combining neurological and psychological research on reported experiences of intensified hypnagogic and hallucinogenic states with archaeological and ethnographic research on cave art and its role in small-scale shamanic societies, Lewis-Williams argues that early humans had the same spectrum of conscious experiences as do contemporary humans. In modern societies, however, many people are less likely than our ancestors to readily accept the relevance and normative significance of altered states such as dreams, visions and induced hallucinations.

For Lewis-Williams, "the body provides raw material for what, in a variety of social contexts, is accepted as some sort of trafficking with supernatural forces or beings."[8] All humans have neurologically generated experiences, but they have to be shared and regulated in a particular way before they can become the foundation for a religion. As the regulation of the alleged interaction between supernatural agents and the members of a coalition was increasingly taken over by shamans, social differentiation became more marked. In human evolution, "religion and social discrimination went hand in hand" (58). Lewis-Williams suggests that as human societies became more complex, so did the modes of regulation and control, which eventually led to clerical and political oppression. In other words, the same altered states and social discrimination are at work in the priestly hierarchies that came to characterize later complex literate societies, including our own. This means "the 'origin' of religion is always with us" (138).

Two of the most commonly reported experiences in the intensified trajectory of consciousness as it moves toward the introverted end of the spectrum and into vivid participatory hallucinations are flying (or floating) above the

---

8   Lewis-Williams, *Conceiving God: The Cognitive Origin and Evolution of Religion* (London: Thames & Hudson, 2010), 149. Page numbers in the following paragraphs refer to this book, unless otherwise noted.

world and descending (or being drawn) through a tunnel or vortex into the depths of the world. Lewis-Williams suggests that these neurologically generated experiences of magical flight and vortex travel led to the belief in a tiered cosmos, which is the most common view of the structure of the cosmos in pre-modern cultures. In an earlier book with David Pearce, Lewis-Williams illustrates the "near universality" of this cosmological pattern with examples from a wide variety of places and times, across every inhabited continent and from the upper Paleolithic to the present. The pre-Columbian inhabitants of Mesoamerica, for example, believed that the world was organized around a great tree that grew out of a mountain, whose branches connected to an upper world and whose roots went down to the underworld. Some Mayan temples were built in correspondence with the idea of tiered cosmos.[9] The holy texts of the Abrahamic religions also reflect cosmologies that presuppose an upper world (sky, heaven, paradise) and an under world (sheol, hades, hell), both filled with supernatural agents.

Lewis-Williams' hypothesis is relatively straightforward. First, the reason this belief in a tiered cosmos is so common is that it is neurologically generated; early humans were simply exploring their internal states of consciousness and found supernatural agents (promoting anthropomorphic promiscuity). Second, some members of early human groups came to be perceived as having special experiences of or access to these other tiers, and acquired a social power that led to discriminatory structures within the community (promoting sociographic prudery). Shamanism played a regulative role in the evolution of both sorts of theogonic mechanism. The social capital of the shaman was derived from his or her capacity to intentionally enter altered states in order to travel the cosmos and interact with supernatural agents, which people believed enabled him or her to control the movements of animals and the weather (ensuring food), to heal the sick, etc.

When Lewis-Williams suggests that everything we call "religion" can be traced to natural neurological activity in the brain, specifically to the mistaken acceptance of hallucinations during altered states of consciousness as perceptions of a supernatural realm, he does not mean that religion is reducible to this experiential domain. Within every culture humans interpret such experiences differently, leading to a variety of formal and material expressions of what he calls the "belief" and "practice" domains of religion. However, he does mean that these beliefs and practices are built upon false interpretations of brain states, incorrectly postulating a supernatural realm that somehow has direct consequential and normative bearing on the natural world. In Lewis-Williams'

---

9 See Lewis-Williams and Pearce, *Inside the Neolithic Mind*, 66–69.

view, as long as religion maintains ritual practices based upon such beliefs about neurological experiences, no reconciliation between it and science is possible.[10]

The Fracture of an Illusion

Pascal Boyer comes to a similar conclusion, although it is based on a different set of empirical findings and disciplinary reflections. Perhaps most well known for his popular book *Religion Explained*,[11] Boyer was one of the first scholars to produce a relatively complete theory of religion based on cognitive science.[12] In his earlier books, he drew heavily upon research in cognitive psychology as well as cultural anthropology to support his claim that the recurrence of patterns in religious belief and behavior across cultures can be explained by an aggregate of complex and independently evolved cognitive mechanisms and social strategies, each of which contributed to early human survival.

In *The Fracture of an Illusion: Science and the Dissolution of Religion* (2010), Boyer summarizes many of his earlier themes but weaves them together in his most explicit challenge to the very notion of "religion." In *Religion Explained*, he had described religion as the commodification of parasitic knowledge of "airy nothing" by priestly guilds. In his newer book, he insists there is "no such thing" as religion; it is an imaginary object, a "package" that exists only as a postulation or "marketing ploy" of particular religious institutions and office holders. This charge has obvious political, as well as psychological implications, to which we will return, but first let us point to some of the evidence that leads Boyer to the verdict that what we call "religions" are only collections of fragments of a variety of mental capacities and tendencies that evolved as adaptations with other (non-religious) adaptive purposes.

For Boyer, supernatural (religious) concepts are parasitic on natural concepts, which are formed and regulated by cognitive modules that evolved as human beings adapted to their natural environment, producing an "intuitive ontology" that supports implicit and easily accessible assumptions about natural objects and processes. Human cognition automatically activates expectations

---

10   Lewis-Williams, *Conceiving God*, 2010, 288.

11   Boyer, *Religion Explained: The Evolutionary Origins of Religious Thought*, (New York, Basic Books, 2002).

12   Boyer, *The Naturalness of Religious Ideas: A Cognitive Theory of Religion*, (Berkeley: University of California Press, 1994).

and inferences about the latter when they are categorized into particular domains. For example, when we imaginatively place an object into the category or domain Person, we automatically (usually without conscious awareness) infer that the object was born, eats, has a body, pursues goals, etc. We assume that an Artifact was made, and that a Plant grows, but that neither is intentional. One way in which "supernatural" concepts are special, argues Boyer, is that they minimally violate our intuitive expectations about objects in an ontological domain; e.g., a ghost that goes through walls, a statue that bleeds, a tree that listens to people's conversations.[13]

Concepts of supernatural agents, which Boyer insists are the most significant of what we call "religious" concepts, minimally violate domain-level intuitions about the category Person – especially the natural inference that Persons are limited by their physicality or embodiment. However, other inferences from non-violated intuitive assumptions about objects identified as belonging to that domain continue to flow naturally. Once a ghost or spirit is "detected," cognitive mechanisms are immediately activated, leading to the assumption that it has goals, could be interested in what we are doing, and may even want to eat. It may initially seem odd that people would maintain belief in such counterintuitive ideas and transmit them from generation to generation. In fact, however, it is precisely their (minimal) counter-intuitiveness that makes them easier to remember and transmit.

This promiscuous searching for supernatural agents in the natural environment is reinforced by the role they are intuitively assumed to have in the inscription of the social and moral life of the coalition. Mickey Mouse, for example, does not qualify as a god (despite his being minimally counter-intuitive) because he is not imaginatively engaged as an existentially relevant social agent capable of enforcing or enhancing prudent behavior within a coalition. Boyer observes that supernatural agents are often concerned about (and watching) the moral behavior of members of a group; moreover, they typically have the power to bring some misfortune or blessing. Rituals of various kinds the world over involve trying to connect or manage the relation between the natural and the supernatural members of more or less coherently defined coalitions.

Even though religious specialists usually describe the relation between gods and ethical behavior in terms of legislation or exemplarity, psychological studies suggest that people tend to default quickly to what Boyer calls the "interested party" notion of how superhuman agents are connected to morality.

---

13   Boyer, *The Fracture of An Illusion: Science And The Dissolution Of Religion. Frankfurt Templeton Lectures 2008*, ed. Grab-Schmidt et al., (Göttingen: Vandenhoeck & Ruprecht, 2010), 29. Unless otherwise noted, page numbers in the following paragraphs refer to this book.

Experiments show that when people are pressed to answer questions about the attributes of gods portrayed in practical scenarios, most automatically defer to descriptions of minimally counter-intuitive discarnate human-like agents who may be watching them and desire to help or hurt them, even if they are able to give "orthodox" definitions of gods (which are often maximally counterintuitive; e.g., God is omniscient and impassible) in other contexts (54). Boyer also briefly treats the application of costly signaling theory to religion, but Atran, and Rossano both deal with this in more detail and so I will return to it below. The salient point for the purposes of this review is the way in which Boyer appeals to a variety of mechanisms that contribute to a mutual reinforcement of sociographic prudery and anthropomorphic promiscuity.

Boyer argues that human beings evolved within a "cognitive niche," which makes them more dependent (in comparison with other species) on obtaining and maintaining social information about other human beings. Given the computational constraints of social interaction, he suggests that thinking about absent agents is both necessary and useful. Humans develop a catalog of possible interaction scenarios constructed when other agents are not around, and this makes actual inference in interaction work better. Boyer points to research indicating how frequently children (perhaps even a majority of children) create imaginary friends. Computing the reactions of the imaginary friend helps to train the mind for actual human engagement, building social capacities for coherent interaction with real, embodied friends (32–34).

In their latest books, both Lewis-Williams and Boyer are more explicit than ever about the need for frank and open discussions about the psychological and political significance of these new scientific insights into the mechanisms by which gods have been and continue to be born(e) within increasingly complexifying human societies. Lewis-Williams suggests it is virtually inevitable that the growth of science will continue to change people's minds about supernatural agents no matter how much "God's empire strikes back."[14] Boyer, on the other hand, believes that the evolutionary mechanisms that have led to the by-product of "religion" are so pervasive that they will probably always have to be accommodated somehow within future secular civilizations, although the health and survival of the latter will require active resistance to theocratic societies that are "versions of Hell on earth."[15] Our next author is also an atheist and is equally dismissive of the truth value of religious claims about supernatural agents, but his theoretical approach to (and the practical implications he draws from) the scientific study of religious others differs in significant ways.

---

14   Lewis-Williams, *Conceiving God*, 2010, 257.

15   Boyer, *The Fracture of An Illusion*, 97.

Talking to the Enemy

Like Lewis-Williams and Boyer, Scott Atran can be considered one of the founding figures of the multidisciplinary "bio-cultural" approach to religion that has emerged in the last few decades. His special expertise is in anthropology and cognitive science, but he has also done significant work in evolutionary biology, psychology, and public policy studies. Atran's earlier books include *Cognitive Foundations of Natural History*,[16] and *In Gods We Trust: The Evolutionary Landscape of Religion*.[17] In his more recent (2010) *Talking to the Enemy: Faith, Brotherhood and the (Un)Making of Terrorists*, Atran incorporates many aspects of his earlier work, but focuses on detailing his recent anthropological research "in the wild" – in contemporary conflict areas such as Palestine, Afghanistan and southeast Asia, where he has interviewed people who knew, or who hope to become, suicide bombers.

Why do people kill and die for "the Cause," i.e., the belief that the world was intended for "our" committed community? Atran observes that this killing and dying is usually not merely for an abstract Cause but for the cause of a specific group, an "imagined family of genetic strangers," whether brotherhood or fatherland, tribe or team. He argues that it is small-group dynamics, such as raising families or playing soccer together, that trump almost everything else as people move through life.[18] Atran insists that (most) fundamentalists and jihadists are not naïve or sociopathic – like the rest of us, they are typical human actors who are motivated by the evolved need to feel emotionally good and physically safe in small-scale groups. Religion and "quasi-religious" forms of devotion like patriotism and even love for humanity, however, can play a powerful role in mobilizing and orienting this natural motivation.

Like our first two authors, Atran argues that "all religions" involve beliefs in "bodiless but sentient souls and spirits that act intentionally." Also like them he does not hesitate to criticize such beliefs themselves as rationally inscrutable, immune to falsification, and even absurd.

---

16 Atran, *Cognitive Foundations of Natural History: Towards an Anthropology of Science*, Reprint edition (Cambridge: Cambridge University Press, 1993).
17 Atran, *In Gods We Trust: The Evolutionary Landscape of Religion* (Oxford: Oxford University Press, 2002).
18 Atran, *Talking to the Enemy: Faith, Brotherhood, and the (Un)Making of Terrorists* (Harper-Collins, 2010), 33. Unless otherwise noted, page numbers in the following paragraphs refer to this book.

> Imagine creatures who consistently believed that the dead live on and
> the weak are advantaged over the strong, or that you can arbitrarily
> suspend the known physical and biological laws of the universe with a
> prayer. If people literally applied such prescriptions to factual navigation
> of everyday life they likely would be either dead or in the hereafter in
> short order – too short for most individuals to reproduce and the species
> to survive (432).

It seems that ideas of supernatural agents would be counterproductive to survival. So how are the gods conceived and why do they stay around? Atran's
answer weaves together the two theogonic mechanisms that I have called
anthropomorphic promiscuity and sociographic prudery.

More than Lewis-Williams and Boyer, however, his portrayal of the pattern
involves detailed attention to the affective and collective security that religion
can provide for people. In his view, this is the key to understanding why religion survives and even thrives more easily than science and secular reasoning
in many modern contexts. Atran argues that religion is not naturally selected
as an adaptation to the environment; it is not innate to human beings. Rather,
it is the result of a "tricking and tweaking" of naturally evolved mental mechanisms and universal sensibilities in our species – a process that "creates religion from cognition." He briefly traces the historical spiral toward ever larger
and more complex human polities that was nurtured and sustained by this
cultural tricking and tweaking of "various aspects of our biologically evolved
cognition in order to cope with a self-generating epidemic of warfare between
expanding populations" (39).

Atran points to studies that show how both children and adults sponta-
neously interpret even the contingent movement of dots and triangles on a
computer screen as interacting agents, detecting intentionality and invent-
ing narratives to make sense of their movement. Humans have evolved an
automatic and "hair-triggered" detection of agency in ambiguous contexts as
a response "to potential threats (and opportunities) by intelligent predators
(and protectors)." This also helps to explain why the majority of supernatural
agents worldwide tend to be either malevolent deities (on the model of predator) or benevolent deities (on the model of protector). This easily trip-wired
tendency to detect agents that might hurt or help us has been manipulated to
serve cultural ends that are far from the adaptive tasks for which they evolved,
including the collective engagement of existential needs for affective security
and anxieties about deception and death (438–440).

These powerful motivations make it an easy move from imagining invisible agents to believing in their actual existence. Once conceived, the gods are

easily borne within human groups. Atran explains that although, or actually because, these concepts "have no consistent logical or empirical connection to everyday reality" (430), they can provide a mechanism by which ordinary cognitive processes can be exploited to produce passionate displays of commitment to the group. Costly displays of commitment to "preposterous beliefs" are more powerful signals of one's solidarity than commitment to a mundane belief that is verifiable (or falsifiable). Such displays and signals often take the form of participating in (otherwise non-productive) rituals, painful initiation ceremonies and sometimes even lead to a willingness to sacrifice one's life for the group. He suggests "collective commitment to the absurd is the greatest demonstration of group love that humans have devised" (450).

For Atran, religion can be roughly described as a community's costly and hard-to-fake commitment to a counterintuitive world of supernatural causes and beings. Why must the commitment be hard-to-fake? The human capacity to imagine and communicate to others about counterfactual worlds brings with it a greater potential for deception and defection. The latter weaken the ability of a community to survive, especially as it gets larger and more complex. "Clerics, rulers and elders can only intermittently monitor peasants, workers and youth to verify that commitments to God, country and authority are kept" (144). A shared belief in supernatural predators and protectors who can be imaginatively engaged in ways that activate deep human affections, and who are always monitoring the behavior of the coalition's members, provides a powerful solution to this emergent cultural problem. Those who actually believe in such agents (generally) give the most genuine signals of commitment, which strengthens the belief of others and reinforces the cohesion of the community.

Unfortunately, the dark side of in-group cohesion is out-group violence. Ironically, and tragically, the most common reactions to terrorists have simply reinforced (and illustrated) the spiraling effect of these mechanisms that helped our ancestors survive but now threaten our long-term affective and collective security.

Supernatural Selection

Matt Rossano's *Supernatural Selection: How Religion Evolved* (2010) is his first book-length offering to the bio-cultural study of religion, but its scope and integrative coherence merit its inclusion in this review. The title might lead one to expect a treatise on intelligent design but as the sub-title suggests, Rossano acknowledges that the evolution of religion can be explained without appeal to the actual existence of supernatural agents. Unlike the other three authors, however, he prefers to leave open the issue of the reality of the gods, suggesting that religion is a matter of subjective relationship and so outside the objective realm of science.

I will return to this issue below, but first let us set out his version of the integrated theogonic mechanisms by which the gods are born(e). Rossano provides a concise summary of his hypothesis about the role of supernatural agents in the evolution of religion:

> Sometime between the disappearance of Homo sapiens from the Levant (about 100,000 ybp [years before present]) and the Upper Paleolithic in Europe (about 35,000 ybp) some of our ancestors thought up the idea of a supernatural world. The world they envisioned was inhabited by spirits, ancestors, and gods who kept a close eye on them, ready to pounce on the first signs of deviance. The idea was an evolutionary winner. Groups who had it fanned out across the globe and quickly overwhelmed those who didn't.[19]

Why did these groups "win"? Once again we find that components of anthropomorphic promiscuity and sociographic prudery are at work. Our ancestors survived (and passed on their genes and cognitive traits to us) because their social and moral life was "supernaturalized," which enhanced the cohesion and health of the coalitions in which they lived in a variety of ways, many of which we have already observed above.

Part of what makes Rossano's contribution so helpful is the clarity with which he tells the story as he weaves together insights from cognitive science, moral psychology, archaeology and other fields. In what follows, I offer a brief summary of his reconstruction of the evolution of religion. Like the other three authors, Rossano argues that humans were moral and social long before they were "religious." Belief in and ritual interaction with supernatural agents was added to human social life relatively recently. By around 500,000 ybp, our hominin ancestors were already socially bonding by singing and dancing around campfires. By around 100,000 ybp, anatomically modern humans (Homo sapiens sapiens) moved out of Africa and into the Levant, i.e., the western part of what we now call the Middle East.

This first time that our ancestors went "out of Africa" they were confronted by groups of Neandertals with whom they had to compete for resources; this

---

19   Rossano, *Supernatural Selection*, 60. Unless otherwise noted, page numbers in the following paragraphs refer to this book.

and other factors such as adverse environmental conditions led them to return to Africa by around 90,000 ybp. Rossano calls the 30,000 years that followed the "African Interregnum." During this period, our genetic ancestors faced more ecological crises and resource shortages and nearly went extinct. Those small bands of humans that learned new strategies for dealing with inter-group cooperation and competition survived. Slowly, human coalitional forms transitioned from hunter-gatherers in smaller and more egalitarian groups to larger and more socially stratified groups. These larger coalitions required more complex social bonding rituals, which played a crucial role in holding them together.

More than the other authors, Rossano highlights childhood imaginative capacity as a main spring for the conception of supernatural agents. As social life became more complex, human imaginative capacity was enhanced through natural selection. New rituals would have taxed attention and working memory, and as social interaction became more demanding, the fitness advantage would have been with imaginative children. Rossano suggests that as these children grew into more socially intelligent adults, they further enhanced the affective and collective intensity of rituals by incorporating engagement with imagined supernatural agents. With the addition of "the supernatural" to social life, rituals were expanded from simple singing and dancing to more complex rites that involved initiations, trust-building and eventually shamanistic healing.

Being more imaginative may also have been correlated to greater susceptibility to the health-enhancing placebo effects of such rituals. In this sense, religion would have contributed to the physical health of the group. Adding supernatural agents not only made people healthier, argues Rossano, it also made them "nicer." The imagined presence of gods (punitive ghosts, spirits, etc.) helped to motivate people to give up some of their self-interest, and comply with the social norms of the group. It also motivated them to punish those who did not comply, or even those whose costly signals of commitment were not sufficiently convincing.[20] The presence of supernatural agents who were always watching, and who were social players who had the power to bring misfortune, went a long way toward solving this difficulty of dealing with potential defectors and deceivers.

---

20   For an introduction to the idea of religion as a form of "costly signaling," see Sosis, "Religious Behaviors, Badges, and Bans: Signaling Theory and the Evolution of Religion," In *Where God and Science Meet,* ed. McNamara and Wildman (New York: Praeger, 2006). 1(2006).

Such adaptive advantages gave "religious" hominin groups a decisive edge over "secular" groups, who "eventually went extinct" (198). Around 60,000 ybp these thriving groups of anatomically modern and robustly religious humans once again left Africa and entered the Levant. This time they outcompeted all other hominin groups. Between 50–30,000 ybp, the supernaturalization of their social life was further complexified by the emergence of ancestor worship, which was reinforced by religious narratives in which the living and the dead were construed as part of a larger coalition. Such narratives provided justification for the increasingly stratified societies that emerged during the Neolithic and beyond. Rossano's account stops here, but of course the story of the evolution of religion continues, through the emergence of complex literate states and the rise of the major religious traditions of the axial age in east, south and west Asia. Indeed, the story continues, and all of us – scientists engaged in the bio-cultural study of religion included – are inexorably emplotted within it.

The Philosophical, Psychological, and Political Significance of the Bio-Cultural Study of Religion

All four of the authors just reviewed are active scientists who have gathered empirical data and formulated complex hypotheses about the evolution of religion, cognition, and culture. All four have also explicitly addressed the relevance of their research for some of the major challenges facing contemporary *Homo sapiens*. In fact, this is increasingly common in the literature in which scientific disciplines overlap in their study of religion.[21] In my view this is a laudable development. In this concluding section, I briefly point out some of the similarities and differences in the way in which these scientists attempt to bring their work to bear on broader public and academic discourse on significant issues in philosophy, psychology, and politics.

First, it is quite clear that the bio-cultural study of religion has an important role to play in wider debates in the philosophy of science, especially insofar as it confirms the inherently embodied and contextual nature of all human knowledge. It is commonly agreed that science is and ought to be guided by *methodological* commitments that embrace anthropomorphic prudery and sociographic promiscuity. In other words, scientific experiments and theories

---

21   See, e.g., Bulbulia et al., eds. *The Evolution of Religion: Studies, Theories, & Critiques* (Santa Margarita, Calif: Collins Foundation Press, 2008); Schaller et al., eds., *Evolution, Culture, and the Human Mind* (New York: Psychology Press, 2009).

should avoid appealing to supernatural agents or authorities in their causal explanations.

Whether scientists should have a *metaphysical* commitment to naturalism is more controversial. Our first three authors are relatively straightforward about their metaphysical commitments, which they find reinforced by their empirical and conceptual work. It is implausible to believe that human cognition is somehow conditioned by intercourse with disembodied and intentional wielders of causal power, as the vast majority of human beings have believed, including most philosophers of science from Plato to Descartes.

Rossano stands out from the others in this regard. Although he agrees with the other three that the evolution of religion can be explained naturalistically by the integration of what I have called theogonic mechanisms, and without recourse to appeals to supernatural agency, he is hesitant to deny the existence of the latter. Because science deals only with the "objective realm" and gods are in the "subjective realm" of relational experience, science cannot decide the question of their existence. He insists "you cannot tell someone they don't have a relationship that they are convinced they have ...(or) force someone into a relationship they are not interested in."[22] From the point of view of the philosophy of science, such comments appear to be remnants of early modern and positivist dichotomies such as subject/object and value/fact. Moreover, in my view, arguing about the possibility or even probability of the existence of supernatural agents is a red herring; the discussion should focus on the plausibility of their hypothesized existence.[23]

Second, each of our authors touches on the psychological relevance of the findings of the bio-cultural study of religion. Although he recognizes some of the deleterious effects of belief in gods, Rossano emphasizes, more than the others, the physical and psychological health benefits of religion. As we have seen, Atran also points to the importance of affective security in small-group dynamics, which can indeed be reinforced by shared imaginative engagement with supernatural agents. However, he pays more attention to the negative psychological distress and emotional hostility toward members of out-groups that such engagement can intensify. Boyer rarely points to the potential positive effects of belief in gods. Although he resists simplistic "religion is the sleep of reason" hypotheses, he expresses concern about the "double consciousness"

---

22   Rossano, *Supernatural Selection*, 27.

23   This is spelled out in more detail in Shults, "The Problem of Good (and Evil): Arguing about Axiological Conditions in Science and Religion," in *Science and the World's Religions, Volume I: Origins and Destinies*, ed. Wildman and McNamara (New York: Praeger, 2012), 39–68. See also Chapter 5 below.

evident in so many contemporary religious people in the secular west. They must express absolute commitment to their beliefs within the context of their religious coalition, but live in the public sphere as though such doctrines are optional or irrelevant.

Lewis-Williams is less tolerant of calls to "respect" those who believe in supernatural agents, referring to members of religious coalitions as "the gullible" rather than "the faithful." He repeatedly points to ways in which taking beliefs in gods seriously can be maladaptive in contemporary life, and to examples of the sanctioning and convoluted justification of "morally repugnant nonsense" that religious elites try to force upon their followers. Unfortunately, this kind of rhetoric can all too easily make things worse, amplifying psychological mechanisms and religious biases that lead people to defend the coalitions with which they identify.

I find Atran's rhetorical approach more promising. More so than Lewis-Williams and Boyer, he attempts to empathize deeply with religious people's affective needs. Unlike Rossano, however, Atran then clearly explains why he believes that gods are merely a result of the "tragedy of cognition," a mental by-product of a strategy for dealing with existential anxiety about deception and death. This suggests the need for developing new strategies for managing anxiety within and across religious "families of origin."[24] Other recent examples of treatments of the psychological significance of the bio-cultural study of religion that do not denigrate the human passion for experiences of intensity include Wildman's *Religious and Spiritual Experiences*[25] and Strozier's *The Fundamentalist Mindset*.[26]

Finally, the political relevance of the bio-cultural study of religion is also explicitly thematized by each of our four authors. For reasons partially connected to his approach to the subject matter, Atran is the most explicit. Throughout his book, he urges public policy makers to take seriously the need for studying "the sacred," which he believes must include ethnographic research and interviews on the ground by qualified scientists. In other words, we must "talk to the enemy" to understand and facilitate healthier ways of inscribing our social lives in a globalizing world. Rossano deals with these concerns briefly at the end of his book, wondering whether it is possible to have the benefits of religion without the intergroup competition out of which it emerged. Boyer

24   Shults, "Transforming Religious Plurality," *Studies in Interreligious Dialogue* 20, no. 2 (2010).

25   Wildman, *Religious and Spiritual Experiences* (Cambridge University Press, 2014).

26   Strozier et al., eds., *The Fundamentalist Mindset: Psychological Perspectives on Religion, Violence, and History* (New York: Oxford University Press, 2010).

and Lewis-Williams are more expansive in their treatment than Rossano and more expressive in their negative evaluation of the political effects of religion than Atran.

Another recent contribution to the bio-cultural study of religion that bears on this theme is John Teehan's *In the Name of God: The Evolutionary Origins of Religious Ethics and Violence*.[27] Although it too was published in 2010 I did not review it above alongside the other four books because it takes a quite different approach that focuses on the study of holy texts, specifically those of the Abrahamic religions. Teehan demonstrates how the ethical codes of these monotheistic traditions are expressions and extensions of, rather than exceptions to, the natural moral intuitions that emerged through the evolution of human brains in social groups. For example, moral commands that seem to elude evolutionary explanations because of their apparent irrelevance for survival, or their radical calls for altruism beyond kin groups, can be interpreted in light of costly signaling theory, which we described briefly above. Teehan also makes a compelling case for the claim that although religion produces powerful examples of altruistic behavior, it also amplifies the mental and social strategies that foster in-group/out-group violence.

Throughout most of this chapter, I focused on illustrating the explanatory power of hypotheses about the role of various components of anthropomorphic promiscuity and sociographic prudery as god-bearing mechanisms in the origin and evolution of religion. Here in the conclusion I have identified another trend within the bio-cultural study of religion: the increasing willingness of scientists to bring their research into wider academic and public discussions about normative concerns. This is a tricky business – as perilous as it is promising. It may not be a trend that all readers of *Religion, Brain & Behavior* will want to follow or even applaud. However, I hope that its pages will sometimes include forays into these lively and important conversations. In this way, the journal may be able to contribute in its own way to some of the greatest adaptive challenges facing humanity in its current natural (ecological) and social (global) environment.

The following chapters represent some of my own attempts to show how bringing the insights of the bio-cultural study of religion to bear on these concerns can help us learn to think and to act together in psychologically and politically healthy coalitions – without bearing gods.

---

27   Teehan, *In the Name of God: The Evolutionary Origins of Religious Ethics and Violence*, (Malden, MA: Wiley-Blackwell, 2010).

CHAPTER 3

Excavating Theogonies

Where do *babies* come from? Archaeologists do not need to dig around for an answer to this question as they attempt to understand and explain the empirical data uncovered at sites in the Neolithic or elsewhere.[1] They certainly need to search for plausible hypotheses to illuminate the vital kinship structures, pregnancy rituals, birthing practices, and neonatal health care policies of any specific community. However, if the community was composed of anatomically modern *Homo sapiens*, archaeologists can appropriately assume that infants appeared within the population as a result of the same basic procedures that produce them today, when ... well, you know.

Where do *gods* come from? In this chapter I will argue that archaeologists (as well as other scientists, philosophers, and theologians) can now also appropriately assume that the reproduction of supernatural agents in ancient civilizations occurred in much the same way that it does today. Although we have known where babies come from for several millennia, only within the last few decades have we come to understand more fully why gods appear (and are cared for) in human populations. As with the process of bearing children, one finds an astonishing variety of ways of ritually surrounding and socially manipulating the process of bearing supernatural agents. Beliefs about and behaviors toward the latter are regulated and transmitted differently in the major religious traditions that were forged within complex literate states during the axial age and now dominate the global landscape. Nevertheless, all members of our species share a phylogenetic heritage that includes sets of cognitive and coalitional tendencies, whose interactions help to explain why gods are so easily born(e) in minds across cultures.

This is the first purpose of this chapter: to offer another brief reconstruction of some relevant theoretical advances in the bio-cultural study of religion, pointing to the convergence of insights from a variety of disciplines around the two conceptual attractors that I call *theogonic* mechanisms (anthropomorphic promiscuity and sociographic prudery). I demonstrate the potential illuminative power of these advances in relation to the empirical findings at

---

1 This chapter is an adapted version of "Excavating Theogonies: Anthropomorphic Promiscuity and Sociographic Prudery in the Neolithic and Now," originally published in *Religion at Work in a Neolithic Society: Vital Matters*, ed. Ian Hodder (Cambridge: Cambridge University Press, 2014).

© F. LERON SHULTS, 2018 | DOI 10.1163/9789004360952_004

Çatalhöyük, a Neolithic archaeological site in southern Turkey. In the field of archaeology I am an amateur (in both senses of the word), and will not pretend to offer expert analyses of the data. And so this chapter has a second, more philosophical and perhaps even more daring, purpose: to explore some of the implications of the unearthing of these theogonic mechanisms for the shared global future of the human race, attending to the adaptive challenges and opportunities that must be faced today in light of our new understanding of the impact that bearing gods has on our mental and social well-being.

In an earlier analysis of the material, social, and spiritual entanglement at Çatalhöyük, I used the term "religion" in a broad sense to indicate the way in which humans symbolically engage what they take to be of ultimate value.[2] In the current context, however, I am focusing more narrowly on a particular feature of human life that also appears across cultures: *shared imaginative engagement with axiologically supernatural agents*. In this sense, "religion" was entangled within and developed alongside all of the other vital matters that shaped human evolution. In the conclusion, I will emphasize the philosophical, psychological, and political significance of the unveiling of the mystery of god-bearing mechanisms, which, like Girard's scapegoat mechanism, only work well when they are hidden.

The bulk of the chapter is a conceptual excavation and reconstruction of ways in which some components of anthropomorphic promiscuity and sociographic prudery may have operated at Çatalhöyük. How was their imaginative interaction with supernatural agents (such as the spirits of ancestors and aurochs) vitally entangled with the material and social dimensions of their lives? To what extent were their production of food and artifacts and their regulation of communal property shaped by their perception of the causal power and social relevance of such agents? It seems to me that bulls, burials, and proprietary production were all mixed together at Çatalhöyük. The first step, however, is to provide another brief summary of the general conceptual framework that will guide my archaeological and philosophical observations.

"Bearing Gods" in the Neolithic

It should be clear enough that my use of the term *bearing* plays a double function, indicating the naturally evolved processes by which gods are *born* in human cognition (by the over-active detection of agency) and *borne* in human

---

2 Shults, "Spiritual Entanglement: Transforming Religious Symbols at Çatalhöyük" in *Religion in the Emergence of Civilization*, ed. Hodder (Cambridge University Press, 2010).

culture (by the over-active protection of coalitions). Here I will continue using the term *god* as basically synonymous with *supernatural* agent; that is, a disembodied (dis-embodi-able, or at least ontologically confused) intentional force that is imagined to have some interest in and causal power over the members of a religious in-group. Despite – or because of – their ontologically confused status, these *agents* are believed to be capable of playing some kind of constitutive and/or regulative role in the social life of a particular human coalition. In this sense, the monotheistic idea of "God" also falls within this category, although it has distinctive features that need to be parsed out in other contexts for different reasons.[3]

My interest here is in uncovering some of the general mechanisms that condition all kinds of *theogonies*. I am using this latter term not in the narrow sense of popular literary accounts of the genesis of the gods, such as Hesiod's graphic portrayal of Cronos' swallowing of divine offspring and mutilation of titanic genitals, but more broadly as a way of referring to any narrative imaginative engagement that reinforces the detection and protection of a particular supernatural agent coalition.

Accepting the risk of blurring still other important distinctions within and across disciplines, I offer a heuristic model of the integration of these mechanisms, which is based on my reconstructive reading of recent empirical findings and theoretical reflections across a variety of fields including archaeology, cognitive science, evolutionary neurobiology, moral psychology, history, social anthropology, economics, and political science. As we have seen, and will see in more detail below, several trends within these and other disciplines converge in supporting the general hypothesis that gods are born(e) as a result of evolved human tendencies to over-detect agents in the natural environment and to over-protect coalitions in the social environment. These cognitive and cultural strategies contributed to the survival of hominid groups before, during, and after the Neolithic. The key question today is whether they are still healthy strategies for adapting in our rapidly changing, pluralistic environment. Are there other directions we could or should pursue?

The conceptual framework depicted in *Figure 1* (see Chapter 1, page 3) can help clarify the options. The level of generality at which human tendencies are depicted on this grid does not allow us to capture all of the nuances within the various theories on offer within the many disciplines we will explore. However, it does capture precisely what is needed to accomplish the general purpose for which the framework has been constructed: clarifying the relation between

---

3   See, e.g., Shults, "The Problem of Good (and Evil): Arguing about Axiological Conditions in Science and Religion," and Chapters 6 and 7 below.

two basic tendencies found among *Homo sapiens,* the integration of which leads to the reproduction of shared imaginative engagement with axiologically relevant supernatural agents.[4]

To review, the horizontal line represents a spectrum on which we can mark the tendency of persons to guess "human-like supernatural force" when confronted with ambiguous phenomena in the natural environment. The anthropomorphically *promiscuous* are always on the lookout, jumping at any opportunity to postulate such agents as causal explanations even – or especially – when these interpretations must appeal to disembodied intentionality. The anthropomorphically *prudish*, on the other hand, are suspicious about such appeals. They tend to reflect more carefully before giving into their intuitive desire to grab at explanations that refer to discarnate intentional entities.

The spectrum represented by the vertical line registers the extent to which a person inflexibly holds on to the modes of inscribing the social field favored by his or her religious in-group; i.e., to the proscriptions and prescriptions that regulate the evaluative practices and boundaries of the supernatural coalition(s) with which he or she primarily identifies. Sociographic *prudes* are strongly committed to the authorized social norms of their coalition, following and protecting them even at great cost to themselves. They are more likely to be suspicious of out-groups and to accept claims or demands that appeal to the supernatural authorities within their own coalition. The sociographic *promiscuity* of those at the other end of the spectrum, on the other hand, leads them to be more open to intercourse with out-groups about alternate normativities and to the pursuit of new modes of creative social inscription. Such persons are also less likely to accept restrictions or assertions that are based only or primarily on appeals to religious conventions.

Most human beings today are intuitively drawn toward anthropomorphic promiscuity and sociographic prudery, which are integrated in the lower left quadrant of *Figure 1*. Why? This is due, in part, to the inheritance of traits that were naturally selected (or by-products of other adaptations) that evolved in early ancestral environments in which survival advantage went to hominids

---

4   In the original version of this essay, I utilized definitions of anthropomorphic promiscuity and sociographic prudery that were broader than those stipulated in Chapter 1 above. In that earlier context, I did not tie those mechanisms so explicitly to *religious* biases. Instead, I referred to them as general cognitive and coalitional tendencies that could (and most often did) contribute to the detection of supernatural agents and the protection of supernatural coalitions. This was also my approach in the original versions of some of the other central chapters in this book. In each case, my revision of these essays has included altering these definitions so that they correspond to the hypotheses of theogonic reproduction theory as articulated in Chapter 1.

whose cognitive capacities enabled them to quickly detect relevant agents (such as predators, prey, protectors and partners) in the *natural* environment, and whose groups were adequately protected from the dissolution that could result from too many defectors and cheaters in the *social* environment.

A growing body of evidence suggests that the chance of survival would have been increased for those small groups of *Homo sapiens* who developed beliefs and rituals related to supernatural agents around 90–70,000 years ago. The integration of these mutually reinforcing theogonic mechanisms was highly adaptive. Sometime around 60,000 years ago it appears that some of these "god-bearing" groups left Africa, out-competing all other hominid species and spreading out across the Levant and into Europe and Asia. All living humans are the genetic offspring of these groups, and so share a suite of inherited traits that support the tendency to infer the presence of supernatural agents and prefer the social norms of their own supernatural coalitions. These naturally evolved traits were tweaked differently in various contexts, which led to the diversity of manifestations of religious life today. Supernatural agent conceptions are never immaculate; the particular features of our gods betray our religious family of origin.

The Neolithic is a particularly fertile time period for the purposes of excavating theogonies. Ancestor worship (or at least imaginative engagement with dead ancestors) had probably already emerged sometime between 50–30,000 years ago, and the upper Paleolithic was characterized by an explosion of innovations in tool-making, art, and burial elaboration. However, the Neolithic was "revolutionary" in many ways, most notably the shift toward sedentism and the domestication of plants and animals. For most of the 20th century these developments were interpreted as the result of human responses to environmental changes and new modes of controlling material production and social organization, which in turn provided the conditions for the emergence of religion. Today, however, theoretical reflections on empirical research from a variety of disciplines have converged to suggest that imaginative engagement with supernatural agents played a generative and regulative role in the Neolithic "revolution."

Published at the turn of the century, Jacques Cauvin's *The Birth of the Gods and the Origins of Agriculture*[5] provides a illuminative example of this trend. He includes an assessment of Çatalhöyük within a broad overview and analysis of a variety of finds from earlier in the Natufian to the diffusion and eastern spread of agriculture later in the Neolithic. Cauvin argues that the key

---

5 Cauvin, *The Birth of the Gods and the Origins of Agriculture*, trans. T. Watkins (Cambridge University Press, 2000).

to the transformation was the development of symbolic imagination and a mythical interpretation of the natural world. In other words, religion was not an after-effect of changes in managing the material world (as in some Marxist hypotheses), but ingredient to the transformation of the human mind that made the Neolithic revolution possible. To put it bluntly, in some sense "religion" was a causal factor in the rise of domestication and sedentism, or at least, as Ian Hodder would put it, "entangled" within the rise of these developments.

Despite his provocative title, which fits nicely with the metaphor guiding the reflections of this chapter, Cauvin's analysis does not really deal with the origin of the *gods* per se, but with the emergence of symbolic and mythical interpretations of the world. I agree that religion played a creative role in this revolution, but want to suggest that closer attention to the actual mechanisms by which *supernatural agents* are born(e) can complement these broader reflections. Given my interest in the fertility of the mental and social fields of Çatalhöyük during the Neolithic, one might expect me to focus on the well-known imagery of the "goddess" so often associated with that site. Unlike Cauvin, however, I do not find the "Goddess and Bull" mythology inspired by Mellaart's earlier interpretations compelling.

As Hodder and Meskell argue, it seems more likely that Çatalhöyük was characterized more by the kind of phallocentrism typical of other sites in the region such as Gobekli Tepe.[6] My interest is not primarily in the perceived sexual antics performed – or the alleged reproductive assistance provided – by any female or male divinities in the mythological "spirit world" of Çatalhöyük, but in the mechanisms by which the god(desse)s themselves were born(e) within the embodied and encultured cognition of its inhabitants. Moreover, that interest itself is driven not only by a fascination with the original revolutionaries of the Neolithic but also with the *current* "inhabitants" of Çatalhöyük, that is, the interdisciplinary and international team of researchers who ritually descend upon it every summer.

Unlike those whose bones, belongings, and abodes they study, for the most part these scientists are anthropomorphically prudish (suspicious of causal explanations that appeal to supernatural agents) and sociographically promiscuous (seeking out the insights of other disciplinary coalitions). In the conclusion of this chapter I will return to the conceptual apparatus of *Figure 1* and examine the tension this creates among the various shareholders interested in the revelations of those who dig the site Now. In the next three sections,

6 Hodder and Meskell, "The Symbolism of Çatalhoyuk in Its Regional Context," in *Religion in the Emergence of Civilization: Catalhoyuk as a Case Study*, ed. Hodder (New York: Cambridge University Press, 2010).

however, my focus will be on the current team's research questions related to power, production, and property in Çatalhöyük during the Neolithic, exploring the extent to which these dynamics may have been entangled with the theogonic mechanisms outlined briefly above.

Burials, Bulls, and Proprietary Production

Perhaps there were no "goddesses" at Çatalhöyük , but there is strong evidence in the data that supports the claim that its original inhabitants engaged in behaviors intended to engage supernatural agents whom they considered to be socially and causally relevant to their coalition. In his analysis of the site in *The Leopard's Tale,* Ian Hodder suggests that "as people, society and crafted materials increasingly became entangled and codependent, so the codependent material agents were further enlisted and engaged in a social world in which spirits were involved."[7] He also explicitly proposes a link between "control of knowledge about and the objects of the spirit world" and the acquisition and maintenance of rights, resources, status and prestige in the community (250). In this context, I take terms like "spirits" and "spirit world" to be roughly synonymous with what I have been calling supernatural agents and coalitions.

My concern is with the role played by theogonic mechanisms in this entanglement. We could point to many instances of material objects that indicate shared imaginative engagement with "gods" in the broadest sense. For example, the polishing and caching of obsidian mirrors may have been conceived as a way of seeing or divining the "spirit world" (229, 239). More speculatively, one might argue that the making of figurines could be an expression of a growing awareness of and interest in detecting or even controlling human-like agents, i.e., anthropomorphic promiscuity. Moreover, the repetitive patterns of architecture and art within houses and across levels suggest a rather prudish sociography. At any rate, I want to focus here on the two types of supernatural agents that seem particularly prevalent within the socius: aurochs and ancestors.

I see dead people. But like most of my cosmopolitan colleagues, I see them rarely, briefly, mostly at funerals, usually once, and only one at a time. The original inhabitants of Çatalhöyük, on the other hand, saw dead people much more often. Indeed, one of the most distinctive features of the "town" is the burial of some of the dead within the houses, often immediately under the

---

7 Hodder, *Catalhöyük: The Leopard's Tale – Revealing the Mysteries of Turkey's Ancient "Town"* (New York: Thames & Hudson, 2006), 195. Unless otherwise noted, page numbers in the following paragraphs refer to this book.

main sleeping area. It is hard to know whether this was comforting or as creepy to them as it is to us. Whatever the case, the removal, burial, and long-term retrieval of skulls as well as other forms of treating the skeletons indicates that their manipulation was perceived as an engagement with supernatural agents whom they imagined played some sort of causal role in the coalition. But human ancestors were not the only, and perhaps not even the most important, disembodied intentional entities by which the thoughts and actions of their daily lives were herded.

Holy cow. What is one to make of all the bull at Çatalhöyük? There are many types of dangerous animals represented in the painting and decoration of the houses, including bears and leopards. However, bulls (aurochs) seem to have played a particularly significant role. It would be anachronistic to call them "sacred" cows, but it appears that the aurochs were indeed set apart as dominant agents within the shared imagination of the coalition. Their buchrania in particular were a prominent part of the décor. Moreover, their positioning sometimes suggests that they bear some relevant relation to the human burials within the homes. We will return below to the possible connection between aurochs and ancestors within what Ian Hodder calls the "prowess–animal spirit–hunting–feasting" nexus that seems so important in life at Çatalhöyük.

At this point I want to emphasize two things. First, as Hodder points out, both of these types of disembodied agents within the spirit world appear to have played a special role in mediating power related to proprietary production, i.e., in providing access to the fruit of communal labor. He hypothesizes that the power of dominant groups such as elders or shamans may have been based partly on their capacity to intercede in relation to *wild animals and ancestors* (204). There may have been competition between forms of power based on the control of knowledge about ancestral ties and auroch behavior and forms of power based on domestic production and accumulation, but Hodder suggests all of these dimensions (material, social, spiritual) were entangled and mutually conditioned one another. "It seems most likely that much of the variation in elaboration of buildings, and in the number of burials, relates to the ability of household members (perhaps especially elders) to mobilize ritual, symbolism, revelation and their performance, even though exchange and production played their part" (183).

Second, the remains of both ancestors and aurochs played a special role in the ongoing process of hiding and revealing that characterized so much of the ritual behavior at Çatalhöyük. The sharp parts of animals (including bulls) were placed in walls, covered and uncovered over and over again. Human skulls (as well as sculptures and other artifacts) were buried, dug up, kept over time, and re-buried. Hodder proposes that this process of material circulation played a role in maintaining social continuity within the houses and had some bearing on status and power. "Things are hidden and then revealed. And often they are hidden in places where the *ancestors and animal spirits* are – beneath floors and behind walls. So when things return, revealed, they bring with them an aura from that other world. They have been magnified in their hidden journey" (170, emphasis added). These repetitive and apparently ritual (un)covering processes appear to be linked to both ancestry and exchange.

In what follows I hope to contribute to the ongoing unearthing of the mysteries of the Çatalhöyük community by exploring ways in which our current knowledge of the naturally evolved human tendencies toward anthropomorphic promiscuity and sociographic prudery, which together help to explain why and how gods are born(e) in the mental and social space of human life, could lead to new hypotheses about the role of supernatural agents within their daily lives in general and proprietary production in particular. Both dead ancestors and aurochs clearly qualify as *supernatural agents*, in the sense defined above. The way in which their remains were engaged indicates shared beliefs about the causal power and intentionality of these disembodied entities within the coalition. What role did the integration of theogonic mechanisms play at Çatalhöyük?

Anthropomorphic Promiscuity at Çatalhöyük

As we have seen, there is a massive and rapidly growing literature supporting the claim that most human beings have a naturally evolved tendency to over-detect agency, intentionality, and purposiveness. Neurological, psychological, and ethnographic research across cultures has demonstrated that human cognition somewhat automatically seeks out (more or less human-like) agents. Even the random movement of dots on a computer screen can easily be interpreted as "intentional" and, especially under stress, most subjects will immediately guess "agent" with little or no priming when confronted with this sort of ambiguous phenomena. Humans seem to have evolved with a "hair-triggered" cognitive mechanism for detecting agents.

Stewart Guthrie's 1993 book *Faces in the Clouds* is still the best place to start for an introduction to the issue of anthropomorphism. Although a great deal of scientific research in the intervening decades has clarified the cognitive and cultural mechanisms involved, Guthrie's book provides a clear exposition of its prevalence and a daring philosophical assessment of its importance. His cognitive theory of religion is built on the reasonable hypothesis that the survival of early humans depended on their ability to perceive any other agents – especially other people – who might be around.[8] Hypersensitivity to human-like agents leads to many false alarms (e.g., seeing faces in the clouds) but it also makes it more likely that hidden agents will be perceived when it is really important (e.g., a camouflaged enemy). For Guthrie, anthropomorphism is by definition the failure of a naturally evolved perceptual strategy, and religion is systematized anthropomorphism.

Religious anthropomorphism is often understood as consisting of the attribution of humanity to gods, but Guthrie turns this around: "gods consist of attributing humanity to the world."[9] In this sense all religions have gods or a god; they all involve "ostensible communication with humanlike, yet nonhuman, beings through some form of symbolic action."[10] The example of Buddhism is often raised as a counterexample, but although some philosophical streams of that tradition resist anthropomorphism (as do minority streams within all the axial age traditions), the vast majority of Buddhists are deeply entangled in shared imaginative engagement with all kinds of supernatural agents such as devas, bodhisattvas and, of course, Buddhas. The question before us is what communication with "gods" such as the spirits of aurochs and ancestors may have looked like at Çatalhöyük.

To my knowledge the most explicit application of the research on (what I am calling) anthropomorphic promiscuity to Çatalhöyük is in the work of David Lewis-Williams and David Pearce.[11] As we noted in Chapter 2, these authors emphasize the *neurological* basis of belief in supernatural agents, which they interpret as the result of mistakenly attributing reality to iconic hallucinations experienced during altered states of consciousness. Based on neurological studies and ethnographic work in many contemporary small-scale societies, they argue that religion evolved as those who were particularly susceptible to or adept at having such experiences (shamans) came to be understood as capable of mediating between the human coalition and the spirit world.

Building on Lewis-Williams' earlier analysis of upper Paleolithic cave art,[12] they argue that the similarity between images found in the latter, such as handprints and geometric designs reminiscent of entopic phenomena in altered states, and the images found at Çatalhöyük suggest that the houses were a "built cosmos" replacing caves as the *axis mundi* within which mediation with spirit

---

8   Guthrie., "A Cognitive Theory of Religion [and Comments and Reply]," *Current Anthropology* (1980).

9   Guthrie, *Faces in the Clouds*, 3–4.

10   Ibid., 197.

11   Lewis-Williams and Pearce, *Inside the Neolithic Mind*.

12   Lewis-Williams, *The Mind in the Cave*.

worlds can occur. Long before sedentism, human groups were participating in shared engagement with supernatural agents, especially human-animal hybrids perceived during hypnagogic states as hovering above or emerging from other worlds below. "The domestication of animals was already conceptually embedded in the worldview and socio-ritual complex we have described before people began actually herding the aurochs."[13] Here we have another example of the claim noted above, defended in different ways by Cauvin and Hodder, that symbolic imaginative engagement with "spirits" contributed to the revolution(s) of the Neolithic.

This interpretation of Çatalhöyük, whatever its other weaknesses or strengths, does not adequately incorporate some of the other popular hypotheses surrounding the phenomena of anthropomorphic promiscuity. Without downplaying the neurological basis of the over-active perception of agency, we should also note that one of the most important reasons for the hypersensitivity of this cognitive mechanism is the adaptive value of quickly detecting *predators and prey*. The predominance of dangerous animals in the art and décor of Çatalhöyük, including reliefs of leopards and bears as well as the teasing and hunting of bulls, suggests that they were particularly interested in perceptual strategies related to these agents.

It makes sense, then, that the spirits of powerful predators would have been attributed power in the spirit world as well, and that the inhabitants of Çatalhöyük would have been primed to detect them. Whatever the details, we can plausibly conclude that their evolved hypersensitivity to seeking out agents was operative in their growing attempts to find, control, respond to, and manipulate dangerous agents even (or especially) in the spirit world. Shared engagement with such imagined agents may well have led to new strategies for finding and controlling actual agents in the natural environment and contributed to domestication and sedentism.

But of course predators and prey are not the only agents that are important to detect; for the species to survive, humans also need to find *protectors and partners*. Here is where the ancestors come in. *Homo sapiens* in the Neolithic, like their forebears and descendants, were born with a tendency to seek out protectors (usually parents) and developed an interest in seeking out partners (potential sexual mates). Our attachment to these embodied human agents does not suddenly disappear when they are not around; even after their death we feel emotionally connected to them. The powerful cognitive mechanics of anthropomorphic promiscuity continue to grind away and, given the significance of our attachment to such care-giving figures within the working models

---

13   Lewis-Williams and Pearce, *Inside the Neolithic Mind*, 141.

by which we navigate life, it is easy to understand why we remain predisposed to perceive their presence.

In the "attachment theory" developed by John Bowlby and others,[14] the dynamics of the human behavioral system are described in explicitly evolutionary terms. Systems in which infants actively sought attachment with care-givers, and care-givers quickly detected and responded to the needs of infants, were naturally selected, and such dispositions became stronger over time. More recently, researchers have demonstrated that the attachment styles developed in infancy and childhood continue to affect adult life, especially in close and romantic relationships.[15] Moreover, this deeply embedded drive for attachment also shapes people's relation to their perceived divine attachment figures, at least in the case of images of "God."[16]

In other words, our naturally evolved hyperactive longing for attachment with embodied human agents easily spills over into a promiscuous seeking for and imaginative engagement with supernatural partners and protectors. Of course gods are not always (or even usually) nice and supernatural agent coalitions are just as (if not more) likely to include fearsome predators as they are potential caregivers. For our purposes, however, the main point is that the data at Çatalhöyük can be illuminated in light of such theories. In my view, the most compelling example is the well-known case of the skeleton of a woman buried embracing a plastered skull. However, we could also point to the burials of infants, figurines or even animals as examples of evidence that our Neolithic ancestors continued to detect the presence of their own dead ancestors (and others) with whom they had developed significant attachments.

We should be suspicious, as always, of overly speculative interpretations of the archaeological evidence, but if the inhabitants of Çatalhöyük were anatomically modern humans with the same basic cognitive mechanisms that we have today, we have good reasons to suspect that the data is at least susceptible to such explanations. However, this is not the whole story. Supernatural agents may be born through overly-sensitive cognitive mechanisms for detecting relevant intentional forces but this does not explain why human families

---

14   Bowlby, *Attachment* (New York: Basic Books, 2008), John Bowlby, *A Secure Base: Clinical Applications of Attachment Theory* (London: Taylor & Francis, 2005).

15   See, e.g., Mikulincer, *Attachment in Adulthood: Structure, Dynamics, and Change* (Guilford Publications, 2007); Rholes and Simpson, eds., *Adult Attachment: Theory, Research, and Clinical Implications*, (New York: The Guilford Press, 2004).

16   Kirkpatrick, *Attachment, Evolution, and the Psychology of Religion* (New York: The Guilford Press, 2004).

continue to bear responsibility for taking care for them. It takes the overly-sensitive protection of a village to raise a god. Or, in the case of Çatalhöyük it took a Neolithic "town."

Sociographic Prudery at Çatalhöyük

The claim that religion plays a role in holding together human groups is hardly new. In fact, for much of the 20th century theories (like Durkheim's) that posited a social function at the root of religion were more influential than views (like Tylor's) that posited belief in "spiritual beings" as its essential characteristic. In the last few decades, however, empirical findings and theoretical reflection across the disciplines that study religion have contributed to an integration of these intuitions. The cognitive mechanisms that give rise to belief in supernatural agents and the coalitional mechanisms that hold groups together are mutually reinforcing. While one finds a natural and healthy competition among scientific hypotheses on these topics, they are for the most part complementary and even convergent.

But how have these developments affected the interpretation of Çatalhöyük? Lewis-Williams and Pearce have proposed that shamans played a key role in regulating the "social contract" of its inhabitants as a way of dealing with their shared interpretations of the introverted end of the consciousness spectrum.[17] In other words, like all societies, they had to develop a "consciousness contract" as a way of dealing with their experiences of altered states of consciousness. The presence of entopic patterns and other representations in the art and architecture of Çatalhöyük that are reminiscent of shamanic cultures suggests to Lewis-Williams and Pearce that their social organization was structurally similar to that of others across the world. However plausible this may be, it does not go very far in explaining the actual mechanisms that produced and maintained the sociographic prudery of this particular Neolithic town.

Whitehouse and Hodder have pressed further by applying Whitehouse's "modes of religiosity" theory to Çatalhöyük.[18] They point to evidence that suggests a slow shift from primarily "imagistic" toward more "doctrinal" modes

---

17   Lewis-Williams and Pearce, *Inside the Neolithic Mind.*
18   For an introduction to this theory, see Whitehouse, *Modes of Religiosity: A Cognitive Theory of Religious Transmission* (Walnut Creek, CA: AltaMira Press, 2004), and Whitehouse and McCauley, eds., *Mind and Religion: Psychological and Cognitive Foundations of Religion* (Walnut Creek, CA: AltaMira Press, 2005).

of transmission during the 1,600 year settlement.[19] The former mode is characterized, among other things, by emotionally intense rituals with low frequency, while the latter mode involves more frequent but less intense rituals. They take the transition from extensive use of buchrania in the lower levels to a growing presence of stamp seals and pictorial narratives in the upper levels to indicate that the coalition became increasingly "doctrinal" over the centuries.

They also suggest that as shamans (loosely defined) came to develop discursive and narrative strategies for transmission the settlement became characterized by more standardized ways of engaging and more authoritative interpretations of the spirit world. This highlights the importance of Çatalhöyük as a transitional site, which may have "paved the way for more centralized, large-scale and hierarchical patterns of political association."[20] One of the values of this theory is the way in which it illuminates the link between cognitive and coalitional structures at Çatalhöyük. The social morphology of the town became more complex as a result of the relation between divergent modalities of ritual transmission. Examining the archaeological data in this light will likely lead to new insights about the impetus for such change, identifying patterns that would otherwise have been missed.

It seems to me that there are at least two other types of theories related to sociographic prudery that could complement this proposal and lead to additional insights about the actual mechanisms at work at Çatalhöyük. The first type has to do with the role of supernatural agents in *moral evolution*.[21] If organisms survive by taking care of themselves, why do we find apparently altruistic behavior in human life, such as actions in which an individual sacrifices her needs for the group? Some of the most popular scientific answers to this question these days are variants of the claim that the imagined presence of "gods" helped to solve the problem of cooperation within coalitions. It is often in the best interest of an individual to defect or cheat, especially if they can do so without being caught. However, if one is convinced that supernatural agents, who have fuller access to knowledge about socially relevant human actions and the power to bring or hinder misfortune, may be (or always are) watching, one is more likely to follow the rules that hold the group together.

---

19   Whitehouse and Hodder, "Modes of Religiosity at Çatalhöyük," in *Religion in the Emergence of Civilization: Çatalhöyük as a Case Study* ed. Hodder (Cambridge: Cambridge University Press, 2010).

20   Ibid., 142.

21   See, e.g., Hauser, *Moral Minds: The Nature of Right and Wrong*, (New York, NY: Harper Perennial, 2007), and Pyysiäinen and Hauser, "The Origins of Religion." *Trends in Cognitive Sciences* 14, no. 3 (2010).

To my knowledge, such hypotheses have not been extensively applied to the Çatalhöyük material. This may be due in part to the fact that so little is known (or knowable) about the ethical codes or norms that guided their daily lives. Nevertheless, we do know that once supernatural agents have emerged within the human imagination, they are automatically attributed qualities that are common to the category Person, such as thoughts, intentions and desires. We also know that the way in which they minimally violate intuitions about this category, such as embodiment, actually makes them easier to remember, which is why ideas of gods are such socially contagious concepts. Given our shared phylogenetic heritage, which includes such cognitive tendencies, it is reasonable to postulate that shared belief in "spirits" played a similar role in enhancing cooperation among the inhabitants of this Neolithic community. Like all other groups of *Homo sapiens*, the members of the Çatalhöyük coalition would have naturally and automatically wondered what their supernatural agents thought or desired.

What do goddesses want? For that matter, what did any of the gods of Çatalhöyük – animal spirits or deceased ancestors of either gender – want? At least during the period of the lower levels, the inhabitants of the houses would have been constantly confronted with images of buchrania (and other symbols) that would have activated the idea that animal spirits were watching them. Repeated burials and re-burials within the houses would also have reinforced a sense that their ancestors may be listening. What are the animal spirits thinking about what they see? What are the ancestors feeling about what they hear? Whatever the specific answers, it is plausible to assume that this general sort of question had the effect of solidifying a willingness to cooperate in the ongoing rituals and moral conventions of Çatalhöyük. The ambiguity produced by the repeated hiding and revealing of bones and other artifacts would only have reinforced the mechanisms of religious reproduction.

A second set of promising hypotheses, linked to what I have called sociographic prudery, that have not (to my knowledge) been applied to Çatalhöyük are those related to *costly signaling theory*. Here too the issue is explaining what appear to be anomalies within evolution, such as extravagant peacock tails, which require a high percentage of metabolic energy and weaken the capacity to evade predators. Such tails, however, are a signal to peahens that their carrier has genes strong enough to survive. In the case of religion, the phenomena to be explained include behaviors that are costly in terms of time and energy, often painful and also without any clear survival value. Richard Sosis, for example, points to the rituals of the Ilahita Arapesh, in which adult males dressed like boars pin down three year old boys and rub their genitals forcefully with stinging nettles.[22] The descriptions of the molestations that must be suffered by males in this coalition, which continue in various forms throughout their lives, would make Hesiod blush.

Sosis argues that these and other religious behaviors, including the pursuit of badges and the acceptance of bans, neither of which provide (but often reduce) adaptive advantage, are actually forms of costly signaling. Participation signals commitment to the group, which strengthens the solidarity of the coalition and indirectly benefits the survival of the gene pool. For the most part, and over the long run, the most convincing displays are by those who are *really* committed to their beliefs and their promises to the coalition. People whose beliefs are internalized are willing to engage in displays of commitment that are (otherwise) so unreasonable that they would be very hard to fake. To cite Atran again: "collective commitment to the absurd is the greatest demonstration of group love that humans have devised."[23] The most reliable signals of this love, which protects the coalition by reinforcing the willingness of its members to cooperate, coordinate, and compete are by those who truly believe that they are in coalition with supernatural agents.

How might this apply to Çatalhöyük? There is no strong evidence of excessively violent rituals, but the art depicts the teasing of wild animals, which appears dangerous indeed. Moreover, there does not seem to be any survival advantage in having extremely sharp bulls' horns protruding from the inner walls of one's house. Using cognitive energy to remember how deeply an ancestor is buried, and physical energy to dig them up and rebury them, have no obvious adaptive value. Given the explanatory power of costly signaling theory in so many other contexts, when *we* dig up something that indicates a widespread form of behavior that does not enhance fitness it seems reasonable to ask whether we have found evidence for shared imaginative engagement with supernatural agents. Participating in the various frequent rituals within the houses, as well as the less frequent rituals connected with hunting, baiting, and feasting on bulls, would have been way of signaling commitment to coalition members, including those in the spirit world. This in turn would have strengthened the cohesion of the Çatalhöyük community.

Unfortunately, the other side of in-group cohesion is often out-group antagonism. Especially under difficult conditions, commitment to one's own group can reinforce discrimination against other groups. This is a natural evolutionary mechanism but, as John Teehan points out, religion can intensify

---

22   Sosis, "Religious Behaviors, Badges, and Bans," 61.

23   Atran, *Talking to the Enemy*, 450. See also Bulbulia, "Religious Costs as Adaptations That Signal Altruistic Intention," *Evolution and Cognition* 10, no. 1 (2004).

this discrimination by giving the moral differentiation between groups a divine sanction and raising the stakes of commitment to cosmic proportions.[24] Teehan's analysis focuses on the Abrahamic religions, demonstrating how the ethical codes of these monotheistic traditions are expressions and extensions of, rather than exceptions to, the natural moral intuitions that emerged through the evolution of human brains in social groups.

The people of Çatalhöyük were clearly not monotheists, and there is not yet any clear evidence of violence toward in-group defectors or out-groups. This lack of evidence is itself one of the most fascinating features of the site, and its further exploration may shed light on the conditions that give rise to religious (or other) violence. In the meantime, there is much to ponder about our own propensity toward anthropomorphic promiscuity and sociographic prudery, which, on this side of the rise of the axial age religions and in this ever more complex space of global pluralism, may no longer have adaptive value for the human race. Seeking out and protecting supernatural agent coalitions with dead ancestors and aurochs may have helped keep the inhabitants of Çatalhöyük alive by providing a kind of emotional and social entangling force that held together their modes of proprietary production. But following similar strategies today could end up killing us.

Theolytic Mechanisms in Science and Theology

Digging at Çatalhöyük has revealed a great deal about the complex revolutionary changes that occurred in the evolution of humanity during the Neolithic. But can its physical and conceptual excavation disclose anything important about the difficult task our species faces today in learning to adapt to ever more complex intellectual and social challenges? If placed within the broader context of the phylogenetic emergence and psychological and political effects of god-bearing mechanisms I believe it can. In this context I only have space for a few provocative suggestions.

As I indicated above, my primary interest is in how insights into the lives of the Neolithic inhabitants of Çatalhöyük can help us understand ourselves as we struggle to understand them. We too are entangled in particular ways of tending to proprietary production, but most of us live a quasi- or post-sedentary life in which not just plants and animals but the natural environment itself is largely domesticated. We have become digital nomads, information huntergatherers roaming around a virtual global socius. The cognitive and coalitional

---

24   Teehan, *In the Name of God*, 174.

tools that we still use in our navigation of this world, however, evolved in an environment with very different natural and social challenges. Tendencies toward anthropomorphic promiscuity and sociographic prudery served (some of) our ancestors well – but how are they working out for us?

Earlier I alluded briefly to the fact that scientists tend to be anthropomorphically prudish and sociographically promiscuous; let me now explain what I mean. Most scientists resist explanations of phenomena that appeal to the causal influence of disembodied intentional agents like the spirits of animals or ancestors. Regardless of academic discipline, such ideas are not welcome in the logical chains of their arguments or in the planning of their empirical experiments.

For example, one will not find in archeological journals any explanations of the data at Çatalhöyük that appeal to the actual causal efficacy of bull spirits (or goddesses) within the coalition. Most archaeologists would also be suspicious of any claims from their colleagues that their knowledge about the site had been revealed to them through angels or astrology. "The religious world increases the number and influence of intentional agents while science ultimately aims to minimize both by seeking alternative accounts of affairs in terms of underlying, predictable, non-intentional mechanisms."[25] In other words, scientists are (or try to be) anthropomorphic prudes.

Scientists are also suspicious of claims that appeal primarily or solely to authority or convention. Certainly they operate within a particular disciplinary tradition, and must take much of what is handed down to them in trust, but scientists raise their eyebrows when a particular argument is immunized from critique because of the reputation of its source or the longevity of its popularity. For example, for many years the consensus of the archeological community was that early hominin groups were violent. For the members of the Hodder team with whom I have interacted, actively seeking out new ways to organize the inquiry of their disciplinary socius in light of the new data at Çatalhöyük is more important than fidelity to such dominant conventions. Like most scientists, they pursue novel ways of inscribing the academic field with new hypotheses. In the case of the current (summer) inhabitants of Çatalhöyük, this also involves openness to intercourse with other disciplines – even philosophy and theology.

We can think of these two reciprocally reinforcing tendencies among scientists as *theolytic* mechanisms because of the way in which they loosen or dissolve (*lysis*) the hold of supernatural agents or gods (*theōn*) on human minds

25   Lawson and McCauley, *Rethinking Religion: Connecting Cognition and Culture* (Cambridge: Cambridge University Press, 1993), 162.

and coalitions. The relationship between these mechanisms are depicted in *Figure 2* (see Chapter 1, p. 64). We do not have space here to deal with the other two (upper left and lower right) quadrants of this diagram, which represent quite different ways of integrating the mechanisms; suffice it to say, they lead to approaches that are either too prodigal or too penurious to work.[26] For our purposes here, the important point is the diametrical opposition between the direction sponsored by theogonic mechanisms (in the lower left quadrant) and the direction represented in the upper right quadrant of *Figure 2*.

The integration of *theolytic* mechanisms unveils the hidden mechanisms by which supernatural agent coalitions are born(e). My rhetorical use of the phrase "unveiling theogonic mechanisms" is inspired by Rene Girard's well-known concept of the scapegoat mechanism.[27] One might argue that the scapegoat mechanism is simply one (important) example of the integration of theogonic mechanisms, insofar as the former involves the detection of an ambiguous intentional force that must be dealt with in order to maintain the social cohesion of and psychological stability of an in-group. In this context, however, it is important to note some similarities and differences between scapegoating and the integrated theogonic mechanisms we have been discussing. Here is perhaps the basic difference:

· The *scapegoat* mechanism creates weak victims, more or less vulnerable, who must be cursed, sent away, or destroyed in order to rid the community of violence, sin or evil.
· The *theogonic* mechanisms, on the other hand, create powerful perpetrators, more or less in-vulnerable, who must be appeased in some way, in order to avoid misfortune or acquire blessing.

Both theogonic and scapegoating mechanisms "work," in the sense that persons within the in-group often feel better and their communities often survive longer because of them. One of the basic similarities between these mechanisms is that their "working" can actually make things worse.

· Removing or destroying *scapegoats* reinforces the powerful belief that our problems can be solved by more violence.

---

26   I discuss these two quadrants in Chapter 7 below ("Theology After Pandora"), and in Chapter 6 of Shults, *Theology after the Birth of God: Atheist Conceptions in Cognition and Culture* (Palgrave Macmillan, 2014).

27   Girard, *Violence and the Sacred* (Johns Hopkins University Press, 1977); Girard, *The Scapegoat*, (Baltimore: Johns Hopkins University Press, 1986).

· Detecting and protecting *gods* reinforces the powerful belief that our problems can be solved by our coalition with supernatural agents.

Another important similarity is that the very process of unveiling any of these mechanisms *weakens* their power. This is because they only work well when they are hidden to those within whom or upon whom they are operating.

When we begin to recognize what we are doing to scapegoats, and what our scapegoating is doing to us, the process no longer automatically has the effect of (temporarily) calming us psychologically and politically. Similarly, as we begin to see how gods are born(e) in human cognition and culture, such conceptions can more easily become the objects of our critical reflection rather than surreptitiously shaping our subjectivity. For the reasons outlined in previous sections, evolution has predisposed many of us to think, act, and feel in ways that keep these mechanisms hidden.

This is why exposing the reproductive processes of god-bearing can be so difficult and even frightening. We have evolved not to challenge beliefs in the things hidden in the walls and foundations of our coalitions. Challenging the beliefs or practices related to the supernatural agents of *other* traditions easily leads to conflicts that can quickly escalate and become dangerous. And so we avoid this too. This may not have been much of an issue at Çatalhöyük, with a relatively small and homogenous population. There would be no reason to challenge the efficacy of engaging the spirit world by hiding and revealing skulls, the sharp parts of animals, and other artifacts. During the Neolithic, the ambiguity and mystery surrounding these processes would only have further activated the natural tendencies to detect human-like agency and to protect one's place in the collective by costly signaling of commitment.

Now, however, we face very different adaptive challenges in an increasingly pluralistic and interconnected global environment, in which we rely ever more deeply and are affected ever more intensely by scientific modes of inquiry and technological developments. During times of crisis, the same theogonic mechanisms that produce and maintain the supernatural agent coalition of a particular in-group also intensify it's members' anxiety about and discrimination toward out-groups. Allowing these procreative urges to run wild in our current context is no longer productive. We can no longer afford to romanticize the human search for gods to protect and partner with us; it may once have been a harmless (or even helpful) romantic distraction, but today it is distracting us from the urgent task of developing new strategies for surviving in a rapidly changing environment.

Facilitating and strengthening theolytic mechanisms will not be easy. It will require great psychological and political sensitivity in relation to a wide array of stakeholders, including not only scientists from diverse disciplines, but also laypeople and leaders associated with religious institutions, museums, educational groups and funding agencies. Nevertheless, it is important to realize that, whether or not we mean it to, digging up Çatalhöyük has a theolytic effect. In my view it is best not to hide these mechanisms; we should bring them out in the open for analysis and evaluation. Engaging in anthropomorphically prudish and sociographically promiscuous reflection on the processes by which gods are born(e) increases our capacity for adapting *intentionally*. Although this is not the job of archaeologists, anthropologists, or psychologists *qua* scientists, the kind of space created by international and interdisciplinary projects like the one at Çatalhöyük provide an excellent opportunity for such intentional dialogue.

But what role could *theology* possible play in this endeavor? The vast majority of theologians have operated within the context of one of the religious traditions that can trace its roots to the axial age; for the most part, they have reinforced the detection and protection of particular supernatural agent coalitions. This direction within theology, which I call its *sacerdotal* trajectory, has by far been the most dominant and obviously contributes to theogonic reproduction. However, there have always been streams of dissent within these traditions, forces that push against anthropomorphic conceptions of the divine and push toward modes of sociography that do not inscribe harsh boundaries between groups. Pressing in this direction, which I call the *iconoclastic* trajectory of theology, has the effect of breaking the power of particular images of gods whose detection protects a conventional way of inscribing the socius.[28]

Theologians (or a-theologians) who follow this latter course today are in a unique position to collaborate with scientists in the unveiling of theogonic mechanisms. For example, demonstrating the logical incoherence of the notion of *an* infinite Supernatural Person, for which there are a wealth of resources even within the monotheistic traditions, can complement scientific challenges to the plausibility of appeals to the intervention of finite supernatural agents in the natural world.[29] In Chapter 5 below I will argue that the best strategy here is not to attempt to disprove the *possibility* or even to weaken the *probability* of such causation through deductive or inductive arguments, but to challenge the *plausibility* of such religious hypotheses by offering abductive arguments that more adequately explain the phenomena without appealing to

28   This is discussed in more detail in Chapters 6 and 7 below, and in Shults, *Theology after the Birth of God*.

29   For a fuller argument for this claim, see Shults, "Science and Religious Supremacy: Toward a Naturalist Theology of Religions," and Shults, *Iconoclastic Theology*.

hidden revelations, thereby making it easier to disentangle the gods from the material and social dimensions of our proprietary production.

To an extent hardly imaginable even a century ago, we have gained significant control over the processes of childbirth. Developing effective means of divine birth control may prove to be much more difficult. It may turn out to be impossible. Perhaps theogonic mechanisms are so deeply embedded in our phylogeny that we can never escape the ontogenetic delivery of the gods. If so, the supernatural population will continue to grow within the natural mental and social space of *Homo sapiens*. The planet already feels overcrowded – physically, emotionally, cognitively.

Given the potentially destructive effects of psychological strategies that are based on detecting divine attachment figures and political strategies that are driven by protecting supernatural coalitions, it seems to me that we can no longer avoid the challenge all parents must face. Some ways of caring for our offspring can become addictive and unhealthy; holding on to them too long is not good for us. Can we learn to let go of our supernatural progeny? It may become easier as the processes behind their mysterious arrival within our families of religious origin are increasingly unveiled.

CHAPTER 4

How to Survive the Anthropocence

In one sense – and, indeed, in the one sense that is perhaps most obvious and yet at the same time easiest to disavow – none of us are going to survive the Anthropocene.[1] We are all going to die. This is just the way it goes with organisms. Nevertheless, many of us, quite naturally, would prefer it if at least a selection of human offspring could continue multiplying and filling the earth. This preference itself is a result of natural selection. The problem, of course, is that the earth is already too full and it is not at all clear how long it can sustain expanding multitudes of *Homo sapiens*. For several decades, a growing number of scientists, policy-makers, and cultural commentators have been trying to draw our attention to imminent ecological crises, explaining their causes and estimating their effects on the survival of humanity and other sentient species.[2]

Unfortunately, the vast majority of people find such explanations and estimations all too easy to ignore. Rather than offering yet another summary of reasons to be alarmed, I want to focus here on some of the reasons why so many people – especially religious people – show surprisingly little alarm when given information about the deteriorating environment of the Anthropocene. My goal is to unveil some of the naturally evolved *religious biases* whose covert operation facilitates the repression or rejection of warnings about the consequences of extreme climate change and excessive capitalist consumption. Learning how to contest these phylogenetically inherited and culturally fortified biases is an important condition – indeed, perhaps a necessary, if not a sufficient, condition – for adapting to and altering our current natural and social environments in ways that will enhance the chances for the survival (and flourishing) of our offspring.

I do not mean to downplay the significance of other factors (nutritional, pedagogical, political, economic, etc.) that must figure into our pragmatic adaptive calculations. My point is that highlighting and warning people about these issues will do little good if we fail to deal with deeply embedded religious biases that surreptitiously shape the reactions of the majority of the

---

1   This chapter is an adapted version of "How to Survive the Anthropocene: Adaptive Atheism and the Evolution of *Homo Deiparensis*," which originally appeared in *Religions* 6, no. 1 (2015): 1–18.

2   See, e.g., Klein, *This Changes Everything: Capitalism vs. the Climate* (Allen Lane, 2014). Kolbert, *The Sixth Extinction: An Unnatural History* (Henry Holt and Co, 2014).

© F. LERON SHULTS, 2018 | DOI 10.1163/9789004360952_005

human population, enabling most of us to immediately dismiss or disregard such calculations. As we will see below, the evolved tendencies that are most relevant for our purposes have to do with mental *credulity* toward religious content and with social *conformity* in religious contexts. These cognitive and coalitional biases are reciprocally reinforcing, and predispose most of us toward believing claims about the manifestation of gods (revelation) especially when engaging with other in-group members in the manipulation of gods (ritual). In other words, superstitious inferences based on the detection of alleged supernatural agents activate segregative preferences based on the protection of allied supernatural groups (and *vice versa*).

The special issue of the journal in which an earlier version of this chapter appeared was dedicated to exploring the role that "religions" might have played – and might continue to play – in exacerbating or easing the current ecological crises that characterize the Anthropocene. In a recent interview in the *New Scientist* about his book *The Meaning of Human Existence,* E.O. Wilson said that he thought religion is "dragging us down," and that, "for the sake of progress the best thing we could possibly do would be to diminish, to the point of eliminating, religious faiths."[3]

In the final section of this chapter, I return to Wilson's concerns and discuss the role that what I call "adaptive atheism" might play in responding to the crises of the Anthropocene. The second and third sections use the conceptual framework introduced in Chapter 1 to help clarify why and how gods are imaginatively conceived and nurtured by believers ritually engaged in religious sects. First, however, when reflecting on claims like Wilson's, which seem to be increasing in frequency in recent years, it is important to be as clear as we can about our use of terms like "religious faiths" and the reasons why we think they might be "dragging us down."

Climate Change, Cultural Cognition and "Religion"

As we have seen, the label "religion" is highly contested within and across the many academic disciplines that study the various phenomena to which the term is commonly applied. In this context, I continue to use the word *religion* to designate *shared imaginative engagement with axiologically relevant supernatural agents.* I further clarify and defend this stipulated definition in more detail below. Of course, this aggregate of traits does not capture everything that

---

3   Wilson, *The Meaning of Human Existence* (Liveright, 2014).

can be said about "religion," but this sort of definition is relatively common in fields such as the cognitive science of religion.[4] Critics of Wilson – like critics of the new atheists – are quick to point out that "religions" have helped hold societies together, provided people with a sense of meaning in life, fostered the production of great works of music and art (and so on). This is no doubt true, and I am all for cohesive societies, meaningful lives, and aesthetic productivity (and so on). But do any of these things depend on widespread belief in and ritual interaction with disembodied intentional forces that are watching over particular in-groups?

We should be happy to discover that they do not. Why? Because whatever else "religions" may produce, they also reinforce evolved biases that consistently lead to mistaken interpretations of natural phenomena and foster antagonism toward out-groups. Moreover, as we will see below, learning to contest these biases can help produce more plausible explanations of causal forces in the world and more feasible social strategies in pluralistic contexts. It is indeed true that other traits often found among people associated with a particular religion, such as a concern for justice or a sense of wonder, may indeed help to encourage creative interventions in socio-cultural practices and political economic systems. In what follows, however, my focus is on the reproduction of religion (in the sense stipulated above) in human minds and cultures, and the extent to which it aggravates the crises of the Anthropocene.

Experts on climate change, as well as public-policy makers concerned with mediating their findings to the general public, have often expressed astonishment at the resistance so many people have toward accepting the scientific consensus. Equally troubling is the lack of environmentally sensitive behavior even among those who are well educated about and explicitly accept that consensus. These concerns have been the focus of a growing number of studies during the last decade or so. In a 2002 article in *Environmental Education Research*, Kollmuss and Agyeman tackled the issue from the point of view of environmental education, summarizing much of the extant literature and arguing for a more complex model of "pro-environmental consciousness" that takes into account both external factors (e.g., demographics) and internal factors (e.g., values).[5] In 2011 a special issue of *American Psychologist* was devoted to exploring ways in which scholars and practioners in the discipline

---

4   See, e.g., Tremlin, *Minds and Gods*. Pyysiäinen, *Supernatural Agents*.

5   Kollmuss and Agyeman, "Mind the Gap: Why Do People Act Environmentally and What Are the Barriers to pro-Environmental Behavior?" *Environmental Education Research* 8, no. 3 (2002).

of psychology might help foster healthy modes of coping with the crisis and contribute to a better understanding of the barriers to action.[6]

One of the most important recent developments in this multi-disciplinary analysis of human responses to climate change has been the incorporation of insights from the cognitive and evolutionary sciences. In a recent review of the "foundational processes" that influence beliefs about climate change, Brownlee and colleagues examined the way in which cognitive dissonance, biased assimilation, confirmation bias, loss aversion, illusions of optimism – and a host of other cognitive biases – shape people's attitudes and beliefs about climate change, and identified ways in which this knowledge could lead to new strategies in environmental education and research.[7] In a similar study on various methodological scenario approaches to climate change research in *Synthese*, Lloyd and Schweizer showed how even some of the most popular models among scientists and policy-makers can be impacted by many of the heuristic biases that have been identified by cognitive psychology, such as availability, overconfidence, and groupthink.[8]

In his article "Why do people misunderstand climate change? Heuristics, mental models and ontological assumptions" in *Climatic Change*, Chen pointed out the special role of "object bias" in skewing interpretations of ecological crises.[9] The static mental models associated with the most common pattern matching heuristics that shape human perception and interpretation work extremely well for dealing with objects. Climate change, however, is not an object; it is a dynamic process. Unfortunately, when people uncritically use their implicit ontological assumptions about objects to try and make sense of (or predict) changes in complex dynamic systems, they consistently and profoundly fail. Chen notes that some physics teachers have developed a radical pedagogical approach to deal with this bias: they *begin* with detailed discussions of ontology before trying to teach novice students about physical processes (like electricity). He suggests that a similarly revolutionary approach might be necessary for making progress in altering people's attitudes toward the processes of climate change.

---

6 Reser and Swim, "Adapting to and Coping With the Threat and Impacts of Climate Change," *American Psychologist* 66, no. 4 (2011).

7 Brownlee, et al., "A Review of the Foundational Processes That Influence Beliefs in Climate Change: Opportunities for Environmental Education Research," *Environmental Education Research*, no. 1 (2013).

8 Lloyd and Schweizer, "Objectivity and a Comparison of Methodological Scenario Approaches for Climate Change research," *Synthese: An International Journal for Epistemology, Methodology and Philosophy of Science* 191, no. 10 (2014).

9 Chen, "Why Do People Misunderstand Climate Change? Heuristics, Mental Models and Ontological Assumptions," *Climatic Change* 108, no. 1 (2011).

These sorts of insights have major implications for understanding the science/policy divide and the common sorts of misjudgments that shape reactions to policy proposals. As Norman and Delfin have pointed out in a recent issue of *Politics & Policy*, cognitive biases are easily activated under conditions of uncertainty and when individuals are trying to assess threats to their survival. They illustrate some of the ways in which cognitive biases like anchoring, framing, false representativeness, availability, attention to intentionality, and affective forecasting lead to systematic errors in judgment, and bad policy decisions.[10] In a similar study in *Mitigation and Adaptation Strategies for Global Change*, Preston and colleagues examined the role of heuristic biases in "adaptation discourse," which all too easily lead to ways of framing the problems, and to policy proposals for solving them, that rely on affective (quick, innate) reasoning processes rather than analytic (slow, methodical) reasoning.[11] Increasingly, scholars in these fields are recognizing the extent to which cognitive biases help to explain resistance to the scientific consensus about climate change and the relative lack of success in policies aimed at promoting pro-environmental behavior.

In their analyses of these issues, Kahan and colleagues utilized the phrase "cultural cognition," by which they mean to refer to "the psychological disposition of persons to conform their factual beliefs about the instrumental efficacy (or perversity) of law to their cultural evaluations of the activities subject to regulation."[12] Cultural cognition is driven by implicit mechanisms like naïve realism and reactive devaluation, which reinforce people's tendencies to immediately judge new information as unreliable when it goes against their culturally congenial beliefs or to dismiss the persuasiveness of evidence when it is offered by members of an out-group. Human beings do not naturally think in terms of Bayesian probabilities; rather, they are prone to process information in ways that confirm their affective orientation, based on prior estimations of risk perception, which makes it all too easy to ignore "experts" whose claims raise challenges to their sense of identity and idealized form of social ordering.[13]

---

10   Norman and Delfin, "Wizards under Uncertainty: Cognitive Biases, Threat Assessment, and Misjudgments in Policy Making," *Politics and Policy* 40, no. 3 (2012).

11   Preston et al., "Climate Adaptation Heuristics and the Science/policy Divide," *Mitigation and Adaptation Strategies for Global Change* 20, no. 3 (2015).

12   Kahan and Braman, "Cultural Cognition and Public Policy," ssrn *Scholarly Paper* (Rochester, NY: Social Science Research Network, August 2, 2005) 149–150.

13   Kahan et al., "Cultural Cognition of Scientific Consensus," *Journal of Risk Research* 14, no. 2 (2011): 168. See also, Kahan, "Cultural Cognition as a Conception of the Cultural Theory of Risk," in *Handbook of Risk Theory*, ed. Roeser (Springer, 2011).

What does any of this have to do with "religion"? In recent years, scholars of climate change have increasingly turned their attention to the relationship between religiosity, religious affiliation, and even specific religious ideas, on the one hand, and attitudes toward (and behavior in response to) reports about ecological challenges and dangers, on the other. In a 2009 article in *Global Environmental Change*, Mortreux and Barnett reported on their interviews with the inhabitants of the islands of Tuvalu, where raising water levels might be considered a good reason to migrate. When people were asked why they did not move, they "consistently referred to the biblical story of Noah as evidence that God would not allow further flooding."[14] In his 2013 report on a case study of the role of religion in the Brazilian Amazon published in *Journal of Rural Studies*, Otsuki concludes that one consequence of the popularity in rural areas of the Pentecostal Church of Assembly of God, which embraces an evangelical Christian message of enjoying earthly prosperity, "was continual conversion of forests into municipalities and promotion of capitalist accumulation."[15]

Experimental evidence suggests that people who are less religious and more analytic tend to be better (on average) at contesting some of the general cognitive and coalitional biases we have been discussing,[16] But are they any different from religious people when it comes to evaluating the scientific consensus on climate change and reacting to the challenges of the Anthropocene? In a recent survey-analysis examining the relation between "place attachment and ideological beliefs" and attitudes toward climate change, Devine-Wright and colleagues found that those with the strongest global attachments were more likely to be "female, younger and self-identify as having no religion."[17] Another recent study compared groups of Christians, Muslims, and secular people in a mixed methods analysis of the impact of religious faith on attitudes toward environmental issues. In their report on this study in a 2014 article published in *Technology in Society*, Hope and Jones attributed the low perception of urgency

14   Mortreux and Barnett, "Climate Change, Migration and Adaptation in Funafuti, Tuvalu," *Global Environmental Change* 19, no. 1 (2009): 100.

15   Otsuki, "Ecological Rationality and Environmental Governance on the Agrarian Frontier: The Role of Religion in the Brazilian Amazon," *Journal of Rural Studies* 32 (2013): 411.

16   Trippas et al., "Better but Still Biased: Analytic Cognitive Style and Belief Bias," *Thinking & Reasoning* (2015). Pennycook et al., "Belief Bias during Reasoning among Religious Believers and Skeptics," *Psychonomic Bulletin & Review* 20, no. 4 (2013); Zuckerman et al., "The Relation Between Intelligence and Religiosity: A Meta-Analysis and Some Proposed Explanations," *Personality and Social Psychology Review* 17, no. 4 (2013).

17   Devine-Wright et al., "My Country or My Planet? Exploring the Influence of Multiple Place Attachments and Ideological Beliefs upon Climate Change Attitudes and Opinions," *Global Environmental Change* 30 (2015).

toward ecological crises among the former two groups as "due to beliefs in an afterlife and divine intervention." Lack of these beliefs among secular participants, on the other hand, contributed to a "focus on human responsibility and the need for action."[18]

But, some defenders of "religion" may object at this stage, is it not the case that some religious people do believe in caring for the earth and do act in environmentally friendly ways? Indeed, research indicates that religiosity can sometimes have a moderating effect on the likeliness to engage in sustainable behaviors. For example, in a cross-cultural comparison of Christian, Atheist and Buddhist consumers in the u.s. and South Korea, Minton and colleagues found that highly religious Buddhists were the most likely to engage in sustainable behaviors.[19] An earlier study by Wardekker and colleagues, which explored the role of "Christian voices" in the United States public debate over climate change, identified three distinct types of narrative: conserving the "garden of God" as it was created, tending to the wilderness so that it becomes the "garden" it should be, and a combination of these two in which God's creation is considered both good and changing. The authors of that study concluded that "religious framings" of climate change could serve as "bridging devices for bipartisan climate-policy initiative" because of the way they resonate with many conservative and progressive members of the electorate.[20]

If our primary concern is dealing with the underlying credulity and conformity biases that impair our ability to respond to contemporary ecological crises, then there are at least two important problems with the sort of analysis represented by these last two studies. By focusing *theoretically* only on institutional "affiliation" or on intensity of "ideological" commitment, this kind of approach ignores those aspects of religion that are most relevant for understanding people's resistance to facing the global challenges of pluralistic human societies in the Anthropocene, namely, shared imaginative engagement with invisible agents who are allegedly invested in upholding the norms and ensuring the survival of a particular in-group. However, this is precisely why cognitive and coalitional biases keep working so well – because they are ignored.

---

18   Hope and Jones, "The Impact of Religious Faith on Attitudes to Environmental Issues and Carbon Capture and Storage (ccs) Technologies: A Mixed Methods Study," *Technology in Society* 38 (2014).

19   Minton et al., "Religion and Motives for Sustainable Behaviors: A Cross-Cultural Comparison and Contrast," *Journal of Business Research*, (2015).

20   Wardekker et al., "Ethics and Public Perception of Climate Change: Exploring the Christian Voices in the us Public Debate," *Global Environmental Change-Human And Policy Dimensions* 19, no. 4 (2009): 512.

And so we also have a *pragmatic* problem. Even if religiously affiliated and committed individuals *explicitly* accept the scientific consensus on climate change and act publically in environmentally friendly ways, insofar as they (and their affines) are also *implicitly* activated by evolved mechanisms that engender conceptions of person-like, coalition-favoring disembodied spirits whose intentions are allegedly relevant for interpreting natural phenomena and normatively inscribing the social field, they are actually *strengthening* the very biases that are contributing to the crises they are trying to solve.

Homo deiparensis

One of the distinctive features of our species is a fascination with naming ourselves. We are, or so we like to claim, *Homo sapiens* (the wise hominid). Of course, wisdom is not the only interesting thing about us; we have also nominated ourselves *Homo faber* (the worker), *Homo ludens* (the laugher) and *Homo economicus* (the shopper). In this context, I want to draw attention to another distinctive feature of our species: our tendency to bear gods. Research in fields as diverse as cognitive science, evolutionary biology, archaeology, experimental psychology, and cultural anthropology has converged in recent years in support of the claim that conceptions of supernatural agents are easily "born" in human minds and "borne" in human groups today as a result of biases that were naturally selected in the early ancestral environment of the upper Paleolithic.

In other words, we are – or we have been for at least the last 70,000 years or so – god-bearing hominids (*Homo deiparensis*). In the sense I am using the term, religion is the result of the integration of inherited cognitive and inculcated coalitional mechanisms that predispose us toward over-detecting human-like forms in the natural environment and over-protecting group-specific norms in complex social environments. The coordinate grid depicted in *Figure 1* (Chapter 1, p. 3) provides a conceptual framework for discussing the possible correlations between – and contestations of – these perceptive and affiliative biases, and their relevance for surviving the Anthropocene.

As noted above, the horizontal line represents a spectrum on which one can mark the tendency of a person to guess "supernatural agent" when confronted with ambiguous phenomena. Anthropomorphically promiscuous individuals jump at any opportunity to postulate discarnate (or otherwise ontologically confused) human-like entities as causal explanations of ambiguous or frightening phenomena. Anthropomorphic prudes, on the other hand, resist superstitious interpretations of nature and hold out for non-intentional explanations.

The vertical line plots the variation among individuals in relation to their tendency to prefer norms authorized by supernatural authorities when evaluating ways to organize the social field. Sociographic prudes are happy to stay home with familiar others and are highly suspicious of the alien values of out-groups. The sociographically promiscuous, on the other hand, are more open to dating other cultures; they tend to resist appeals to conventional religious (and other) authorities that enforce segregative inscriptions of society.

The integration of anthropomorphic promiscuity and sociographic prudery served our ancestors well in an environment where survival depended on quickly perceiving any relevant agents, and consistently defending the resources and values of one's in-group. Shared imaginative engagement with axiologically relevant supernatural agents – religion – powerfully reinforced these biases and gave a survival advantage to hominid groups whose members had this aggregate of traits. The integration of theogonic (god-bearing) mechanisms, represented in the lower left quadrant of *Figure 1* was an evolutionary winner.

In more than one sense, gods were the "best guess" available to our early ancestors. Hypothesizing the presence of a "human-like agent" – even when there was no clear evidence that such an agent existed – was "best" because it provided further motivation to keep trying to detect hidden agents, which was necessary for survival. Given the importance of honing this hypersensitive disposition, it would have been better to keep believing that there might be animal-spirits or ancestor-ghosts in the forest than to guess that the cause of weird noises or movements was simply the wind or shifting shadows. Although these biases regularly triggered false positives, the guesses they produced were cognitively cheap and inferentially rich. Once the human mind thinks it has detected an intentional force, attributions of person-like qualities to the putative agent (e.g., "may be angry" or "wants something") are easily triggered by other cognitive devices like mentalization and teleological reasoning.

So, over-active cognitive dispositions like these led to the emergence of god-concepts, but why did people keep socially entertaining them? Supernatural agents may be easily born in human minds but, as we have seen, it takes a ritually engaged village to raise them. The gods that stick around and become entangled within the communal rituals of religious sects are typically those that serve as "better guards." As human groups get larger, it becomes more difficult to keep an eye on everyone and be sure that they are following the norms of the coalition. When the members of an in-group really believe in the existence and causal relevance of disembodied intentional forces who are interested in their behavior, and who have the power and desire to reward or punish them, they are more likely to follow the rules even if no other embodied human agents are watching.

Especially when resources are low, or under otherwise stressful conditions, the most competitive coalitions are those whose members are able to cooperate and remain committed to the group. It is easy to understand why self-serving tendencies in individual organisms have been naturally selected over time. However, the societies in which individual human beings live, and on which they depend for survival, will fall apart if there are too many self-serving cheaters, freeloaders, or defectors. Research in the bio-cultural sciences of religion suggests that cooperative commitment within some hominid coalitions during the upper Paleolithic was improved by the intensification of shared belief in and ritual engagement with potentially punitive gods.[21] Vindictive supernatural agents would be able to catch misbehavior that natural agents might miss, and could punish not only the miscreants, but also their offspring or even the entire community. Accepting the existence of invisible or ambiguously apparitional "watchers" helps to enhance the motivation to obey conventional regulations and stay committed to the in-group.

So, what does any of this have to do with the Anthropocene, and the ecological and economic crises we face today as a species? The problem is that these religious credulity and conformity biases, which served our ancestors well, are no longer adaptive in some of the contexts within which a growing number of us find ourselves. Most of us do not live in relatively homogenous small-scale groups, hunting and gathering across wide expanses like the African savannah, but in pluralistic, densely-packed, large-scale societies rapidly running out of agricultural resources for supporting our expansive sedentation. The hyper-active detection of supernatural agents and the hyper-active protection of supernatural groups helped earlier human civilizational forms emerge and hold together, but it now seems like we must learn to contest these evolved biases if we are to adapt to (and alter the conditions of) the Anthropocene.

If large numbers of the population interpret natural phenomena like tsunamis and hurricanes as acts of God (literally), they are less likely to pay attention to scientific reports about climate change. Why worry about the planet if the supernatural agent of one's in-group is going to create a new heaven and a new earth anyway? If large numbers of the population are motivated to inscribe the social field in ways that enforce the values putatively revealed by their God to elite members of their coalition, they are less likely to alter their patterns of consumption. Why worry about unequal distribution of resources today if one

---

21   Johnson, "God's Punishment and Public Goods: A Test of the Supernatural Punishment Hypothesis in 186 World Cultures," *Human Nature* 16, no. 4 (2005); Norenzayan, *Big Gods: How Religion Transformed Cooperation and Conflict* (Princeton University Press, 2013).

expects a supernatural agent to return at any moment with eternal rewards for the in-group and eternal punishment for the "wicked"?

At this stage, some defenders of "religion" may once again protest. They do not believe such things, nor do any of their educated friends. They are deeply concerned about climate change and capitalist consumption and so are most of their cosmopolitan colleagues. Their conception of "God" promotes neither superstition nor segregation. Even if the latter were true it would, unfortunately, be irrelevant. Cross-cultural psychological research indicates that no matter what the intellectual elite and priestly class of a religious in-group says, the vast majority of regular believers immediately give in to naturally evolved biases toward detecting person-like, coalition-favoring gods when faced with real-life religious scenarios.[22]

As if this were not bad enough, these evolved cognitive and coalitional mechanisms are so deeply intertwined that mental *credulity* about gods and ritually enhanced social *conformity* constantly strengthen one another, implicitly and somewhat automatically, all too easily obscuring and promoting the powerful biases that skew our readings of (and reactions to) problems like climate change.

The Reciprocity of God-Bearing Biases

In other words, many of the component mechanisms of anthropomorphic promiscuity and sociographic prudery are *reciprocally reinforcing*. As we have seen in earlier chapters, this is one of the central tenets of theogonic reproduction theory. The fact that – and the ways in which – these biases are mutually intensifying continues to be confirmed and clarified by proliferating empirical research and theoretical developments in the many fields that contribute to the bio-cultural study of religion. As we saw in Chapter 1, a variety of studies indicate that activating people's anxiety about their own mortality, or the welfare of their kith and kin, can increase their tendency to interpret ambiguous phenomena as caused by potentially punitive disembodied agents; conversely, priming individuals with thoughts about possible invisible watchers can reinforce their tendency to protect their in-group, and sometimes leads them to express antagonism toward out-group members. In this section, I point to a

---

22   Slone, *Theological Incorrectness: Why Religious People Believe What They Shouldn't* (Oxford University Press, 2007). Barrett, "Dumb Gods, Petitionary Prayer and the Cognitive Science of Religion," in *Current Approaches in the Cognitive Science of Religion*, ed. Pyysiäinen and Anttonen (New York: Continuum, 2002).

few examples from the rapidly growing multi-disciplinary literature that demonstrates the mutual amplification of theogonic (god-bearing) biases.

In a 2012 article in the *Journal of Experimental Social Psychology*, Gervais and Norenzayan presented evidence from three experimental studies for what they call the "supernatural monitoring hypothesis: That thinking of God triggers the same psychological responses as perceived social surveillance."[23] Psychologists have known about the association between socially desirable responding and religiosity for quite some time, but priming experiments provide a way to demonstrate a *causal* relationship between them. Their studies, which used both explicit and implicit methods for priming concepts of God, confirmed their hypothesis that thinking about supernatural agents activates sensitivity to reputational cues that others are watching, and causes an increase in behaviors considered socially acceptable – especially among believers. As the authors point out, these experiments also lend credence to the claim that supernatural agent concepts, once they arise in a culture, may "foster cooperative behavior by making religious believers feel as if they are monitored by their gods."[24]

Cooperative behavior has rather obvious survival benefits, so what is the problem? The problem is that the dark side of in-group cohesion is out-group antagonism. The correlation between religion and prejudice has also been well-known and documented for decades by social psychologists, but more recent experiments in cognitive psychology have shed light on the mechanisms that link them. In a study of Singaporean Christians and Buddhists published in 2014 in *The International Journal for the Psychology of Religion*, Ramsay and colleagues found that participants who were primed with concepts or images related to supernatural agency tended to become more prejudiced and antagonistic toward out-groups. Members of both religious traditions demonstrated more negative pretest to posttest attitude change toward homosexuals when primed with religious in-group words, in comparison with those primed with neutral words. Even when there is no explicit religious value-violation, bias toward culturally relevant out-groups increases when believers are primed with religious concepts. The authors concluded that religion may exert its prejudicial effects "indirectly through activation of associated cultural value systems."[25]

23   Gervais and Norenzayan, "Like a Camera in the Sky? Thinking about God Increases Public
     Self-Awareness and Socially Desirable Responding," *Journal of Experimental Social Psy-
     chology* 48, no. 1 (2012).

24   Ibid., at 302.

25   Ramsay et al., "Rethinking Value Violation: Priming Religion Increases Prejudice in Singa-
     porean Christians and Buddhists," *International Journal for the Psychology of Religion* 24,
     no. 1 (2014): 1.

In other words, some of the mechanisms of anthropomorphic promiscuity promote sociographic prudery (and *vice versa*). In a 2014 article in *Psychological Science*, Neuberg and colleagues used data from the Global Group Relations project to investigate the relation between religion and intergroup conflict among 194 groups in 97 sites across the world. Their goal was to discover the extent to which religious infusion, that is, the extent to which religious rituals and discourse permeate the everyday activities of groups and their members, "moderated the effects of two factors known to increase intergroup conflict: competition for limited resources and incompatibility of values held by potentially conflicting groups." They found that when religion was infused within group life, "groups were especially prejudiced against those groups that held incompatible values, and they were likely to discriminate against such groups."[26] The evolved predilection toward protecting one's own in-group by antagonizing out-groups is easily activated when one's mental and social worlds are filled with messages about and ministrations toward watchful supernatural agents.

In a 2014 study published in *Psychiatry Research*, Reed and Clarke demonstrated the effect of religious context on the content of visual hallucinations in individuals high in religiosity. Perceptual experiences in the absence of external stimuli – that is to say, hallucinations – are usually associated with schizophrenia or similar mental conditions, but they actually occur quite commonly in large parts of the population. Using a subliminal prime methodology (word-detection task), the authors found that "participants measuring high on religiosity were more likely to report false perceptions of a religious type than participants low on religiosity." Both religious and non-religious participants (none of whom were schizophrenic) made false perceptions based on priming, but those who were high in religiosity produced more false perceptions with a religious content; in other words, their hallucinations were more likely to be related (directly or indirectly) to the supernatural agents of their in-group. The authors hypothesize that "context becomes a framework for processing through which context-relevant information or response to stimuli is facilitated and context-irrelevant information is suppressed."[27]

What happens to human brains when they regularly engage in religious rituals? In a 2013 target article in *Religion, Brain & Behavior*, Schjoedt and colleagues explored the ways in which religious interactions tend to deplete

---

26   Neuberg et al., "Religion and Intergroup Conflict," *Psychological Science* 25, no. 1 (2014): 198.

27   Reed and Clarke, "Effect of Religious Context on the Content of Visual Hallucinations in Individuals High in Religiosity," *Psychiatry Research* 215, no. 3 (2014): 597.

cognitive resources. They proposed a "resource model of ritual cognition in which collective rituals limit the cognitive resources available for the individual processing of religious events," and demonstrated the way in which "rituals directly suppress and channel default cognition in order to facilitate the construction of collective memories, meanings, and values among ritual participants."[28] Rituals tend to be characterized by incomprehensible – or at least causally opaque – interactions. The perception of goal-demoted and causally opaque actions in rituals uses up participants' cognitive resources, limiting their capacity to activate the usual executive systems that support critical analysis. Ritual practices deplete cognitive resources in such a way that people become more susceptible to the suggestions and narratives of religious authorities or ritual officers. Other studies suggest that ritual contexts even alter basic assessments about bodily and mental processes.[29]

In a 2014 article in *The Journal of Social Psychology*, Riggio and colleagues described two self-report experiments designed to show how religiosity affects attributions of causality. Participants read a story about a hypothetical man (Chris) who had a heart attack, and then (depending on the version of the story) used either religiously or medically authorized behaviors to improve his health, and either lived or died. When Chris used religious behaviors and lived, highly religious individuals attributed this outcome to God. However, when Chris used the same behaviors and died, these individuals showed a form of excuse-making the authors call a "God-serving bias." Like the cognitive predisposition toward a self-serving bias, such attributions implicitly support the maintenance of strongly held beliefs (especially beliefs related to group identification and belonging) even in the face of contradictory evidence. The authors conclude that religious belief systems, which claim to have a supernatural basis, not only lead to "low-quality thinking but to dangerous thinking, especially because it is purposeful and motivated by emotional processes...such belief systems, in being defended, lead to extremes in thinking and behavior that are *dangerous to all people.*"[30]

In a 2013 article in the *Israel Journal of Ecology & Evolution*, Purzycki and Sosis proposed the idea of an "extended religious phenotype,"[31] incorporating

---

28   Schjoedt et al., "Cognitive Resource Depletion in Religious Interactions." *Religion, Brain & Behavior* 3, no. 1 (2013): 40.

29   Astuti and Harris, "Understanding Mortality and the Life of the Ancestors in Rural Madagascar." *Cognitive Science* 32, no. 4 (2008).

30   Riggio et al., "Unanswered Prayers: Religiosity and the God-Serving Bias," *The Journal of Social Psychology* 154, no. 6 (2014): 491–514. Emphasis added.

31   Purzycki and Sosis, "The Extended Religious Phenotype and the Adaptive Coupling of Ritual and Belief." *Israel Journal of Ecology & Evolution* 59, no. 2 (2013): 102.

the two "central features" of religion: "the coupling of ritual behavior and supernatural agency attribution." Belief in supernatural agents is possible because of evolved systems devoted to the detection and attribution of mental states, but this does not explain why people believe in the particular gods of their in-group. They suggest that the predictable variations found in religious *content* across cultures are a result of attempts to deal with particular problems posed by environmental challenges in specific niches. Religious systems evolve in response to the demands of their *context*, socio-ecological niches that they help to construct, using mechanisms such as costly signaling and shared belief in supernatural surveillance to maintain the cohesion of the system. An "adaptive religious system" only survives if its members become and remain emotionally and (in some sense) intellectually committed to it; ongoing ritual engagement plays an important role in fulfilling these conditions. "Ritual behaviors and religious beliefs exist in a feedback loop in which behaviors affect beliefs and beliefs affect behaviors."[32]

In other words, religious credulity and religious conformity biases reinforce one another. Why is this relevant for understanding and responding to climate change? Because these deeply ingressed biases shroud the operation and amplify the effects of the other cognitive and coalitional biases we reviewed above, further distorting interpretations of (and decelerating reactions to) the ecological and economic crises of the Anthropocene. Educational, psychological, and public-policy experts are coming to realize that communicating more (or even better) *explicit* information about these crises is not going to help as long as people's perception of this information is *implicitly* biased. Unveiling and contesting heuristic mechanisms like anchoring, affective forecasting and self-serving bias is likely a necessary condition for the long-term success of any proposed solution to the kind of problems facing pluralistic, globalizing civilizations. However, even that monumental task is not likely to succeed unless and until the reciprocally reinforcing *religious* biases, which in many cases conceal and buttress those other generic mechanisms, are *also* unveiled and contested.

Let us take the example of "solution aversion." Campbell and Kay explored the function of this bias in the context of an analysis of the relation between ideology and motivated disbelief in a 2014 article in the *Journal of Personality and Social Psychology*. They reported on four experiments that studied the role of motivated reasoning (that is, rationalization processes that are implicitly shaped by biases outside of conscious awareness) in people's attitudes or perceptions of climate change. The authors discovered that the source of the motivation to disbelieve scientific reports is not necessarily related to an

---

32   Ibid., at 103.

aversion to the *problem* itself, but to an aversion to *solutions* popularly associated with the problem. The skepticism of many U.S. Republicans toward environmental science, for example, is partly a result of a conflict between ideological values (preference for a free market) and the sorts of solutions typically proposed for dealing with climate change (like regulating the free market). The answer to the problem of skepticism about scientific claims, therefore, "is not to simply present the public with more or better data but to consider other motivating factors."[33]

Biases like solution aversion are fortified and intensified by religious credulity and conformity. People who regularly engage in shared imaginative engagement with supernatural agents will implicitly perceive problems and proposed solutions through the lens of the axiological norms authorized by the supernatural coalitions to which they are committed. Because these norms are reinforced by ritual interactions that can exhaust cognitive resources, promote anxiety about hidden punitive forces, and increase antagonism toward out-groups, it is hardly surprising that individuals strongly committed to religious in-groups sometimes find it difficult to acknowledge problems associated with the Anthropocene, much less to commit themselves to solutions that challenge their superstitious interpretations of natural causes and segregative inscriptions of the social field.

It is encouraging to hear the arguments (and see the actions) of many "religious" people who are explicitly trying to promote the well-being of the environment and a fairer global distribution of wealth.[34] Tragically, however, such efforts may be implicitly undermined by the way in which their participation in *religion* – shared imaginative engagement with axiologically relevant supernatural agents – reinforces deep biases toward anthropomorphic promiscuity and sociographic prudery in their fellow believers, thereby demoting the sort of critical reflections and cultural relations that are needed for surviving the Anthropocene. This is particularly obvious among conservative and evangelical Christians, one of the most significant voting blocs in the U.S.[35] Nevertheless, there are some reasons for optimism.

---

33 Campbell and Kay, "Solution Aversion: On the Relation Between Ideology and Motivated Disbelief," *Journal of Personality and Social Psychology* 107, no. 5 (2014): 811.

34 Unfortunately, ecological sensitivity and economic desire are all too often in conflict with one another. This seems to apply not only to members of large-scale monotheistic coalitions, but to small-scale societies as well. See, e.g., Guen et al., "A Garden Experiment Revisited: Inter-Generational Change in Environmental Perception and Management of the Maya Lowlands, Guatemala," *Journal of the Royal Anthropological Institute* 19, no. 4 (2013).

35 Irwin and Martinez, "The Effects of Protestant Theological Conservatism and Trust on Environmental Cooperation," *Journal for the Scientific Study of Religion* 56, no. 1 (2017);

Adaptive Atheism

The term *atheism* is almost as contentious and contested as *religion*. In this context, I am using the former to designate attempts to make sense of the world and to act sensibly in society without appealing to supernatural agents or authorities (as explained at the end of Chapter 1). This stipulated definition highlights the creative efforts of those who contest the evolved tendency to rely on imaginative engagement with gods (*theōn*) when dealing with socio-ecological challenges. Insofar as it generates new modes of axiological engagement within pluralistic societies that alter the conditions for critical theoretical discourse about – and creative behavioral responses to – threats facing the human race and other sentient species, atheism can be conceived as an adaptation (in the general sense) to a radically interconnected global environment.

In the contemporary academy, and in the daily lives of an increasing number of people, supernatural agents are no longer the "best guess" when it comes to explaining surprising phenomena. Scientists and (non-religious) philosophers are trained to become anthropomorphically prudish, to resist the temptation to automatically attribute intentionality to unknown causes. If something unexpected happens in a test tube during a laboratory experiment, a chemist is not likely to hypothesize that it was a "ghost." If an inferential link seems to be missing in a chain of logical argumentation, a (non-religious) philosopher is not likely to accept the strategy of inserting a "god."

In many pluralistic societies today, supernatural agents no longer serve as "better guards." Scandinavian countries, for example, are among the happiest and most successful in the world and yet are also ranked as the most secular and atheistic. In *Living the Secular Life*, Zuckerman reviews the survey data that demonstrates that when it comes to measuring factors like happiness, valuing motherhood, promoting peace and murder rates, the least theistic states come out far better than the most theistic states.[36] It seems that, at least under some conditions, democratically elected secular governments, whose policy-making procedures are relatively transparent to their people, sponsor cooperative behavior at least as well as shared credulity about supernatural agents – without automatically activating the defense mechanisms of religious conformity biases.

Schwadel and Johnson, "The Religious and Political Origins of Evangelical Protestants' Opposition to Environmental Spending," *Journal for the Scientific Study of Religion* 56, no. 1 (2017).

36   Zuckerman, *Living the Secular Life: New Answers to Old Questions* (Penguin Press, 2014).

Happily, as we will see in more detail in Chapter 12, sociographic promiscuity and anthropomorphic prudery are also reciprocally reinforcing. The integration of these theolytic forces helps to unveil and challenge the evolved religious biases of *Homo deiparensis*. It is important to remember why the tendencies to fantasize about invisible agents and to become fanatical when protecting one's in-group are so common across human cultures – and so difficult to contest. These biologically evolved and socially bolstered tendencies are widely distributed in the current human population because they provided survival advantage to our early ancestors during the upper Paleolithic, enabling them to outcompete other hominid coalitions.

Like racist, sexist, and classist biases, *theist* biases have helped hold together increasingly complex human societies throughout the Neolithic, axial and modern ages. Today, however, in the diverse, cosmopolitan niches in which most of us live, these attitudes and behaviors have become maladaptive. Moreover, they are contributing to the degradation of the global environment in which all of us live.

As we noted at the beginning of this article, E.O. Wilson has recently claimed that religions are "dragging us down," and so "for the sake of progress the best thing we could possibly do would be to diminish, to the point of eliminating, religious faiths." In *The Meaning of Human Existence*, about which he was being interviewed when he made these comments, Wilson argued that:

> Human existence may be simpler than we thought. There is no predestination, no unfathomed mystery of life. Demons and gods do not vie for our allegiance. Instead, we are self-made, independent, alone, and fragile, a biological species adapted to live in a biological world. What counts for long-term survival is intelligent self-understanding, based upon a greater independence of thought than that tolerated today even in our most advanced democratic societies.[37]

This is the sort of anthropomorphic prudery we have come to expect in reflective, scientific analysis, and the sort of sociographic promiscuity we have come to hope for in policy proposals for pluralistic contexts.

But what would happen if "religious faiths" were eliminated? Is that even possible – or desirable? It seems more likely that shared imaginative engagement with supernatural agents (*religion*) will slowly dissolve as new generations find little or no use for this ancient adaptive strategy. But what will happen then? Of course, these are not the sorts of questions one can answer

---

37   Wilson, *The Meaning of Human Existence*, 26.

definitively in advance. We will have to figure it out together as we go along. If the analysis in this chapter is correct, however, then it may be that one of the conditions for surviving the Anthropocene is figuring out relatively soon how to facilitate the contestation of theistic credulity and conformity biases.

As I have argued elsewhere, there is a sense in which the academic discipline of *theology* can play a helpful role in this process.[38] This claim may surprise many readers, because most people are only familiar with the *sacerdotal* trajectory of theology, which in fact has dominated discourse within and among the west Asian monotheistic traditions. This trajectory is quite clearly compromised by evolved religious biases, having pressed anthropomorphic promiscuity to infinity and sociographic prudery to eternity with the conception of a supernatural Agent whose norms are the grounds for punishing (or rewarding) all Groups whatsoever.

However, if we think of theology in the broadest sense as *the critique and construction of hypotheses about the existential conditions for axiological engagement*, then it is easier to discern this ancient discipline's *iconoclastic* trajectory. The latter has certainly been the minority report in theology, but its proponents have consistently pressed toward anthropomorphic prudery and/ or sociographic promiscuity (see *Figure 2*, Chapter 1, p. 64 above), challenging the logical coherence and/or practical implications of the idea of an infinite intentional Being – without giving up on the existential *intensity* of *intentional* engagement with natural *infinities*. Liberating these iconoclastic forces from· the bio-cultural gravitational pull of religious biases is a good place to start. I discuss these two theological trajectories in more detail in Chapter 7 below.

Shared imaginative intercourse with supernatural agents emerged over time as evolved hyper-sensitive cognitive tendencies led to mistaken perceptions, which slowly became entangled within erroneous collective judgments about the number of potentially punitive agents in the social field. Allowing the covert operation of these evolved biases to continue unchecked reinforces commitment to favored in-group superstitions and antagonistic out-group segregations. Of all the tasks that face humanity as we try to adapt to (and alter) the Anthropocene, one of the most difficult will be un-learning these deeply embedded, reciprocally reinforcing heuristic habits.

This is why it is so important to talk openly about the mechanisms of religious reproduction – especially with the younger *Homo sapiens* among us. When it comes to explaining where babies come from, and how much effort is required to take care of them, we know that waiting too long can have devastating effects. The behaviors that lead to sexual and religious reproduction can

---

38   See especially, Shults, *Iconoclastic Theology,* and Shults, *Theology after the Birth of God.*

feel sensational to our bodies, but most people become quite sensitive when asked to bare their souls and talk about these feelings. All of this is completely natural. When discussing such intimate issues, it is important to be delicate – but it is also important to be direct. Having "the talk" about *religious* reproduction should involve more than simply explaining how "it" works. It is equally important to explain the socio-ecological consequences of "doing it."

We are not likely to find solutions to the global ecological and economic crises of the Anthropocene unless and until we learn how to accept our finitude and axiologically engage one another – intentionally and intensely – without bearing gods.

CHAPTER 5

Can Theism be Defeated?

For all of the reasons we have been discussing, god-bearing individuals usually find it extremely difficult to contest their religious credulity and conformity biases. This applies not only to laypeople, but to scholars who are affiliated with religious in-groups as well. Philosophy can help. It is no coincidence that the majority of professional philosophers embrace (or lean toward) atheism. Unfortunately, the road most travelled by unbelieving philosophers and scientists with a predilection for engaging the intellectual elite of culturally dominant monotheistic coalitions is marked by deep ruts carved out by centuries of debate over the same tired arguments for and against the probability (or possibility) of the existence of an omnipotent, omniscient, omnibenevolent God.

Occasionally one encounters theistic dialogue partners who are willing to budge a bit on issues like divine predestination or foreknowledge. When pressed on the coherence or plausibility of the very idea of a mysterious intentional force whose thoughts and desires (and even presence) can only be detected by in-group members ritually engaged in religious sects, however, progressive defenders of the faith make the same sort of predictable evasive hermeneutical maneuvers as their more traditionalist colleagues. How can we escape this disputatious cul-de-sac within which apologists and (too many) atheists have been circling for so long?

Instead of focusing primarily on the logical incoherence and empirical intractability of religious faith, which to many non-believers seem so obvious, I suggest we pay more attention to the way in which ritual interaction with imagined supernatural forces is engendered by an aggregate of evolved biases, to which many believers seem so oblivious. If we stay only at the level of explicit claims about gods (or God) we do not notice the implicit motivational reasoning that immunizes faith from critique and so easily activates believers' defensive reactions to (and even perceptions of) challenges to their supernatural beliefs and behaviors.

The apparent futility, monotony, and interminability of theoretical debates with religious apologists might lead us to conclude that an atheist's energy would be better spent identifying and implementing practical solutions to concrete problems. I'm all for the latter.[1] However, one good reason for

---

1 See, e.g., the discussion of policy-oriented computer modeling and simulation methodologies at the end of Chapter 12.

© F. LERON SHULTS, 2018 | DOI 10.1163/9789004360952_006

staying engaged in a debate over the metaphysical implications of scientific discoveries about the mechanisms of religious reproduction is that the arguments of apologetic philosophers of religion and other religiously affiliated scholars can play an important role in reinforcing the attitudes and actions of religious laypeople, who are all too often resistant to any pragmatic solution that is forbidden (or not authorized) by the supernatural agents whom they think are watching over them.

On the one hand, many believers spend an inordinate amount of time imaginatively engaging secretive and punitive supernatural agents who supposedly have designs for redeeming those who are part of their religious alliance. On the other hand, they all too quickly become belligerent when confronted with the religious beliefs and behaviors of those allied with other supernatural coalitions, which they find rather decidedly bizarre. Atheists are often astonished that theists fail to notice this double standard. Perhaps even more surprising is the apparent lack of concern expressed by believers even after this partisan bias is pointed out. At least religious apologists are concerned enough to be defensive about it! As students of the history of philosophy and theology, they realize that the rise of scientific naturalism and the spread of pluralistic secularism present new and serious challenges to the credibility and cohesion of their religious coalitions.

Insofar as apologetic arguments provide cover for the flourishing of theistic credulity and conformity biases, thereby exacerbating the global tensions we all face, it is worthwhile to take the time (and space) to challenge them. Debunking the theological claims of their religious colleagues is one way that godless philosophers can promote an adaptive atheism. In this chapter, I briefly describe three steps that may help us break the bad habits that have for so long characterized discourse about the extent to which science, and especially the science of religion, debunks theism. These steps are woven into a conversation with a recent book by Helen De Cruz and Johan De Smedt titled *A Natural History of Natural Theology: The Cognitive Science of Religion and Philosophy of Religion.*[2]

De Cruz and De Smedt have clearly demonstrated the extent to which – and the way in which – evolved cognitive tendencies play a role in the emergence of theistic ideas about God and in the formulation of theistic arguments meant to defend belief in his existence. Their contributions to the "naturalization"

---

2   De Cruz and De Smedt (Cambridge, MA: The MIT Press, 2014). Unless otherwise noted, page numbers throughout this chapter refer to this book. An earlier version of this essay was originally published as a commentary on their book; see Shults, "Can Theism Be Defeated? CSR and the Debunking of Supernatural Agent Abductions," *Religion, Brain & Behavior* 6, no. 4 (2015).

of theology and (theistic) philosophy of religion are a gift to the field. It is tempting to praise one chapter after another, but instead I will focus primarily on the one aspect of the book I found disappointing: their claim, stated early and often, but most forcefully in the final chapter, that the cognitive science of religion (CSR) "cannot straightforwardly provide a debunking account of natural theology and religion." The main warrant they offer for this claim is the difficulty they find "in choosing an appropriate level of explanation of how cognitive capacities generate religious beliefs" (179).

Their lack of confidence in the debunking force of empirical findings and theoretical developments in CSR and related disciplines may be due, in part, to the way in which they focus primarily on the *content* biases that *generate* ideas about gods (including God), and the relative lack of attention they give to the *context* biases that *nurture* them. This is why, in my view, their "naturalization" does not go far enough. The resilience of natural theology is indeed bolstered by the covert operation of hyper-sensitive *cognitive* biases that engender facile concepts of hidden, purposive agents. However, these god-conceptions only have staying power because of the facility of *coalitional* biases that reinforce shared imaginative engagement with such agents.

For the sake of this conversation, I accept their definition of *theism* as belief "in the existence of an omniscient, omnipresent, eternal morally perfect being who created the world and sustains its existence continuously" (xiii). I accept their definition of *religions* as "practices and beliefs that bind communities of people and that link them to a supernatural realm… features of human behavior that are regarded as religious (e.g., belief in supernatural beings, engaging in rituals) are present in all human cultures…" (12). And, finally, I accept their definition of *debunking* as the provision of arguments that "examine the causal history of a particular belief in a ways that undermines that belief" (2).

Can theism – and religious belief in supernatural agents in general – be debunked by philosophical reflection grounded in the findings of CSR? De Cruz and De Smedt are not alone in answering "no," or at least "not easily." It is no surprise that this is the unanimous answer among theistic apologists. But it is somewhat surprising how popular this answer is among agnostic (and even some atheist) scientists. Although there are several notable exceptions, several of which we have reviewed in earlier chapters, most CSR scholars seem hesitant to draw any explicit conclusions about the actual existence of the gods (or God) conceived in the human minds they study.

As the theoretical integration of biological *and* cultural factors continues to gain momentum, new light is being shed on the way in which religious beliefs, attitudes and practices rely on and reinforce anxiety about in-group cohesion. This is one reason I prefer to refer to the multi-disciplinary "field" in which most of these scholars work (or at least sometimes play) as the *bio-cultural* study of religion. Finding defeaters for religion may not exactly be "straight-forward," but if we pause to ask for directions – to reflect on the inferential directionality of our argumentation – it becomes relatively easy to spot theistic bunk.

Step 1. Leave the Cul-de-sac of Deductive/Inductive Arguments

Like most scholars on either side of (or on the fence in) the debate over the rationality of theism, De Cruz and De Smedt focus almost entirely on arguments that rely on inductive or deductive modes of inference. In other words, their discussions of the existence of God center around two sorts of question: whether it can be *proven* by valid reasoning from appropriate premises, and whether it can be rendered *probable* based on the evaluation of appropriate evidence. This is evident, for example, in their analysis of the "two types of possible defeaters" of theism, namely, those that rebut and those that undercut religious belief. "A *rebutting* defeater gives us reason to think the conclusion must be false. By contrast, an *undercutting* defeater does not challenge the conclusion directly, but makes us doubt that the evidence supports the hypothesis" (183).

The important thing to notice here is the (only?) two possible sorts of defeaters the authors treat rely on deduction or induction, eclipsing abduction and retroduction (to which I will return below). A rebutting defeater relies primarily on *deductive* inference. De Cruz and De Smedt do not think this sort of defeater works because theists and atheists disagree on the premises. "One's *prior assumptions* about the existence of God mediate to an important extent the perceived reliability of cognitive faculties that are involved in the formulation of natural theological arguments – this holds for both debunkers and vindicators" (198, emphasis added). As a side note, I should point out that this seems inconsistent with their claim elsewhere that "natural theology, unlike most other forms of theology, does not explicitly presuppose the existence of God …(and) should be intelligible regardless of one's metaphysical assumptions by appealing to observations and intuitions shared by all" (11). The more important point here, however, is the one on which we agree: focusing on deductive arguments gets us nowhere.

An undercutting defeater relies on *induction*, challenging some aspect of the evidence that theists offer in their attempts to lend credibility to belief in God. This is especially prevalent in the discussion around teleological arguments. At the end of their chapter on the argument from design, the authors insist that claims about theistic evolution (for example) cannot be evaluated purely by the empirical evidence. The rationality of the design argument "relies on the *prior probability* one places on the existence of God.... The reason why some find the design argument compelling and others do not lies not in any intrinsic differences in assessing design in nature but rather in the *prior probability* they assign to complexity being produced by chance events or by a creator" (84, emphases added). I think the differences in assessing design are also shaped by the extent to which a person has contested the evolved disposition toward guessing "an idiosyncratic hidden agent interested in my in-group" when confronted with ambiguous phenomena, but at this stage I want to emphasize my agreement: focusing on inductive arguments gets us nowhere.

Apologists and atheists have driven around in circles in this inferential cul-de-sac for centuries, and so it is no surprise the interlocutors keep meeting each other at the same old impasses. Despite its astonishing fecundity in so many other arenas of discourse, the "cognitive turn" in science (and philosophy) has not yet altered the course of (a)theological debate, which all too often follows the ruts carved out by the longstanding attempts to (dis)prove God through deduction or render God (im)probable through induction. It is time to explore other avenues. Yes, theists and atheists have quite different "prior assumptions" and "prior probabilities," but where did *these* come from? We cannot answer this question simply by appealing to the cognitive generation of the *content* of such (dis)belief. We must also ask about the coalitional *contexts* within which ideas about the gods (and God) are kept alive.

Step 2. Start at the Site of Alleged Religious Abductions

C.S. Peirce used the term *abduction* to refer to the way in which we develop conjectures that are intended to make sense of ambiguous phenomena. I observe a surprising fact (C). But then I reflect – or intuit – that if (A) were true, (C) would be a matter of course. This gives the hypothesis (A) an initial plausibility. In everyday life, we usually go with this "best guess" unless and until we encounter some challenge to it. In scholarly life, however, we are encouraged to overcome our confirmation bias, to reflect critically on our own idiosyncratic interpretations, and to invite others to challenge our hypotheses. Although abductive inferences may be based on earlier observations and utilized in later logical formalizations, they are not validated (as in induction) or proven (as in deduction); rather, they are rendered more or less theoretically *plausible* within a particular context in which they are evaluated as more or less pragmatically *feasible*.

Religious ideas about animal-spirits, ancestor-ghosts or gods are not the result of deduction or induction, but of abduction. For example, a Christian does not think the Eucharistic wafer has turned into the body of the risen Christ because she has observed its transmutation multiple times, nor because she has deduced this conclusion from theologically correct premises about the hypostatic union of the two natures of Christ. Rather, she finds herself confronted by a highly ambiguous phenomenon, the "surprising fact" that everyone around her participating in this ritual seems to be detecting the (real) presence of a supernatural agent (C). If the ritual officer (priest) really belonged to a social category of persons who were divinely imbued with a special power (A), then (C) would be a "matter of course." Hypotheses like (A) "work" in the sense that they hold together believers ritually engaged in religious sects.[3]

Ideas about counterintuitive supernatural agents are relatively immune to *inductive* challenges because they can live forever in the meta-representational limbo created by regular participation in causally opaque rituals; such "prior probabilities" can never be empirically falsified through observation. God-concepts are also relatively immune to *deductive* challenges because the symbolic representational contexts of religious in-groups are so open-textured that an endless array of quasi-propositions can be generated to qualify and protect the "prior assumptions" within them. *Abductive* challenges, on the other hand, press those who think they have detected a mysterious contingently-embodied intentional force to reflect carefully on the way in which they might be unconsciously immunizing such hypotheses from serious critique because they have a conflict of interest in maintaining the idiosyncratic beliefs of their own religious coalition.

An emphasis on the importance of *context* biases as well as *content* biases for understanding the origin and development of religion was already present in many of the early founding texts of the bio-cultural study of religion,[4] but as we saw in Chapter 1 above, it seems to have been intensifying in more recent literature. Extensive cross-cultural psychological experimentation and ethnographic research suggest that one of the main reasons that "punitive supernatural agent" (e.g., the son of a god who is "coming to judge the living and the dead") continues to feel like the "best guess" to people in religious contexts is that these sorts of agents function as "better guards." Shared belief that there

---

3 I discuss the importance of abductive (as well as retroductive) inferences for understanding the "phylogenetic fallacies" of theistic argumentation in more detail in Chapter 4 of Shults, *Theology after the Birth of God*.

4 See, e.g., Lawson and McCauley, *Rethinking Religion*; Boyer, *The Naturalness of Religious Ideas*.

are invisible "watchers" who have the capacity and desire to reward or punish those who do not follow the norms of the in-group tends to increase cooperation within and commitment to religious coalitions.

In an earlier article on the epistemic status of *scientific* beliefs, De Cruz and De Smedt do emphasize the role that the context of a large scientific community plays in the acquisition of truth-approximating knowledge under a broad range of conditions. Using the Price equation to assess the effects of cultural transmission and cognitive biases in scientific progress, they show that even low fidelity of transmission and substantial *cognitive* bias can be offset "by the tendency to make many different scientific inferences and by a large scientific *community*."[5] I would argue that this is partially due to the way in which academic contexts press individuals to reflect critically on the extent to which their hypotheses (abductive inferences) are open to critique from other individuals outside their own in-group (or research team) and might be surreptitiously shaped by conflict of interest.

In *A Natural History of Natural Theology*, the authors indirectly approach the issue of abduction when they discuss the way in which "inference to the best explanation" plays a role in arguments from design (64). My interest, however, is in the *primal* abductions about supernatural agents that have become deeply ingressed within a theist's interpretive scheme long before he or she gets around to dealing with abstract theological hypotheses about the best explanation for apparent design. It is precisely religious abductions of this sort – those that flow naturally from the evolved bias toward guessing that a hidden, person-like, coalition-favoring force is the cause of ambiguous phenomena – that scientific and philosophical training encourages one to *challenge*. Supernatural agent abductions are not simply prior "assumptions" or "probabilities," but biased hypotheses powerfully protected from critique by ongoing participation in the shared imaginative engagement of a particular religious coalition, wherein one is constantly required to send credible and costly signals of commitment to other in-group members.

How do scientists, (non-religious) philosophers, and most educated people in general, respond when they hear claims about UFO abductions, the detection of spirit-guides at a séance, celestial forces fulfilling astrological predictions, or the presence of trolls in the Norwegian forest? They consider them bunk. It is not always clear why "gods" are given a pass. De Cruz and De Smedt devote considerable time to "Reformed" epistemology, without asking why this should be given more credence than "Lutheran," "Mormon" or "Hindu"

---

5   De Cruz and De Smedt, "Evolved Cognitive Biases and the Epistemic Status of Scientific Beliefs," *Philosophical Studies* 157, no. 3 (2012): 427.

epistemology. Alvin Plantinga claims that Christians have access to more evidence than naturalists (and members of other out-groups) because of their special experience of divine revelation,[6] and Justin Barrett appeals to the biblical myth of the "fall" of Adam and Eve in his explanation of the failure of nonbelievers to detect God.[7] Normally this sort of special pleading would never be allowed to stand in serious, academic discourse.

Appealing to the noetic effects of sin, or some other flaw appraised by a punitive supernatural agent, to discredit the hypotheses of one's opponents, is surely one of the most appalling of noetic sins, and yet most scholars in the bio-cultural study of religion just let such claims slide. Why? The response "because these claims occur in the context of religious and theological discourse, which deals with spiritual realities beyond the boundaries of science" simply begs the question: why would anyone think that spiritual forces are real in the first place? The scientific study of religion has provided a really good answer to this question: such hypotheses are the result of abductive inferences covertly guided by implicit cognitive and coalitional biases. At the end of their book, De Cruz and De Smedt suggest that "one of the challenges for the metaphysical naturalistic worldview is to explain why such beliefs are widespread if their referents (supernatural entities) do not exist" (198). In light of the bio-cultural study of religion, it is not at all challenging to explain why beliefs in UFO abductions are widespread although their referents (probing aliens) do not exist. Why hesitate to make similar claims about gods (or God)?

Step 3. Don't be Afraid to Pursue Retroductive Destinations

"Retroduction," a term also introduced by Peirce, refers to inferences that lead to claims about what makes a phenomenon possible or, better, the conditions for its actualization. Like abduction, it involves the formulation of hypotheses, but *retroductive* conjectures are about the conditions without which a phenomenon could not be (or become) as it is. They "lead back" (*retro-ducere*) from more or less plausible and stable abductions to that which determines the existence of the phenomenon itself. Now, scientists make retroductive

---

6   Plantinga, "Games Scientists Play," in *The Believing Primate: Scientific, Philosophical and Theological Reflections on the Origin of Religion*, ed. Schloss and Murray (Oxford University Press, 2009), 167.

7   Barrett, "Cognitive Science, Religion, and Theology," in *The Believing Primate: Scientific, Philosophical and Theological Essays on the Origin of Religion*, ed. Schloss and Murray (New York: Oxford University Press, 2009), 97–98.

inferences all the time. They have no difficulty claiming that bodies, brains, ritual behaviors, cultural artifacts, etc., all exist and have varying conditioning effects upon one another.

Yet, when it comes to supernatural agents, many go out of their way to emphasize that CSR has no bearing whatsoever on their existence. Perhaps this is due, in part, to the constant reminders they hear from theists that one cannot "prove" a negative like "God does not exist" (a discussion of *reductio ad absurdum* arguments is beyond the scope of this essay). As should be clear by now, however, I am not interested here in proof (or probability), but in *plausibility*. In fact, scientists quite often *deny* the existence of something in their hypothesizing about causal relations. Most physicians in the 18th century believed in the existence of humours in the blood and other bodily fluids that affected human temperament and disease. Most physicists in the 19th century believed in the existence of ether, an unseen medium through which light allegedly travelled. Today, scholars in these fields have no qualms about claiming that ether and humours *do not exist*.

To take an example more directly relevant to the bio-cultural study of religion, Paul Bloom devotes most of his book *Descartes' Baby* to explaining the evolved mechanisms that generate conceptions of "immaterial" entities like a "soul" that can be separated from a "body," biases that so easily mislead children (and adults) into accepting Cartesian dualism. Bloom explicitly claims: "Descartes was mistaken ... We do not have immaterial souls."[8] However, he concludes by reassuring religious people that all of this is "logically separate from the question of whether God exists."[9] He has no problem retroductively inferring that there is no Cartesian "ghost" in the machine. Why, then, the reticence to straightforwardly reject hypotheses that appeal to the "Holy Ghost" in the hearts of Christian believers or "ancestor-ghosts" in the heart of the forest?

Like virtually all other scientists and (non-religious) philosophers, De Cruz and De Smedt explicitly accept *methodological* naturalism. "Thus when investigating the cognitive basis of intuitions in natural theology, we will not adopt metaphysical naturalism, which holds that there are not supernatural entities, nor will we assume a metaphysical theism, which takes the existence of God as a given. Moderate naturalism is *neutral with respect to metaphysical assumptions*" (59, emphasis added). It is certainly methodologically virtuous not to "assume" that something (say, a UFO, a troll or a god) does not exist, but moderate naturalism quite naturally leads to moderated metaphysical (retroductive)

---

8   Bloom, *Descartes' Baby: How the Science of Child Development Explains What Makes Us Human*, (New York: Basic Books, 2005), xii.
9   Ibid., 216.

claims about the existential conditions of causally complex phenomena – as illustrated throughout *A Natural History of Natural Theology*.

Still, one might be tempted to simply leave (apologetic) theologians and philosophers of religion to their imaginative engagement with supernatural agents, and get on with the work of critiquing and constructing scientific hypotheses. It is certainly not the job of scientists *qua* scientists to address the theoretical and practical problems associated with religiously-salient biases. But scientists are human too, and I see know reason why they should feel compelled *not* to point out the maladaptive effects that theism has on our species when it comes to (for example) dealing with the challenges of climate change and the injustices of consumer capitalism. The cognitive and coalitional biases that reproduce religion were naturally selected in the upper Paleolithic because these traits granted survival advantage to individuals in groups where they were widely distributed in the population.

In the complex, large-scale, pluralistic societies in which most of us live, however, these biases are no longer adaptive – at least if our concern is with the (temporal) adaptation of the race as a whole, as well as other sentient species, and not simply with the (eternal) survival of a particular religious in-group. The problem is that the cognitive biases that generate superstitious interpretations of nature and the coalitional biases that exacerbate segregative inscriptions of society are reciprocally reinforcing. As we saw in Chapter 1, priming people to think about supernatural agents can activate anxiety about out-groups, and participating in religious rituals can enhance credulity toward supernatural authorities. Theism, which follows out the hyper-sensitive tendency to detect supernatural agents to infinity and presses the hyper-sensitive tendency to protect supernatural coalitions to eternity, *intensifies* the psychological, political, and philosophical problems associated with religious belief and behavior.

Can theism be defeated? There is really no need to try and "defeat" it. If the plausibility of naturalistic explanations of causal forces in the cosmos and the feasibility of secularist inscriptions of society in pluralistic contexts continue to capture the imagination of new generations of *Homo sapiens*, then theism, like every other form of bias that loses its ritual hold on young minds, may just slowly fade away into cultural irrelevance.

In their response to my original commentary in *Religion, Brain & Behavior*, De Cruz and De Smedt said that they disagreed "with the claim that natural theological arguments are mainly written with the aim of signaling commitment to in-group members." As evidence against this claim (a claim which I did not make), De Cruz and De Smedt point out that "many natural theological arguments are formulated in a context of intellectual diversity, in particular one where naturalistic worldviews are on the rise… [and can be seen] as a countermovement to the rising influence of naturalism in philosophy and everyday life."[10]

The way in which they phrase this claim, as well as the evidence they offer to refute it, betrays a misunderstanding of costly signaling theories of religion. Costly signals are not something one "aims" to send. They are sent automatically and somewhat unconsciously as deliverances of the theistic credulity and conformity biases described above. Moreover, one would expect costly signals of the theological type to be sent in *exactly* the sort of context De Cruz and De Smedt describe: pluralistic environments in which the religious worldview of one's in-group is being robustly challenged.[11] Signals of one's commitment to theism through costly scholarship are (unconsciously) directed not toward one's unbelieving opponents, but toward the religious in-group crowd watching the match.

Theoretical analyses of literature *in* the philosophy of religion and empirical research *on* philosophers of religion have led to increasingly serious warnings about the pervasive theistic biases that permeate this academic discipline.[12] In the last few years, several scholars have argued that empirical findings and theoretical developments in the cognitive science of religion have provided powerful debunking arguments against religious beliefs.[13] This has elicited a

---

10   De Cruz and De Smedt, "Naturalizing Natural Theology," *Religion, Brain & Behavior*, 6, no. 4 (2015): 22.

11   Mahoney, "The Evolutionary Psychology of Theology," in *The Attraction of Religion: A New Evolutionary Psychology of Religion*, ed. Slone and van Slyke (London: Bloomsbury Academic, 2015).

12   See, e.g., Tobia, "Does Religious Belief Infect Philosophical Analysis?" *Religion, Brain & Behavior* 6, no. 1 (2016); Draper and Nichols, "Diagnosing Bias in Philosophy of Religion." *The Monist* 96, no. 3 (2013): 420; Nola, "Do Naturalistic Explanations of Religious Beliefs Debunk Religion?" in *A New Science of Religion*, ed. Dawes and Maclaurin (London: Routledge, 2003).

13   Wilkins et al., "Evolutionary Debunking Arguments in Three Domains" in *A New Science of Religion*, ed. Dawes and Maclaurin (London: Routledge, 2013) Griffiths and Wilkins, "Crossing the Milvian Bridge: When Do Evolutionary Explanations of Belief Debunk Belief?" in *Darwin in the Twenty-First Century: Nature, Humanity, God*, ed. Sloan et al., (Notre Dame, IL: University of Notre Dame Press, 2015); Teehan, "The Cognitive Bases of the Problem of evil," *The Monist* 96, no. 3 (2013): 325; Teehan, "Cognitive Science and the Limits of Theology," in *The Roots of Religion: Exploring the Cognitive Science of Religion*, ed. Trigg and Barrett (Surrey, UK: Ashgate, 2014); Schaffer, "Cognitive Science and Metaphysics: Partners in Debunking," in *Goldman and His Critics*, ed. Korblith and Maclaughlin (New York: Wiley-Blackwell, 2016); van Eyghen, "Two Types of 'Explaining Away' Arguments in the Cognitive Science of Religion," *Zygon* 51, no. 4 (2016).

host of responses from apologetic (or agnostic) philosophers, eager to point out that none of this *entails* that religious beliefs are false, and that CSR *alone* is insufficient as a defeater of theism.[14] However, I have argued that focusing on entailment relations is a red herring. By definition, negative logical assertions cannot be proven; no set of premises will ever entail that gods and ghosts (or unicorns for that matter) do not exist. The more relevant question is whether belief in such beings is *plausible*.

Moreover, as we have seen in earlier chapters, CSR is *not* in fact alone. While this discipline focuses on explaining the evolved mechanisms that shape belief, controlled experimental designs in several other disciplines have demonstrated the way in which manipulating religious beliefs increases errors and misattributions. CSR itself may not be able to comment on the ultimate reliability of religious beliefs, but

> experimental psychology has a great deal of evidence supporting their unreliablity and malleability... Not only is it impossible to prove a negative, these types of arguments also ignore abductive standards of evidence, which prefer the most likely explanation for a phenomenon while requiring the fewest additional assumptions ... [CSR] must be viewed in the light of findings from experimental psychology indicating that religious and spiritual intuitions are unreliable in shaping beliefs (e.g., the over-detection of animacy), and that rational processes [used to defend them] likewise represent motivated biases. A range of processes occurring at multiple levels (neural, cognitive, personality, social) have been shown by experimentation to produce misattributed RS [religious/spiritual] thoughts and experiences indistinguishable from spontaneous or "genuine" ones. This has relevance to meta-physical claims because it indicates that these experiences, often refered to as sui generis or evidential of the supernatural can be accounted for by naturalistic mechanisms, thus constituting "false positives."[15]

---

14   Leech and Visala, "The Cognitive Science of Religion: A Modified Theist Response," *Religious Studies* 47, no. 3 (2011); Johnson, et al., "The Elephant in the Room: Do Evolutionary Accounts of Religion Entail the Falsity of Religious Belief?" *Philosophy, Theology and the Sciences* 1, no. 2 (2014); Barrett and Trigg, "Cognitive and Evolutionary Studies of Religion," in *The Roots of Religion: Exploring the Cognitive Science of Religion*, ed. Trigg and Barrett (Surrey, UK: Ashgate, 2014); Visala, *Naturalism, Theism and the Cognitive Study of Religion: Religion Explained?* (Routledge, 2016).

15   Galen, "Overlapping Mental Magisteria: Implications of Experimental Psychology for a Theory of Religious Belief as Misattribution," *Method and Theory in the Study of Religion*, 29, no. 3 (2017), 27, 33.

All of the scientific disciplines that contribute to the bio-cultural study of religion facilitate the debunking of theism. Naturally, quite naturally, theistic biases will continue to shape the reactions to (and perceptions of) this research among individuals for whom engaging in religious sects has become a central part of their narrative identity and a primal source of meaning for those they love.

This applies to laypeople and scholars alike. Attacking specific religious beliefs directly through arguments involving proof and probability can all too easily make things worse, amplifying precisely those biases that are blocking the embrace of a robust naturalist and secularist worldview in the first place. This is why I recommend the strategy of (more or less gently) continuing to point out the role that evolved content *and context* biases play in engendering and nurturing their conceptions of the idiosyncratic supernatural agents imaginatively engaged in the rituals of their in-group.

But it is important to remember that science is not the only anaphrodisiac when it comes to religious sects. Insofar as it promotes analytical and critical reflection, philosophy can also jolt people out of the mood. To reiterate, I am not suggesting that atheist philosophers (or unbelievers in general) are not biased. There is no doubt that some non-religious folks struggle with sexism, classism, racism, and other biases. By (my) definition, however, an atheist has learned, or is in the process of learning, how to overcome the credulity and conformity biases associated with theism (or, through some combination of personality and contextual factors, never came under the seductive influence of these biases in the first place). Insofar as theism bolsters sexism, classism, and racism (at least at the population level), those concerned about the latter ought to embrace the task of undermining the former. Godless philosophy can help. In my view, the philosophical resources found in the work of Gilles Deleuze are among the most useful for fostering an adaptive atheism.

CHAPTER 6

The Atheist Machine

My focus in this chapter is on what I will call *the atheist machine*, the multiple uses and effects of which are expressed throughout the productions, registrations, and consumptions of the literary corpus of the philosopher Gilles Deleuze.[1] In their first co-authored book, *Anti-Oedipus*, Deleuze and his frequent collaborator Felix Guattari challenged the psychoanalytic idealization and capitalist appropriation of Oedipus, and set out a plan in which – or a plane on which – a new set of questions could be productively engaged: "Given a certain effect, what machine is capable of producing it? And given a certain machine, what can it be used for?"[2] Using the language of *A Thousand Plateaus*, we could say that the abstract machine of adaptive atheism produces rhizomic lines of flight whose absolute deterritorialization molecularizes the transcendent pretenses of monotheistic molarities. I will argue that the atheist machine is always at work wherever schizoanalysis (or rhizomatics, micropolitics, pragmatics, etc.) proceeds, as long as it proceeds.

In their last co-authored book, *What is Philosophy?*, Deleuze and Guattari argued that "Wherever there is transcendence, vertical Being, imperial State in the sky or on earth, there is *religion*; and there is Philosophy only where there is immanence ... only friends can set out a plane of immanence as *a ground from which idols have been cleared*."[3] When it comes to dealing with priestly erections of arborescent icons within a religious Imaginarium, the schizoanalytic task of the Deleuzian Friend is definitely destructive. "Destroy, destroy. The whole task of schizoanalysis goes by way of destruction..."[4] As the last few sections of *Anti-Oedipus* make clear, however, this destruction is inextricably linked to the positive and creative tasks of schizoanalysis.

---

1   This chapter is an adapted version of "The Atheist Machine," which was originally published in Powell-Jones and Shults, eds., *Gilles Deleuze and the Schizoanalysis of Religion* (London: Bloomsbury Academic, 2016). Like some of the earlier essays in the current book, the original version of this chapter utilized a more generic definition of anthropomorphic promiscuity and sociographic prudery. I have reworked this version significantly to render it consistent with the stipulated (and fractionable) definitions outlined in Chapter 1.

2   Deleuze and Guattari, *Anti-Oedipus: Capitalism and Schizophrenia* (Minneapolis: University of Minnesota Press, 1983), 3.

3   Deleuze and Guattari, *What Is Philosophy?* (Columbia University Press, 1996), 43. Emphasis added.

4   Deleuze and Guattari, *Anti-Oedipus*, 342.

After a brief review of the relation between atheism and schizonalysis in Deleuze's work, I return to the conceptual framework introduced in Chapter 1. As we have seen, insights derived from empirical findings and theoretical developments within the bio-cultural sciences of religion can help us understand how and why gods are so easily born(e) in human minds and groups. We also need to refresh our memories about the historical contingencies surrounding the emergence of the (western) monotheistic idea of "God" – an infinite supernatural Agent who has a special plan for a particular Group. In the second section, I briefly explain how the advent of this conception, which turned out to be logically, psychologically, and politically unbearable, contributed to the assemblage of the atheist machine during the axial age.

Next, I utilize the conceptual framework of theogonic reproduction theory as a heuristic model for clarifying the dynamics at work within and among the four main social-machines treated in the *Capitalism and Schizophrenia* project (i.e., the territorial, despotic, capitalist, and war machines). As we will see, the atheist machine plays a special role in the creative production of the (revolutionary) war machine. Finally, I will explore the implications of the integration of these machines, memories, and models for the productive task of *becoming-atheist*, that is, for the experimental construction of bodies without organs on the plane of immanence without any recourse to transcendent religious Figures imaginatively engaged by subjugated groups whose rituals allegedly mediate divine revelation.

Elsewhere I have spelled out the connections between theogonic reproduction theory and Deleuzian philosophy in more detail.[5] In the current context, I limit myself to a broad outline of the theory, demonstrating its usefulness for abstracting a Deleuzian atheist machine, and extracting its revolutionary force for the schizoanalysis of religion.

Atheism and Schizoanalysis

The goal of schizoanalysis is "to analyze the specific nature of the libidinal investments in the economic and political spheres, and thereby to show how, in the subject who desires, desire can be made to desire its own repression ... All this happens, not in ideology, but well beneath it ..."[6] One of the goals of *theological* schizoanalysis, I suggest, is to show how subjects come to desire their own *religious* repression. "All this" does indeed occur "well beneath" the

---

5   Shults, *Iconoclastic Theology: Gilles Deleuze and the Secretion of Atheism.*
6   Deleuze and Guattari, *Anti-Oedipus*, 115.

surface of priestly ideology. As we will see in the next section, evolved cognitive and coalitional mechanisms surreptitiously regulate desiring-production by engendering god-conceptions in human minds and cultures. At this stage, however, our focus is on the way in which schizoanalysis works to challenge the striations and segmentations of the socius effected by priestly figures, whether psychoanalytic or religious.[7]

Deleuze expresses astonishment that so many philosophers still find the death of God tragic. "Atheism," he insists, "is not a drama but the philosopher's *serenity* and philosophy's *achievement.*" The dissolution of God is not a problem. "Problems begin only afterward, when the *atheism* of the concept has been attained."[8] Why, then, would the Deleuzian Friend continue to devote attention to religious ideas, such as concepts of God that hold up monotheistic molarities? First of all, chipping away at such repressive representations is valuable in and of itself. But Deleuze suggests another motivation for poking around religious and theological edifices. "Religions," he argues, "are worth much less than the nobility and the courage of the atheisms that they inspire."[9]

Some of Deleuze's most inspiring pages are those in which he attends to sacerdotal stratifications; this makes sense in light of his claim that "there is always an atheism to be extracted from religion." In fact, Deleuze singles out Christianity as that religion that *secretes atheism* "more than any other religion."[10] This helps to explain his frequent criticism of that long-dominant monotheistic Coalition.

However, Deleuze explicitly separates *all* religion from philosophy, art, and science. The latter three require more than the making of "opinions," which are attempts to protect ourselves from chaos based on the invocation of "dynasties of gods, or the epiphany or a single god, in order to paint a firmament on the umbrella, like the figures of an Urdoxa from which opinions stem." Art, science, and philosophy "cast planes over the chaos... (they) want us to tear open the firmament and plunge into the chaos. We defeat it only at this price."[11] Each of these struggles with chaos in its own way, "bringing back" varieties (art), variables (science) or variations (philosophy). Efforts within all three disciplines are always and already bound up in the struggle against *opinion* – especially opinions woven into sacred canopies defended by religious hierarchies.

---

7   Deleuze and Guattari, *A Thousand Plateaus: Capitalism and Schizophrenia* (New York: Continuum, 2004), 171.

8   Deleuze and Guattari, *What Is Philosophy?* 92. Emphasis added.

9   Deleuze, *Two Regimes of Madness: Texts and Interviews 1975–1995* (Semiotext, 2007), 364.

10 Deleuze and Guattari, *What Is Philosophy?* 92.

11 Ibid., 202.

What does any of this have to do with schizoanalysis? Does Deleuze really link schizoanalysis (and rhizomatics, micropolitics, pragmatism, etc.) to *atheism*? Indeed he does. In *Anti-Oedipus*, Deleuze notes that denying God is only a "secondary thing," and accomplishes nothing if "man" is straight away set in God's place. The person who realizes that "man" is no more central than "God" does not even entertain the question of "an alien being, a being placed above man and nature." Such a person, he observes, no longer needs "to go by way of this mediation – the negation of the existence of God – since he has attained those regions of an auto-production of the unconscious where the unconscious is no less atheist than orphan – immediately atheist, immediately orphan."[12] For the schizoanalyst, the unconscious is not mediated by Oedipus or Christ (or any other religious Figure): it is *immediately* orphan *and* atheist.

In his critique of psychoanalysis Deleuze identifies three errors concerning desire: lack, law, and the signifier. These are in fact the same error, an "idealism that forms a pious conception of the unconscious." But where did these errors come from? "These notions cannot be prevented from dragging their *theological* cortege behind – insufficiency of being, guilt, signification... But what water will cleanse these concepts of their background, their previous existences – *religiosity*?"[13]

In *A Thousand Plateaus* these notions are explicitly linked to the triple curse cast on desire by "the priest," the most recent figure of which is the psychoanalyst: "the negative law, the extrinsic rule, and the transcendental ideal."[14] The similarity between traditional interpretations of the Genesis myth as a "Fall" and models of the Oedipal conflict that rely on privative, punitive, and palliative categories is hard to miss: both understand desire in terms of loss, guilt, and idealization – as under the curse of anxiety, prohibition, and displacement from a desexualized paradise.[15]

In the plateau on "Nomadology," Deleuze also explicitly links atheism to the creative war machine that was invented by the nomads. "It may be observed that nomads do not provide a favorable terrain for religion; the man of war is always committing an offense against the priest or the god... The nomads have

---

12   Deleuze and Guattari, *Anti-Oedipus*, 65–66.

13   Ibid., 121. Emphasis added.

14   Deleuze and Guattari, *A Thousand Plateaus*, 171.

15   I explored this connection in more detail in Shults, "De-Oedipalizing Theology: Desire, Difference and Deleuze" (Grand Rapids, MI: Eerdmans, 2011). When I wrote that chapter in 2009, I was still attempting to domesticate Deleuze within a sacerdotal, albeit apophatic, prophetic vision of my religious family of origin (Christian evangelicalism).

a sense of the absolute, but a singularly *atheistic* one."[16] Although the phrase "war machine" does not appear in *Anti-Oedipus*, we do find references there to a "revolutionary machine," and to hunters in nomadic space who follow the flows and escape the "sway of the full body of the earth."[17] Atheism and schizoanalysis cannot be separated. "For the unconscious of schizoanalysis is unaware of persons, aggregates, and laws, and of images, structures, and symbols. It is an orphan, just as it is an anarchist and an *atheist*."[18]

This link between atheism and schizoanalysis will come as no surprise to those familiar with Deleuze's earlier single-authored works of philosophical portraiture, in which he consistently hammers away at religious ressentiment and traditional notions of God, and celebrates the atheistic effects of Nietzsche, Spinoza, Hume, and even Kant (1984).[19] In *Difference and Repetition*, he encourages us not to judge the atheist from the point of view of the belief that supposedly drives him, but to judge the believer "by the *violent atheist* by which he is inhabited, the *Antichrist* eternally given 'once and for all' within grace."[20] In *Logic of Sense*, Deleuze insists that there has only ever been one ethics, the *amor fati* of the humor-actor who is "an anti-God (*contradieu*)" – the Stoic sage who "belongs to the Aion" and opposes the "divine present of Chronos."[21]

Deleuze found atheism a somewhat obvious place to begin. Instead of loitering around the starting line of philosophy, he encouraged us to get moving, to experiment on the plane of immanence by creating concepts. Getting people to the starting line, however, is harder than Deleuze seemed to realize. One of the most important effects (and uses) of an atheist machinic assemblage, I suggest, is the disassembling of the god-bearing machines that reproduce supernatural agents in the human Imaginarium and covertly pressure believers to keep nurturing them through regulated ritual engagement. Unveiling these evolved mechanisms, which operate "well beneath" theological

---

16   Deleuze and Guattari, *A Thousand Plateaus*, 422. Emphasis added.

17   Deleuze and Guattari, *Anti-Oedipus*, 354, 163.

18   Ibid., 342. Emphasis added.

19   Deleuze, *Nietzsche and Philosophy*, (New York, NY: Columbia University Press, 1983); Deleuze, *Expressionism in Philosophy: Spinoza*, (New York: Cambridge, Mass: Zone Books, 1992); Deleuze, *Empiricism and Subjectivity*, (New York: Columbia University Press, 2001); Deleuze, *Kant's Critical Philosophy: The Doctrine of the Faculties* (Minneapolis: Univ Of Minnesota Press, 1985).

20   Deleuze, *Difference and Repetition*, (New York: Columbia University Press, 1995), 96. Emphasis added.

21   Deleuze, *The Logic of Sense* (New York: Continuum, 2004), 170–171.

ideologies, is an important initial step as we begin to have "the talk" about religious reproduction.

Theogonic Mechanisms: How Gods are Born(e)

Where do babies come from? Why do parents keep them around? As I argued in Chapter 3, archaeologists working at sites like Çatalhöyük do not have to dig around for answers to such questions. As they unearth Neolithic skeletons and artifacts, clearing the ground of "idols" (or, at least, of "figurines"), they can confidently assume that the regular arrival and continued nurture of the infants in that community were the result of the same basic sort of coital procedures and mating strategies that were naturally selected during the evolution of *Homo sapiens* in the upper Paleolithic and that continue to replenish the human population today. Although research on these practices in contemporary contexts might yield insight into some interesting variations, cultural anthropologists know enough about human biology and social psychology to explain, without additional field work, where the babies in their field sites come from and why the parents keep them around.

A similar confidence is emerging among scholars in the bio-cultural sciences of religion about the mechanisms by which *gods* are born in human minds and borne in human cultures. As we have seen in previous chapters, in the last quarter century theoretical proposals based on empirical research within a wide variety of fields such as evolutionary biology, archaeology, cognitive science, moral psychology, and cultural anthropology, have been converging around the claim that religious phenomena can be explained by the evolution of *cognitive* processes that over-detect human-like forms in the *natural* world and *coalitional* processes that over-protect culturally-inscribed norms in the *social* world.

The phenomena associated with "religion" are complex and contested (like the term itself), but for the purpose of this interdisciplinary experiment I will continue to use the term to indicate an aggregate of features that have in fact been found in every known culture, past and present, namely, *shared imaginative engagement with axiologically relevant supernatural agents.* Where do conceptions of gods come from, and why do groups keep them around? Belief in supernatural revelations and participation in supernatural rituals are the result of the integration of evolved perceptive and affiliative tendencies that contribute to what I have been calling "anthropomorphic promiscuity" and "sociographic prudery." Here I return to the coordinate grid introduced in *Figure 1* (see Chapter 1, p. 3), which provides a conceptual framework for discussing the possible correlations between these types of cognitive and coalitional dispositions – and their contestation.

Let's begin with a quick review. Why are humans so prone toward *superstition*, that is, to proposing and accepting interpretations of ambiguous (and especially frightening) natural phenomena that are based on false conceptions of causation? Such interpretations are due, in part, to evolved *cognitive* tendencies that pull us toward the left side of the horizontal line in Figure 1. When we encounter some pattern or movement we do not understand, our first guess is likely to involve the attribution of characteristics like mentality and animacy. This over-active predilection helps to explain why we so easily see "faces in the clouds" and worry about hidden supernatural forces that may intend us harm. Moreover, we quite often double down on such guesses and keep scanning for human-like agents even when there is no clear evidence of their presence. This tendency to assume that hard-to-detect agents are the cause of hard-to-understand events served our upper Paleolithic ancestors well; otherwise, we would not be here to write and read about them.

For example, early hominids who developed hyper-sensitive cognitive devices that scanned for agency (intentionality, purposiveness, etc.) were more likely to survive than those who did not. What made that noise in the tall grass? Was it a human enemy or some other animal? Or was it just the wind? Those who quickly guessed "intentional force" and acted accordingly were more likely to avoid being eaten (if the animal was a predator) and more likely to find food (if the animal was a prey). Despite almost constant false positives in the short run, this over-active perceptual strategy would have granted survival advantage in the long run. It would have paid off to keep searching for and believing in such hidden agents. Anxiety about the failure to find an actual agent generates other hypotheses; just because we are paranoid does not mean that an animal-spirit or angry ancestor-ghost was not really lurking in the grass before it mysteriously disappeared.

These and other mechanisms that contribute to anthropomorphic *promiscuity* are distributed in human populations as part of our phylogentic inheritance. Most of us still jump at any opportunity to postulate human-like entities as causal explanations even – or especially – when these interpretations must appeal to counterintuitive disembodied intentional forces, i.e., to "supernatural agents." Of course, it is also possible to contest this sort of evolved bias. Scientists and philosophers, for example, are trained to become anthropomorphically *prudish*. Far more cautious about such appeals, and typically critical of superstition in general, they are more likely to resist ascribing intentionality to unknown causes. If something strange happens in a test tube during an experiment, the chemist will not guess that it was a "ghost." If something seems to be missing in a causal (or logical) chain, the (non-religious) philosopher will not insert a "god."

Why are humans also so easily prone toward *segregation*, that is, to making and reinforcing inscriptions of the social field that protect their own in-groups from contamination or domination by out-groups? Our evolved *coalitional* biases can pull us toward the bottom of the vertical line in Figure 1. This (often vehement and sometimes violent) fortification of boundaries is engendered, in part, by an evolved over-active tendency to embrace and defend supernaturally authorized conventional modes of segmenting and regulating society. This naturally generated prejudice for one's own collective makes it tempting to just stay at home where the proscriptive and prescriptive norms feel most familiar. This tendency is so powerful that people will often engage in costly and painful behaviors in order to follow the rules – and willingly inflict pain on those who do not. It makes sense that such a hyper-sensitive propensity toward protecting one's own coalition would also have served our early *Homo sapiens* ancestors well.

When it comes to competition among small-scale societies, especially when resources are low or under other stressful conditions, those groups that are most likely to survive are those in which the individual members are able to cooperate and remain committed to the group. Natural selection reinforces the tendency of an individual organism to watch out for itself, but if there are too many cheaters, freeloaders, or defectors in a society it will quickly fall apart. Research in the bio-cultural sciences of religion suggests that this problem was solved in some hominid coalitions during the upper Paleolithic by an intensification of shared belief in and ritual engagement with potentially *punitive* supernatural agents (such as animal-spirits or ancestor-ghosts). Such coalition-favoring "gods" could catch misbehavior that regular natural agents might miss and could punish not only the miscreants, but their offspring or even the entire group. Belief in invisible or ambiguously apparitional "watchers" helped to enhance the motivation to follow the rules and stay within the coalition.

Contemporary humans have also inherited this sociographic *prudery*. Most people somewhat automatically follow the authorized social norms of their in-group, or at least put great effort into building up a reputation for doing so. Here too, however, these evolved biases can be contested. Those who are *promiscuous* in their sociography are less likely to accept claims about or demands for the segregation of human groups that are based only (or even primarily) on appeals to authorities within their own coalition. They are more likely to be open to intercourse with out-groups about alternate normativities and to the pursuit of new modes of creative social engagement. In-group bias helped (some of) our ancestors survive in small-scale societies in difficult socio-ecological niches. Today, however, this evolved predisposition does not always serve us well – especially those of us who live in large-scale, urban societies characterized by the pressures of globalization and radical pluralism. A growing number of policy-makers and legislators in such contexts refuse to appeal to "ghosts" or "gods" in their attempts to inscribe the public sphere.

As we have seen, the mechanisms of anthropomorphic promiscuity and sociographic prudery often reinforce one another. Supernatural agents who are cared for and ritually engaged within a coalition then become easy imaginative targets for the hair-triggered agency detection mechanisms of each new generation. In the environment of our early ancestors the selective advantage went to hominids who developed cognitive capacities that quickly *detected* relevant agents in the natural milieu and whose groups were adequately *protected* from the disruption that could result from too many cheaters in the social milieu. These god-bearing traits are distributed in contemporary populations as part of our phylogenetic inheritance, which helps to explain why so many people today are still so easily pulled into the trajectory represented in the lower left quadrant of *Figure 1*.

The explanatory power of the disciplines that contribute to the bio-cultural study of religion challenges the plausibility of belief in ghosts, gods, and other culturally postulated disembodied intentional forces. As noted in Chapter 5 above, scientists and (non-religious) philosophers may not be able to provide deductive logical arguments that *disprove* the existence of supernatural agents or inductive evidence that *invalidates* claims about their causal relevance, but they can offer powerful abductive and retroductive arguments that render their existence *implausible*. The more reasonable hypothesis is that shared imaginative intercourse with supernatural agents emerged over time as naturally evolved hyper-sensitive cognitive tendencies led to mistaken perceptions of intentionality that slowly became entangled within erroneous collective judgments about the extent of the social field.

The (relative) success of science and the (relatively) peaceful cohesion of democratic, pluralistic societies require that those who want to participate in the academic and public spheres learn how to challenge the cognitive and coalitional biases that promote superstition and segregation. But if the biases that lead to shared imaginative engagement with supernatural agents were so deeply woven into the genetic and memetic structures of human life, why and how did they come to be challenged in the first place? Scientific naturalism and political secularism are expanding in many parts of the world. A growing number of us do not think we need gods to make sense of the natural world or to act sensibly in the social world. Where did such "atheistic" ideas come from? They were already gestating during the axial age.

Monotheistic Memories: The Birth of (A)theism

During the tenth millennium BCE, shared belief in local animal-spirits or limited ancestor-ghosts was enough to hold together small-scale societies of hunter-gatherers. Shamanic engagement with such finite supernatural agents even sufficed for the egalitarian sedentary collectives that began to form during the Neolithic. Over the millennia, however, in many contexts across the most fertile areas of east, south, and west Asia, human groups grew in size and complexity and claimed ever-larger plots of land for themselves. So did their gods. As coalitions were amalgamated or assimilated by one another, smarter and more powerful supernatural agents emerged – "high gods" who could monitor the behavior of more human agents and trump the local spirits or ancestral authorities of the newly-merged coalitions. Ever bigger groups evolved ever bigger and ever more punitive gods, which helped to ensure that everyone (or at least a sufficient percentage of the population) cooperated and stayed committed.

During the first millennium BCE, within the largest and most complex literate states across east, south, and west Asia, a new sort of god-concept was born in the minds of intellectual and priestly elites: an all-encompassing Supernatural Agency whose influence was universal and in relation to whom all behavior was punished (or rewarded). The period from approximately 800–200 BCE is commonly called the "axial age" because it represents a turning point, or axis, in the transformation of civilizational forms in human history. The most common ideas about an ultimate Reality that emerged in east and south Asia during this period did not explicitly (or unambiguously) involve the attribution of intentionality to an infinite Force. Dao and Dharma, for example, were supposed to be morally relevant for any and all groups, but many Chinese and Indian religious scholars seriously questioned whether such Realities should be primarily conceived as person-like and coalition-favoring.

There was far less doubt in the monotheist traditions that emerged in the wake of the west Asian axial age: we are made in the *image* of God and God has a special plan for *our* group. The identity of Jewish – and eventually Christian and then Muslim – coalitions was tied to narratives about the creation of Adam and the call of Abraham to a promised land (paradise lost, and found, in west Asia). Theological debates among these religious in-groups center around questions about the extent to which (or even whether) Moses, Jesus, or Muhammad mediate divine law-giving and care-giving. Which group has the definitive revelation of – and ritual access to – the one true God who will personally punish (or reward) everyone for all eternity? Monotheism is anthropomorphic promiscuity and sociographic prudery gone wild – superstition and segregation applied to infinity.

As we will see in the next section, Deleuze often noted a special relation between monotheism and what he called the *despotic* machine. When the coding of flows in the "primitive" territorial socius are overcoded in the despotic socius, then "the ancestor – the master of the mobile and finite blocks – finds himself dismissed by the deity, the immobile organizer of the bricks and their infinite circuit."[22] For Deleuze, the main role of the deity seems to be the inscription of debt into the very existence of the despot's subjects, who now owe their very being to the despot-god. "*There is always a monotheism on the horizon of despotism*: the debt becomes a debt of existence, a debt of the existence of the subjects themselves."[23] Even if the priest (or the prophet) connected to the king-despot does not see the disobedient actions or disrespectful attitudes of the people, the inescapable Eye of God will – and no sinner can hide from his judgmental Voice and punitive Hand.

Among the despot's bureaucrats, the monotheistic priest has a special role: administering the face of God and interpreting His intentions. "A new aspect of deception arises, the deception of the priest: interpretation is carried to infinity and never encounters anything to interpret that is not already itself an interpretation."[24] The revelation that is allegedly encountered in holy texts and engaged in rituals is ambiguous; it can be (and must be) endlessly interpreted in new ways because ideas about counter-intuitive discarnate forces are not empirically constrained. What does the Torah (Bible, Qur'an) *mean*? What does God *want* us to do *now*? The transcendent God of monotheism, Deleuze notes, "would remain empty, or at least *absconditus*, if it were not projected on a plane of immanence of creation where it traces the stages of its theophany." Whether it takes the form of imperial unity or spiritual empire, "this transcendence that is projected on the plane of immanence paves it or populates it with Figures."[25]

On the one hand, the intellectual and priestly elites of monotheistic coalitions insist that their supernatural Agent has appeared and will continue to appear in the finite world. On the other hand, they also insist that His glorious nature is infinitely transcendent and beyond comprehension – even the despot may misinterpret God.[26] This tension has always characterized *theology*, which

---

22   Deleuze and Guattari, *Anti-Oedipus*, 217.
23   Ibid., 215. Emphasis added.
24   Deleuze and Guattari, *A Thousand Plateaus*, 126–128.
25   Deleuze and Guattari, *What Is Philosophy?* 88–89.
26   Eisenstadt et al., *The Origins and Diversity of Axial Age Civilizations*, (State University of New York Press, 1986), Arnason et al., eds., *Axial Civilizations and World History*, (Leiden;

was also born during the axial age. Broadly speaking, theology is the construction and critique of hypotheses about the existential conditions for axiological engagement.[27] What is it that makes possible – or actual – the real, finite human experience of valuing and being valued? In their attempts to answer this sort of question, the majority of theological hypotheses within the monotheistic coalitions that eventually came to dominate most of west Asia and Europe (and much of the rest of the globe) followed the theogonic trajectory depicted in *Figure 1* (Chapter 1, p. 3).

Even among theologians (as well as priests and prophets) who were committed to the sacerdotal regulation of religious minds and groups within particular monotheistic in-groups, however, one can also find minority reports that contest the idea of God conceived as a person-like, coalition-favoring, punitive disembodied Entity. We have already alluded to the first reason the intellectual elite in such religious groups might have for resisting finite images of God as, for example, a "Father" or "Judge": whether material or semiotic, such images (icons) are all too easily taken by regular religious folk as actual representations of an infinitely glorious and holy divine Reality that ought not to be represented. This is (part of) the motivation behind warnings against idolatry and occasional acts of physical iconoclasm. An infinite God *must not* be represented for doxological reasons.

However, God *cannot* be represented for logical, psychological, and political reasons. One of the existential requirements for intentionality is being in relation to something not identical to oneself, that is, to an object of intention. This is the case even if one is intentionally relating to one's imagined, future self – intending, for example, to become a better person. Intentionality presupposes an in-tensional relation to that which one is not, or which one does not yet have. In other words, it requires being-limited, which is the de-finition of finitude. This is why absolute infinity cannot be intentional: to conceive it as such would be to imagine it as *related* to an object that it was not (such as a finite creation), in which case it would not be *absolutely* unlimited. Moreover, cognitive and coalitional biases evolved to engage *finite* supernatural agents, and the pressure exerted by the notion of an all-knowing and all-powerful *infinite* despot-God is simply psychologically and politically unbearable. People may memorize and repeat orthodox doctrinal formulations about God's omniscience, omnipotence, and impassibility but, especially under stress, they

Boston: Brill Academic Pub, 2004); Bellah, *Religion in Human Evolution: From the Paleolithic to the Axial Age* (Cambridge, MA: Harvard University Press, 2011).

27   I defend this definition and demonstrate the sense in which Deleuze is an atheist "theologian" in Shults, *Iconoclastic Theology*.

immediately fall back into their default tendency to imagine finite, temporal gods who are interested in their kith and kin.[28]

The idea of "God" as an infinite disembodied intentional Force was tentatively born(e) in the minds of theologians who pressed the evolved biases toward anthropomorphic promiscuity and sociographic prudery as far as they would go – which turned out to be too far. If God is so transcendent that He cannot be represented, then He cannot be conceived (or perceived) as a human-like agent (or anything else). If God eternally fore-knows and pre-ordains *everything*, then it is hard to understand the point of praying to or ritually engaging Him. Throughout the centuries, monotheistic theologians have worked hard to defend hypotheses about the conditions for axiological engagement that utilize images (icons) of God as a Person who cares about a Group while simultaneously emphasizing that such images must be broken.

Evolved cognitive mechanisms for detecting finite agents crumple under the pressure of trying to think an infinite intentional Entity. Evolved coalitional mechanisms for protecting in-groups implode (or explode) under the stress of trying to live in complex literate states. It is not hard to understand why and how atheism would emerge as an option (albeit rarely, slowly, and tentatively) as monotheism took over within large-scale, pluralistic societies. The abstract, transcendent God described by the priest does not seem to have any relevance for daily life. All these people around me have different views of gods whom they think care about their group. They try to explain the natural world in superstitious ways that make no sense to me. They try to regulate the social world in segregative ways that make it difficult for me and those I love. Perhaps we can make sense of the cosmos and behave sensibly in the socius without bearing God – or any other finite supernatural agents preferred by particular in-groups.

The assemblage of the atheist machine involved the contestation of evolved theogonic mechanisms, which opened up lines of flight that were previously unimaginable. Although its use within and effect on the mental and social fields of the civilizations that emerged out of the west Asian axial age were initially quite limited, the atheist machine began to unveil the implausibility of the various (contradictory) ideas about supernatural causality and the infeasibility of the various (contradictory) ritual strategies for organizing normativity. Even when contesting the relevant cognitive and coalitional biases is not consciously used to clear the ground of religious Icons, it automatically has a somewhat theolytic (god-dissolving) effect.

---

28   Slone, *Theological Incorrectness*.

The intensification and integration of the forces of anthropomorphic prudery and sociographic promiscuity are part of the actualization of the atheist machinic assemblage, which follows the trajectory in the upper right quadrant depicted in *Figure 2* (see Chapter 1, p. 64). The effects of the atheist machine are obviously destructive but, like all schizoanalytic (rhizomatic, pragmatic, micropolitical) proceedings, its uses are also productive.

Its most palpable productions are naturalism and secularism. There are many varieties of *naturalism*, but most share a resistance to appeals to supernatural agency in theoretical explanations of the natural world, especially in the academic sphere. Individual scholars may continue privately to harbor superstitious beliefs, but most are (at least) methodologically naturalistic in the sense that they exclude god-concepts from their scientific hypotheses. There are also many varieties of *secularism*, but most share a resistance to appeals to supernatural authority in practical inscriptions of social worlds, especially in the public sphere. Individual civil leaders in complex, democratic contexts might maintain membership in religious in-groups, but a growing number are (at least) methodologically secularist in the sense that they exclude divine-sanctions from their political proposals.

We do not yet know what naturalist-secularist bodies can do. Whatever they *can* do, hypothesizes the atheist, their axiological engagement is not conditioned by human-like, coalition-favoring gods. Atheism follows out the logic and practices that flow from the integration of the theolytic forces, pressing beyond methodological versions of anthropomorphic prudery and sociographic promiscuity and insisting on *metaphysical* naturalism and secularism. The atheist machine cuts away at superstitious beliefs and segregating behaviors based on shared imaginative engagement with supernatural agents, and constructs pragmatic plan(e)s within socio-ecological niches in which survival no longer depends on the detection and protection of the gods of particular in-groups.

I have argued that the naturally evolved theogonic biases operate "well beneath" monotheistic ideology, reproducing repressive religious representations that fuel the despotic machine. I now want to make more explicit the relation between the theolytic forces and the other three social-machines described in the *Capitalism and Schizophrenia* project.

Deleuzian Social-Machines in Bio-Cultural Perspective

It is important to remember that Deleuze does not think of the social-machines as concrete, historical formations of the socius that were (or will be) realized in a particular order. Rather, they are abstract machines that are actualized in diverse ways within all complex social assemblages, precisely in their intensive mutual interactions. The territorial, despotic, and capitalist machines are all social-productions that "fall back" on desiring-production; each in its own way creates a "full body," a "recording surface" that inscribes lack, law, and idealization on the schiz-flows of the Real, which is pure becoming. As we will see, although the war machine can be captured by the State, in itself it is the creative element or productive force of rhizomic lines of flight that escape repressive representations.

The territorial (or primitive) machine is the "first form of socius, the machine of primitive inscription."[29] A socius is produced whenever there is a coding (inscription) of stock (consumption) that falls back upon the flow (production) of desire. The first mode of representation organizes itself at the surface by the coding of filial flows through alliances, thereby creating a "territory." The unit of alliance is debt, and alliance, suggests Deleuze, is "representation" itself. When it falls back on the desiring-production of human bodies, the territorial machine constitutes a debt system involving "a voice that speaks or intones, a sign marked in bare flesh, an eye that extracts enjoyment from the pain." An element of transcendence (representation of an ideal) is introduced, but it remains "quite close to a desiring machine of eye-hand-voice."[30] The territorial assemblage is declined on the full body of "the earth" through the *coding* of lateral alliances and extended filiations.

The despotic (or barbarian) machine, on the other hand, appears with the force of a "projection that defines paranoia," in which a "subject leaps outside the intersections of alliance-filiation, installs himself at the limit, at the horizon, in the desert, the subject of a deterritorialized knowledge that links him directly to God and connects him to the people."[31] Deleuze describes despotism as the first principle of a paranoiac knowledge that withdraws from life and from the earth, producing a judgment of both. The socius will now be inscribed on a new surface, not the earth, but the full body of "the despot" (or his god). The voice is no longer one of alliance across filiations, but "a fictitious voice from on high." The *overcoding* of the despotic machine (or imperial barbarian formation) is characterized by the mobilization of the categories of *new* alliance and *direct* filiation.

The eyes watching the hands' inscription of bodies are replaced by the Eye and the Hand of the despot, who watches everyone through the eyes of his bureaucrats, officials, and priests, and subordinates graphism to the Voice

---

29   Deleuze and Guattari, *Anti-Oedipus*, 155.

30   Ibid., 207.

31   Ibid., 211.

that "no longer expresses itself except through the writing signs that it emits (revelation)." Now, interpretation becomes all important: "The emperor, the god – what did he mean?"[32] Having claimed a direct and transcendent filiation, the despot appropriates all the forces of production. All alliances are now organized around and oriented toward him. Instead of blocks of mobile and finite debt coded by horizontal alliances, the despot extracts taxes for a vertical tribute that feeds a constantly expanding glorious expenditure.

In *A Thousand Plateaus*, this is also spelled out in relation to the "facialization machine," which effects an overcoding wrought by the signifying despotic Face, irradiating a surveillance that reproduces paranoid faces. The savage system of *cruelty* is replaced by the barbarian system of *terror*. The despotic State, Deleuze insists, is an abstraction that is realized only as an abstraction.[33] As an abstract machine, it can be conceived as "the common horizon" to what comes "before" and what comes "after," that is, as a complex of syntheses that can overcode the territorial machine's coding of break-flows and, in turn, that can become relativized and incorporated within the capitalist machine's axiomatization of decoded break-flows.

This *decoding* of flows that characterizes the capitalist (or civilized) social-machine has also always been present in human populations, even if only as that which was "warded off" by primitive and barbarian social inscriptions (and the nomads). This machine has a deterritorializing effect, but it is only "relative." It immediately reterritorializes the decoded flows on the "full body" of Capital. The surplus value of production, as well as the qualities of alliances, which had been coded through kinship or overcoded through tribute are now decoded, rendered quantitative and relativized in relation to the surplus flux of the market, which registers value on the basis of the potential for earning wages or generating profit. The capitalist machine is fully installed when money begets money, when Capital itself becomes filiative. "It is no longer the age of cruelty or the age of terror, but the age of *cynicism*, accompanied by a strange *piety*..."[34]

What about the war (or revolutionary) machine? Despite its name, the primary use (and effect) of this machine is not war. Only when it is appropriated by the State apparatus of capture does war necessarily become its object. The essential aim of the war machine is "revolutionary movement," escaping the molar organization and conjugation of flows through a becoming-molecular that effects an *absolute* deterritorialization (whether artistic, scientific, or

32   Ibid., 244.
33   Deleuze and Guattari, *A Thousand Plateaus*, 240.
34   Deleuze and Guattari, *Anti-Oedipus*, 245. Emphasis added.

philosophical). Once the capitalist machine has relativized the despotic machine's overcoding of the territorial machine and taken over the socius, every struggle involves the construction of *"revolutionary connections"* in opposition to the *"conjugations of the* [capitalist] *axiomatic."*[35] Resisting facial-ization (and oedipalization), the war machine creates and populates smooth space with "probe-heads" that draw lines of flight, cutting edges of deterritori-alization that become positive and absolute, "forming strange new becomings, new polyvocalities."[36]

How can the conceptual framework of theogonic reproduction theory, derived from bio-cultural scientific models of the origin and evolution of religion, shed light on the repressive (and liberating) functions of the Deleuzian social-machines?[37] Once again, it is important to emphasize that, for Deleuze, *all* of these abstract social-machines are operative in *every* hu-man population – although in each concrete context they are more or less successful in their coding, capturing, axiomatizing, or escaping in relation to one other. This means that we should not think of any particular, historical assemblage as the manifestation of one of these social-machines; rather, each concrete assemblage is characterized by the dynamic interplay among them as they "fall back" upon desiring-machines.

It is also crucial to remember that social-machines do not operate on the same plane as theogonic mechanisms, whose operation was described in detail in Chapter 1. The evolved tendencies that contribute to anthro-pomorphic promiscuity and sociographic prudery are entangled within the desiring-production of culturally embodied human cognitive systems, the desiring-production of real machines that eat, talk, shit, and fuck.[38] The social-machines, on the other hand, are forms of "social production" that include an element of "antiproduction coupled with the process, a full body that func-tions as a *socius*. This socius may be the body of the earth, that of the tyrant, or capital." The god-bearing biases we have been exploring throughout this book, however, operate "well-beneath" the "surface" formed by the machinic ideolo-gies of each socius, where "all production is recorded, whereupon the entire process appears to emanate from this recording surface."[39]

---

35   Deleuze and Guattari, *A Thousand Plateaus*, 522.
36   Ibid., 211.
37   I give a fuller answer to this question in Chapter 5 of my *Iconoclastic Theology: Gilles De-leuze and the Secretion of Atheism*. In the latter context, however, I utilized more general definitions of the theogonic mechanisms. See footnote 1 above.
38   See p. 1 of Deleuze and Guattari, *Anti-Oedipus*.
39   Ibid., 11.

Let us begin with the "primitive" territorial machine. Archaeological evidence and ethnographic analogy suggest that the hominids that flourished in the upper Paleolithic were extremely anthropomorphically promiscuous. They somewhat automatically postulated ambiguously embodied intentionality behind everything – rivers, trees, crystals, the weather, and the earth itself. Early hominids may also have been (relatively) sociographically promiscuous, at least in some contexts. In Europe, for example, it seems that *Homo sapiens* got along surprisingly well with Neanderthals, at least until around 35,000 BCE. In the wide open spaces of Africa and the Levant, interaction with alien hominid groups would have been more rare, and such encounters may not have felt as immediately threatening as they do in environments with scarce resources and denser populations. The integration of these promiscuous tendencies is still prevalent among human coalitions found today in some indigenous cultures and New Age groups.

In this sense, social assemblages strongly shaped by the territorial machine have a special relationship with the upper left quadrant of *Figure 1*. Because desiring- and social-machines operate on different levels, however, one should not confuse the assemblage with an individual whose cognitive and coalitional tendencies could be depicted in that quadrant. Rather, the social-machines should be conceptualized on a different, third dimension relative to the two dimensions of *Figure 1*, or perhaps as dynamic transversal (or Minkowski) planes that intersect that two-dimensional plane in various ways (depending on the parameters that constitute the relevant phase space). At any rate, it makes sense to think that assemblages whose social production primarily falls back upon the "body of the earth" will foster (and be fostered by) the presence of anthropomorphically and sociographically promiscuous individuals.

Social assemblages dominated by the "barbarian" despotic machine seem to have a special relationship with the lower left quadrant of *Figure 1*. As we have seen, the socius that falls back upon the "body of the tyrant" is inextricably linked to monotheism, which is a powerful promoter of sociographic prudery. The one true God (ours) has revealed the norms by which all human groups are to be regulated and judged. All assemblages whatsoever are subject to his rule. There is no point in arguing with or trying to trick an infinite, unchanging despot-God, whose prescriptions and proscriptions are absolute. The despotic (monotheistic) social machine also powerfully promotes anthropomorphic promiscuity. First and foremost, it encourages the detection of a Supernatural Agent who is allegedly everywhere at all times. Imaginative engagement with more limited disembodied spirits such as angels, demons, or saints is usually permitted as long as paranoia about them does not challenge the authority of His priestly bureaucrats, who are watching and waiting to enforce divine judgments.

What about the lower right quadrant of *Figure 1*? This is where we represent individuals who are both anthropomorphically and sociographically prudish. There is an important sense in which the "civilized" capitalist machine intersects with this quadrant. On the one hand, when all production falls back on this socius, it is easier for people to become more anthropomorphically prudish. Despite the claims of televangelists, the accumulation of surplus wealth has nothing to do with pleasing supernatural agents and everything to do with the mechanisms by which money begets money. As the capitalist machine increasingly decodes the qualitative (personal) alliances upon which territorial coding and despotic overcoding relied, individuals have less reason to worry about immaterial anthropomorphic spirits (or a Divine Spirit) and can focus on the means of material production.

On the other hand, the inscriptive prudery of the capitalist machine is absolute: it forces *all* surplus value to fall back on the "full body" of Capital, converting all codes to abstract quantities (Money). It spreads a universal anxiety: everyone must accumulate surplus value for their *own* group (family, corporation, state, military, etc.). As with the other social machines, we should be careful not to equate capitalism with any one of the quadrants of *Figure 1*. There is a sense in which this machine supports sociographic promiscuity by decoding human affiliations and encouraging the multiplication of images. It quite clearly does not directly promote sociographic prudery in the sense articulated in Chapter 1; that is, it does not foster the protection of *supernaturally* authorized norms per se. However, insofar as it is capable of quantifying and relativizing *all* codes (even supernatural images), and assimilating them into the surplus flux of the market, it can facilitate the (doubly) prudish thoughts and behaviors represented in the lower right quadrant of the coordinate grid.

Both the territorial and capitalist machines promote tendencies that at least partially challenge the despotic mode of theogonic reproduction. Their inscriptions inevitably throw wrenches into the monotheistic machine. The war machine, however, fractures the repressive "representations" of all three of the other modes of social-production. It has no time (or place) for the segmentarity of Oedipus, much less for the sedentary arborescence of the transcendent Icons of monotheism. In this sense, it is always consuming, registering, and producing an atheist machine. The monotheistic machine exists only by overcoding territories and resisting the axiomatizations of the immanent capitalist field that relativize its preferred religious Figure. The nomads who invent the "revolutionary" war machine want to "have done" with the judgment of God.[40]

---

40   Deleuze, *Essays Critical And Clinical* (Minneapolis: Univ Of Minnesota Press, 1997), 126.

The forces of the war machine open up lines of flight that promote anthropomorphic prudery and sociographic promiscuity. Escaping the facialization machine and drawing positive and absolute lines of deterritorialization, this machine populates a smooth space with "probe-heads ...that dismantle the strata in their wake, break through the walls of signifiance, pour out of the holes of subjectivity."[41] The nomads refuse the segmentation of sedentary collectives whose striation of the socius finds its center of gravity in the State. The *nomos* of the war machine is a movement and composition of people that cannot be captured in the apparatus of the "law." Its becoming is a celerity that constantly invents tools and weapons that can be used on the move in the encounter with and the production of new modes of social assemblage.

I have suggested that all of these social-machines shape (and are shaped by) underlying cognitive and coalitional tendencies, which are distributed and contested in various ways across human populations. How is the atheistic, schizoanalytic machine functioning in our own contemporary context? What destructive and creative theolytic effects is it having?

Becoming-Atheist

Deleuze has helped clear the ground for revolutionary experimentation by disclosing the repressive power of social-machinic representations. I have tried to show how *theogonic* mechanisms, which integrate and intensify superstitious and segregative tendencies, make this process of clearing far more complicated than it initially appears. The repressive representations they (re)produce are reinforced by naturally evolved biases that all too easily lead to the detection of gods and the protection of in-groups. This is why we also need to pay closer attention to the uses and effects of *theolytic* mechanisms. How can we produce atheistic registrations and consumptions on the field of immanence as we clear the ground of the religious Figures of transcendence that make us anxious and distract us from creating new connections? As Deleuze consistently emphasized, the criteria for answering such questions can only be discovered in the actual, problematic process of schizoanalysis.

Developments in the bio-cultural sciences provide us with conceptual tools that can supplement the insights that arise in the debates among defenders and detractors of psychoanalysis. They help us unveil the secrets of theism, especially the cognitive incoherence and coalitional irrelevance of representations of an infinite personal God. Such prodding exerts a pressure

---

41   Deleuze and Guattari, *A Thousand Plateaus*, 210.

that intensifies the secretion of atheism. But this is not enough; the forces of theogonic reproduction have led to adaptive defenses that continue to hold subjects within religious coalitions. For example, theologians committed to monotheistic in-groups can insist that these "mysteries" are part of what is adorable about the divine nature or part of what is hidden in the divine plan. Appealing to concealed secrets, secrets that are appealing in part because of their concealment, keeps the secretion in check. This is one of the reasons that theology should not be left to theists.

It is important to keep talking about where the god-conceptions within in-groups come from in the first place because unveiling theogonic mechanisms automatically weakens them; they function well only when they are hidden. We cannot know ahead of time what effects the atheist machine will have. The secretion of productive atheism will not solve all our problems and will surely create some new ones. However, insofar as it clears the ground of arborescent religious Icons that reinforce mythical and superstitious interpretations of nature and divide us through supernatural segregations of society, at least it gets us moving.

We do not yet know all that godless bodies can do, but we do know they can move on the surface, liberate lines of flight, construct rhizomes, feel the movement of the pack, and unleash the creative forces of art, science, and philosophy. For obvious reasons, such movements threaten groups whose molarity depends on centralized imaginative engagement with supernatural agents. Like the State apparatus, despotic religious societies treat their secrets with gravity, but inevitably – it is the nature of secrets – something oozes out, something is perceived. The war machine treats secrets with celerity, molecularizing their content and linearizing their form.[42]

This is why the atheist machine feels so dangerous to the monotheistic machine, which uses its massive arsenal to crush or domesticate it. But we nomads have no reason to fear: we have weapons of mass secretion that work just by bringing them into the open.

---

42   Ibid., 320.

CHAPTER 7

Theology after Pandora

In these next three chapters, I explore the more or less anaphrodisiacal effects that different sorts of theology can have on people's desire to engage in religious sects. When it comes to bearing gods, fundamentalist religious groups are far more fertile than liberal religious groups. Demographically, the same is true for bearing babies. This is one of the reasons why mainstream, progressive denominations have declined while many charismatic and conservative denominations have grown, or at least come closer to maintaining their population levels. However, I will argue that liberal permissivism is at least as problematic as conservative fertility (though for different reasons).

In this chapter, I clarify the tension between the sacerdotal and iconoclastic trajectories of theology in the work of one of most progressive theologians in the evangelical tradition. Chapter 8 assesses the "natural theology" of a theologian who has devoted much of her scholarly efforts to reconstructing Christian doctrine in light of challenges from evolutionary biology and climate science. Chapter 9 examines the (relatively) theolytic writings of one of the world's leading liberal theologians committed to a radically pluralistic approach to religion.

Each of these theologians is (more or less) critical of some of the god-conceptions in the Christian tradition and some of the repressive and oppressive effects of the religious coalitions with which they are affiliated. For this, the atheist can be grateful. For reasons I hope to make increasingly clear, however, this kind of "critical" theological reflection all too easily provides cover for the religious reproduction of the sort of superstitious beliefs and segregative behaviors that are exacerbating the global (and local) challenges we all face.

The evangelical theologian Stanley Grenz loved to make connections between theology and popular culture, and he had a special predilection for science fiction – especially Star Trek (Next Generation, of course) and X-Files (like agent Mulder, he believed the truth was "out there").[1] He also knew that our fascination with extra-terrestrials is more about our own alienation, our own strangely anxious and hopeful sentience, than it is about imagined alien

---

1 This chapter is an adapted version of "Theology After Pandora: The Real Scandal of the Evangelical Mind (and Culture)," which was originally published in a posthumous Festschrift for Stan Grenz, ed. Tidball et al., *Revisioning, Renewing, Rediscovering the Triune Center* (Eugene, OR: Wipf and Stock, 2014).

© F. LERON SHULTS, 2018 | DOI 10.1163/9789004360952_008

This is an open access chapter distributed under the terms of the CC BY-NC-ND 4.0 License.

creatures. More than most evangelical theologians, Stan focused on what was happening here and now on planet Earth. Nevertheless, like every other evangelical I know, he also anticipated an eschatological renewal of this world, a new Earth (and a new Heaven), re-created in some sense by the supernatural agency of Christ. In this chapter I borrow one of Stan's well known methodological strategies, reflecting on theological themes in the context of engaging a popular science fiction film. Although he almost certainly would not have agreed with my material proposal, I know he would have welcomed the conversation.

One might think that the reference to "Pandora" in the main title of this chapter was to the Greek myth in which the first woman, modeled of clay by Hephaestus as part of Zeus's punishment of mankind for Prometheus' theft of fire, released evils into the world by opening a box (or jar) given to her by the gods. At the end of this Chapter we will return to poor Pandora, but the reference here is actually to the planet Pandora in the 2009 film *Avatar*. The film portrays the conflict between the Na'vi, the (mostly) friendly natives of Pandora, and the invading human forces of the RDA mining corporation, the (mostly) nasty humans bent on acquiring the aptly named "unobtainium" that lie buried beneath the surface of the planet. I use this fictional account of a conflict of ideas and societies on Pandora as material for reflecting on the possibilities for theology to respond in new ways to the intellectual and political challenges we face here and now on planet Earth.

In his influential book on *The Scandal of the Evangelical Mind*, Mark Noll begins by asserting that "the scandal of the evangelical mind is that there is not much of an evangelical mind... American evangelicals have failed notably in sustaining serious intellectual life."[2] In his more recent *Jesus Christ and the Life of the Mind*, Noll finds some signs for modest optimism but remains "largely unrepentant" of his negative evaluation.[3] Stan Grenz was an obvious exception. His scholarship set a standard of excellence among those who self-identified as part of the North American evangelical subculture. Fundamentalists were often scandalized by his writings, not for Nollian reasons, but because he challenged the *status quo* they were so concerned to protect. Nevertheless, the existence of a few such scholars does not diminish the larger point: Noll is right to decry the lack of intellectual rigor among evangelicals. The problem, however, is much deeper and more serious than Noll realizes.

The real scandal of the evangelical mind, I will argue, cannot be separated from the scandal of the evangelical *culture*, and vice versa. Balancing

---

2 Noll, *The Scandal of the Evangelical Mind* (Grand Rapids, Mich.; Wm. B. Eerdmans Publishing Co., 1995), 3.

3 Noll, *Jesus Christ and the Life of the Mind* (Grand Rapids, Mich.: Wm. B. Eerdmans Publishing Co., 2013), 105.

piety with more appreciation of the Christian intellectual tradition will not solve the problem. Balancing social concern with better scholarship will not solve the problem. Such efforts merely reorganize the chairs on the deck of the sinking Titanic (to allude to another film by James Cameron). The deeper problem facing evangelicalism is one that is shared by other religious coalitions. Although the material details of evangelical belief in and hope for eschatological renewal have their own peculiarities, they are formally structured by the same evolved dynamics that have contributed to the emergence of religion in every known human society: widely shared imaginative engagement with axiologically relevant supernatural agents.

Evangelicals will continue to stumble as long as they cannot see the hidden cognitive and coalitional mechanisms that shape their mental and social life. This chapter is an attempt to unveil these mechanisms by engaging recent scientific discoveries about (and philosophical reflection on) the emergence, evolution, and transmission of human religiosity. Once we can see what we are doing, it will be easier to explore new possibilities for the discipline of theology. In the concluding section of this chapter, I propose a way of doing theology that does not appeal to extra-terrestrials (or supernatural agents) who favor a particular human coalition. For reasons I will try to make clear, I call this the *iconoclastic* trajectory of theology.

The first three sub-sections set out the basic argument of the chapter. First, I provide yet another overview of the conceptual framework of theogonic reproduction theory, which we have been discussing in earlier chapters. Second, I use this framework to analyze the "theological" options portrayed in the movie *Avatar*, as the (increasingly) evil capitalists fight the (initially) naïve tree-huggers for control of the planet Pandora. This then sets the stage for a description of evangelical groups as examples of a particular kind of supernatural agent coalition, typical of those religions that trace their roots to the axial age, which normally follow what I call the *sacerdotal* trajectory of theology. The *real* scandal of the evangelical mind (and culture) is that evolved mechanisms are surreptitiously shaping its theological practice, reinforcing the psychological repression and political oppression that everyone sees but no one is sure what to do about. The last two sub-sections explore the possibility of a quite different approach to doing theology.

Bearing Gods in Mind and Culture

Why are we religious? As we have seen, empirical findings and theoretical reflections across a variety of fields including archaeology, cognitive science, evolutionary neurobiology, moral psychology, social anthropology, and political theory suggest that the contemporary human beliefs, activities, and emotions normally associated with "religion" are shaped by naturally evolved mechanisms that are part of our shared phylogenetic heritage. This section provides another introductory and integrative reading of significant trends within these fields, which coalesce around the general claim that shared ritual interaction with gods naturally emerges in contemporary human minds and cultures as a result of cognitive and coalitional tendencies that helped our early hominid ancestors survive in small-scale groups, granting them a competitive advantage in an upper Paleolithic environment. The question that faces us today is whether these tendencies are still adaptive in our rapidly changing, pluralistic, global environment.

First, let me reiterate the sense in which I am using some key terms. The term *bearing* has a double meaning, indicating the naturally evolved processes by which gods are *born* in human cognition (by the hyper-active detection of agency in the *natural* environment) and *borne* in human culture (by the hyper-active protection of coalitions in the *social* environment). When I refer to these mechanisms as *theogonic* (god-bearing), I am not alluding to literary accounts of the genesis of the gods such as Hesiod's *Theogony*, but to the way in which such bio-cultural mechanisms engender any and all narrative imaginative engagement that reinforces the human desire to participate in religious sects.

As we have seen, scholars in the disciplines that contribute to the bio-cultural study of religion often use the term *god* as shorthand for any culturally-postulated, discarnate intentional force – as synonymous with "supernatural agent" or "superhuman entity." In other words, not only Yahweh, Zeus, or Vishnu, but also ghosts, genies, and goblins would be referred to as *gods*. The differences between these kinds of gods are obviously significant, but for the sake of participating in this interdisciplinary dialogue I will follow this usage in this chapter. At this point, it is important to note that they all share at least two key features: intentionality and contingent embodiment.

A supernatural agent *coalition* is a social nexus that is held together, at least in part, by appeal to the power or authority of gods allegedly watching the group and concerned about its members' evaluative (or axiological) judgments or moral actions.[4] That is to say, the way in which members of the group evaluate one another's (and their own) beliefs, behaviors and attitudes are in some way constituted or regulated by supernatural agents who are taken to be strategic players in the survival of the group. All of this leads to my use of the

---

4   See, e.g., Bulbulia, "Nature's Medicine: Religiosity as an Adaptation for Health and Coopera-
    tion," Sosis, "Religious Behaviors, Badges, and Bans," and the other references from Chapter 1
    in the current volume.

term *religion* in this context to refer to "shared imaginative engagement with axiologically relevant supernatural agents."

With these definitions in place, I now return to another discussion of the conceptual grid we have been exploring (*Figure 1*, Chapter 1, p. 3), but this time focusing on the way in which it can help us differentiate between two trajectories in the discipline of theology. As we have seen, this initial fractionation of theogonic mechanisms (anthropomorphic promiscuity and sociographic prudery) does not capture all the complexity of "religion," but this framework is sufficient for its purpose; namely, as a heuristic device for discussing the interactions between two basic tendencies that are part of the phylogentic inheritance of all *Homo sapiens* (including evangelicals).

Once more (and for the last time in this volume), let's review the basic tenets of theogonic reproduction theory. As we have seen, the integration of mechanisms that engender anthropomorphic promiscuity and sociographic prudery was an evolutionary winner. In the early ancestral environment the selective advantage went to hominids whose cognitive capacities enabled them to quickly *detect* relevant agents (such as predators, prey, protectors, and partners) in the natural environment, and whose groups were adequately *protected* from the dissolution that could result from too many defectors and cheaters in the social environment.

Hyper-sensitive detection often led to false positives; e.g., identifying a noise in the forest as a predator (or prey) when it was really the wind. However, occasionally it really was a predator (or prey) and those whose detective capacities were weak or lazy – it's probably just the wind – got eaten (or failed to eat) and so their genes were not passed on. Hyper-sensitive protection often led to serious punishment of cheaters, the demand for costly signals of commitment from those suspected of considering defection, and willingness to attack and kill members of out-groups. The good news (for the in-group) is that these strategies did in fact lead to stronger (longer-lasting) coalitions.

In fact *over*-sensitive detection and protection (and other components of anthropomorphic promiscuity and sociographic prudery) increased the chance of survival during a critical period of time in human history. In earlier chapters, we reviewed part of the growing body of evidence that suggests that around 90,000–70,000 years ago, some *Homo sapiens* groups developed more complex beliefs and rituals in which they imaginatively engaged supernatural agents they detected in the environment (e.g., spirits attributed responsibility for weather, good hunting, etc). These contingently embodied (or ontologically confused) intentional forces were often believed to have the power to punish cheaters or defectors (or their family members). They also might be watching at any time, which increased group member's motivation to follow social norms. Sometime around 60,000 years ago it appears that some of these "god-bearing" groups left Africa, out-competing all other hominid species and spreading out across the Levant and into Europe and Asia, eventually incorporating other kinds of supernatural agents such as ancestor-ghosts.

All living humans are the genetic offspring of these groups, and so share a suite of inherited traits (differentially distributed across the population) that support the tendency to detect supernatural agents and protect supernatural coalitions. In other words, most human beings today are intuitively and "naturally" drawn into the lower left quadrant of *Figure 1*. These evolved traits were tweaked differently in various contexts, which led to the diversity of manifestations of religious life we see across cultures of gods. Conceptions of gods are never immaculate; the particular features of our supernatural agents give away our religious family of origin.

During the axial age (circa 800–200 B.C.E.), the challenges of pluralism and organizational hierarchy in complex literate states across west, south, and east Asia required more complicated and stronger forms of coalition. In other words, bigger cultures needed bigger gods. In the monotheistic religions that trace their roots to Abraham (Judaism, Christianity, and Islam), this took the form of belief in an infinite person-like Supernatural Agent who has power over all Coalitions whatsoever. These religions are *sacerdotal* insofar as they require their members to signal commitment to the group by costly participation in "priestly" rituals that are intended to mediate the power of the "sacred."

When theology follows this sacerdotal trajectory, it reinforces detection of a particular Supernatural Agent concerned about the well-being (and obedience) of a particular Coalition. In other words, it canalizes and facilitates the unchecked operation and integration of the theogonic mechanisms depicted in the lower left quadrant of *Figure 1*. Such an organizational strategy worked relatively well for centuries, at least from the perspective of those Coalitions whose crusades against and colonization of religious others enhanced their own prosperity.

Avatar Theology

What does any of this have to do with the imaginary planet of Pandora? Like most science fiction, the movie *Avatar* portrays a mixture of dystopian and utopian idealizations projected from the writer's own anxieties or hopes about human society. As far as we can tell, nothing like an axial age had occurred on Pandora; the writers depict the Na'vi (and other indigenous tribes) as an odd combination of shamanic small-scale clans and proto-barbarian despotic states. The RDA corporation is a stereotypical organizational cog within an industrial-military complex, driven by nothing more than a lust for more profit. Some of the scientists hired by RDA, however, want to study (and perhaps even learn from) the Na'vi.

As we saw have seen, most human beings on planet Earth have evolved dispositions that collectively encourage them to bear the gods of their own in-group. The two main groups combating in the movie *Avatar*, the Na'vi and the RDA, can be taken as representatives of social assemblages that have been primarily shaped by two of the machines we discussed in Chapter 6: the territorial and the capitalist social-machines (respectively). Although social-machines cannot be plotted onto the same plane as the cognitive and coalitional tendencies registered on our coordinate grid (*Figure 1*), we might imagine them as operative in a third dimension that creatively interacts with those individual level variables (or perhaps as Gaussian curvatures, or even shifting time-space curvatures within a Riemannian manifold, that intersect with them under various parametric conditions).

At any rate, focusing initially only on the coordinate grid, we can say that most of the Na'vi are anthropomorphically promiscuous: supernatural agency is detected at work in moss, trees, animals, and mountains. Many of the Na'vi are quite open (at least initially) to other modes of inscribing the socius, enthusiastically sending their children to the school run by the RDA scientists. The corporate leaders of RDA, on the other hand, are anthropomorphic prudes, refusing to acknowledge even the human-like agency of the Na'vi, whom they refer to as "blue monkeys." Members of the RDA coalition are also quite prudish in their sociography (although not in the "religious" sense), forcing their own norms upon others, with little patience for anything that challenges their capitalist inscriptions.

What does any of this have to do with *theology*? For the sake of this chapter, let us work with a broad definition of this field of inquiry: theology as the critique and construction of hypotheses about the existential conditions for finite axiological engagement. In this sense, both the Na'vi and RDA have their own "theologies." Each group has its own (more or less explicit) shared hypotheses about that which makes possible their experiences of valuing and being valued by others. Here we are not talking about this or that particular value, but that which generates the conditions for all valuation whatsoever. Now the *sacerdotal* trajectory in theology, by far the most common on Earth since the axial age, develops hypotheses that appeal to particular supernatural agents and their coalitions; "*our* God" is the basis of and judge over all values and actions. I will return below to the upper right quadrant, in which we can trace what I have called the *iconoclastic* trajectory of theology.

The typical ways in which members of the Na'vi tribe and the RDA corporation tend to make sense of the conditions for finite axiological engagement illustrate what I call the prodigal and the penurious trajectories of theology. These theological trajectories are buoyed by the integration of cognitive and coalitional tendecies that are integrated in the upper left and lower right quadrants of *Figure 1* (respectively). We can call the trajectory in the upper left corner *prodigal* because it is promiscuous in relation to *both* the cognitive and coalitional tendencies; i.e., it can lead to an extravagant expenditure of energy on imaginative engagement with supernatural agents (ubiquitous detection of intentionality) and on profligate pursuit of ever new experiences with other groups (inadequate consolidation of sociality).

The theological trajectory that gravitates toward the lower right quadrant of *Figure 1* is *penurious* in the sense that it is stingy in relation to *both* types of evolved mechanisms; i.e., it can lead to a tightfisted refusal to acknowledge members of out-groups (failure to "see" actual, natural intentional agents), and miserly resistance toward sharing with and learning from other cultures (stubborn maintenance and expansion of in-group norms). Of course, the RDA capitalists are not sociographically prudish in the specifically religious sense articulated in Chapter 1; they do not try to validate their norms by appealing to *supernatural* authorities. Nevertheless, we can see how anthropomorphic prudery and at least some of the component mechanisms that undergird sociogaphic prudery are at work in the cognitive interpretations and normative inscriptions of the leaders of the RDA.

The main point here is that these trajectories represented by the Na'vi and the RDA are indeed *theological*. What is it that makes possible (perhaps even originates, orders, and orients) value-laden engagements? The hypothesizing that guides the prodigal trajectory of the Na'vi is characterized by a relatively loose and open interaction with a pervasive field of supernatural agency that (early in the story) is not specifically concerned with protecting a particular coalition. The penurious hypothesizing of the RDA is guided by a strict allegiance to the invisible (yet quite "natural") hand that guides the flow of capital-money, and whose alleged neutrality justifies the behavior of those who learn to control it.

In the movie, the planet Pandora is portrayed as actually infused with the supernatural energy of a mother tree-goddess who (spoiler alert) eventually makes the animals of the planet fight against RDA. Jake Sully (the hero) becomes a kind of warrior-priest who is able to convince her of the evil of RDA. Because of her supernatural intervention, the RDA is thwarted and forced to leave the planet; the Na'vi coalition is saved.

But let's come back to Earth. Clearly there are some groups on our planet too who resemble the RDA and others who live somewhat like the Na'vi.

However, neither of these theological strategies will be adequate for saving *our* planet; in our late modern, pluralistic, globalizing context, we will not be able to *live together* under these conditions. As several cinematic observers pointed out, the movie Avatar is rather obviously intended as a negative commentary on u.s. interventions in the Middle East. The Colonel with a southern accent (George W. Bush?) and the RDA Administrator in charge of doling out contracts (Dick Cheney?) are blind to everything but the "unobtainium" (oil) hidden under the land of indigenous peoples (Iraqis, Afghanis, etc.).

My point here is that although we may well celebrate the movie's denigration of the trajectory represented by RDA, we humans do not actually live on Pandora and so the trajectory represented by the Na'vi is also doomed to fail. If in fact there were tree-goddesses to whom the colonized worldwide could appeal, things would be different; invading forces (military or economic) could be defeated by petitioning such supernatural agents who could harness the powers of nature. But there are not. And they cannot. Earth is not Pandora. I believe that the upper right quadrant is our best "theological" option. Before exploring this possibility, however, let me back up and demonstrate the way in which contemporary evangelicalism illustrates the *sacerdotal* trajectory, which has been the most popular mode of theological hypothesis construction for the last two millennia here on Earth.

Evangelical Supernatural Agent Coalitions

First, what is an "evangelical"? The question is not merely academic, as Stan Grenz, and those who tried to exclude him from this category, knew quite well. For my purposes, it suffices to use a broad definition of the term, referring generally to those who participate in religious coalitions shaped by various attempts in the mid-20th century (by the likes of Billy Graham) to find a middle way between fundamentalism and liberalism in Christianity. Such groups have achieved dominance in many areas in the United States and Britain, and continue to expand in many parts of the world, including my new homeland of Norway. Although this was my own religious "family of origin," I did not know how important this appellation was until I was informed of my "evangelical" identity at college.

Now many (but certainly not all) evangelicals would laugh at the idea of a tree-goddess who controls animals and cares about a small-scale coalition. However, most evangelicals do imaginatively detect a whole host of ambiguously discarnate or contingently embodied intentional forces who are interested in *their* coalition: angels, demons, disembodied ancestors (saints), etc. They also believe in a powerful and wise Supernatural Agent who will punish cheaters and defectors and protect those who remain faithful to an in-group, rewarding them with a place in an everlasting heavenly Coalition. The fact that members of out-groups (other religions, or even other sorts of evangelicals who disagree on some point of polity or biblical interpretation) believe that *their* coalitions will be protected by the gods (or God) that *they* have detected is explained away as the result of demonic delusion or even sin. In my view, this appeal to the noetic effects of sin is one of the most appalling of noetic sins. But I digress.

While most evangelicals find themselves comfortably in the sacerdotal trajectory (the God we detect protects our Coalition), it is interesting to observe how differently their right and left wings typically respond to the other two trajectories so far explored. In my experience, those on the evangelical "right" are usually more worried about New Agers (the prodigals) than they are about capitalist corporations that ravage the poor and the environment (the penurious). On the other hand, those on the evangelical "left" tend to react more harshly to RDA types and are less anxious about the touchy-feely spiritualism that characterizes some recent forms of the ecclesial socius. Which is more important – rejecting the (interpretation of) gods detected by others or expanding our own coalition by loosening social norms?

Wherever evangelicals fall in their answers to such questions, they remain within the sacerdotal trajectory. Here they are in good (or at least plentiful) company. Like the other Abrahamic religions, evangelical coalitions are held together by shared imaginative engagement with particular kinds of contingently embodied intentional forces. These may be explicitly divine figures detected at rituals (the presence of Jesus at the Eucharist, the Holy Spirit at a baptism) or lesser supernatural agents detected in everyday life (an angel when in need of protection, a demon when feeling temptation, a former saintly coalition member when in need of inspiration).

Evangelicals may be well trained in theological doctrine, and give orthodox answers to questions about divine infinity, immutability, aseity, and omniscience, but (like everyone else) they easily fall back into "theologically incorrect" models of God as a human-like intentional entity who is emotionally concerned with the struggles of their coalition in real space and time. This is because the evolved theogonic mechanisms naturally lead people to imagine finite gods who are watching over small-scale groups.[5]

Most professional theologians (at least in America) are paid by institutions that support a particular religious coalition or set of coalitions that follow the sacerdotal trajectory. A great number of these institutions require faculty to sign a "statement of faith," signaling their commitment to the in-group. As one

---

5 See, e.g., Barrett, "Dumb Gods, Petitionary Prayer and the Cognitive Science of Religion," and Slone, *Theological Incorrectness*.

example, let us take the institution with which Mark Noll has been associated for much of his career. Wheaton College demands that the scholars it employs – the "evangelical minds" it hires – assert and re-assert every year that they believe *inter alia* in supernatural agents like Satan, that out-group members will be punished eternally, that a text revealed by a supernatural agent is the final authority on all matters it discusses, and that physical death entered the world when Adam and Eve, the historical parents of the entire human race, disobeyed God.[6]

Similar claims could be culled from other statements of faith imposed by hundreds of similar institutions. But let us set aside for a moment the plausibility of particular assertions within such statements. The very fact that intellectual exploration is policed and restricted by forcing scholars to limit their claims (in any field) to assertions that are consistent with a particular coalition's appeal to supernatural agents (which only they can appropriately detect and properly interpret) is a symptom of the *real* scandal of the evangelical mind (and culture).

Stan Grenz was a leader in reforming, renewing, and revisioning evangelical theology.[7] His efforts were consistently attacked by those who were afraid that his intellectual rigor and engagement with contemporary culture and science were a threat to their own coalitions. I always admired Stan's courage and integrity in setting out his positions. He was the epitome of irenicism and never insisted that someone agree with him before (or after) engaging in serious theological conversation. I'm quite sure he would have resisted the radical proposals that I set out in the next section, but equally sure he would have encouraged me to tell it like I see it.

Iconoclastic Theology for Terrestrials

We do not live on Pandora. There are no tree-goddesses to save us. Those of us who agree that unbridled capitalism requires an infinite expansion of resources, and is rapidly depleting our finite ecological limits, have little faith in the RDAs of planet Earth. Given our evolved tendencies to detect supernatural agents, and our social entrainment within west Asian religious traditions, it is easy to believe that our only and best option is the sacerdotal trajectory. This

---

6   http://www.wheaton.edu/About-Wheaton/Statement-of-Faith-and-Educational-Purpose. Accessed 15 August 2017.

7   His extensive corpus was engaged by other chapters in the Festschrift of which the original version of this chapter was a part. Examples of his efforts at reformation include Grenz, *Renewing the Center: Evangelical Theology in a Post-Theological Era* (Grand Rapids, MI: Baker Academic, 2000); Grenz, *Revisioning Evangelical Theology* (Downers Grove, IL: IVP Academic, 1993).

adaptive strategy may have worked well (in terms of holding together complex social groups) during the axial age but the integration of these cognitive and coalitional mechanisms lead us to misinterpret ambiguous natural events (like tsunamis) or to ignore clear natural events (like global warming), appealing instead to supernatural causes or promises.

The same evolved dynamics that aid in the coalescence and maintenance of relatively large religious coalitions, like evangelicalism, also fuel antagonism toward perceived in-group defectors (like Stan Grenz) and a willingness to sanctify violence against out-groups. The sacerdotal trajectory helped some members of the species to hold together during a difficult period, but the exponential growth and rapidly increasing global connectedness of the human population require new ways of constructing and criticizing hypotheses about the conditions for axiological engagement.

How could evangelicals possibly participate in the *iconoclastic* trajectory of theology, which resists the evolved tendencies to over-detect agency and over-protect groups, and gravitates toward the upper right corner of *Figure 2* (Chapter 1, p. 64)? After all,this trajectory is diametrically opposed to the theogonic forces that have nurtured their traditions for centuries. In fact, many Christian theologians have indeed followed the iconoclastic trajectory – at least sometimes, at least partially. The real question is whether they can follow it consistently. I call this trajectory "iconoclastic" because the integration of theolytic tendencies has a jarring, and potentially destructive, effect on the religious images (icons) shared by members of a religious coalition, weakening their explanatory and cohesive power. When theologians follow this trajectory, they construct and criticize hypotheses about the conditions for axiological engagement without immediately appealing to a particular supernatural revelation or to the rituals and social norms of their in-group. In other words, they become more sociographically promiscuous and anthropomorphically prudish.

Before offering examples of this iconoclastic trajectory in theology, let me illustrate the integration of these theolytic mechanisms in non-theological scholarship. In sciences such as physics and chemistry, for example, scholars resist (or at least try to resist) the tendency to appeal to discarnate intentional forces or to the beliefs of politicized organizations with some investment in the research. If something strange happens in a test tube, the chemist's first guess is not "ghost." If a laboratory heavily funded by a pharmaceutical company announces that the drug produced by that company is more effective than previously thought, other scientists will remain skeptical until the research is repeated by another group. Even in sciences like sociology and political theory, which do indeed need to detect human agents and whose subject matter includes the dynamics of group cohesion, their *scientific* explanations of these phenomena do not appeal to supernatural agents or insights available only through revelation to a particular coalition. If they did, they would not be taken seriously as scholars. Anthropomorphic prudery and sociographic promiscuity are promoted by participation in academic communities.

What about *theologians*? Can someone who is anthropomorphically prudish and sociographically promiscuous still be a theologian? Yes, in a sense. In fact, many Christian theologians (and even some evangelical theologians) do resist the idea of God as a coalition-favoring, contingently-embodied "person." The problem is that movement in the iconoclastic (upper right) direction on the coordinate grid of *Figure 1* is almost always pulled back down (toward the lower left) by the powerful forces of the theogonic mechanisms. The sacerdotal trajectory relatively easily and "naturally" overrides the iconoclastic trajectory because the vast majority of theologians are operating within coalitions whose cohesion depends on shared imaginative engagement with supernatural agents. A first step for theologians interested in pursuing the iconoclastic trajectory – if they dare – is to trace and liberate lines of flight already present within the axial age religions, in which they are expert; for evangelicals, this usually means Christianity.

We can identify at least three different pathways or *modes of intensification*, already present (albeit suppressed) in the Christian tradition, which lead in this direction. The first is what I call the *intellectual* mode, in which the intensification of *conceptual analysis* leads to a recognition of the logical impossibility of the idea of *an* infinite Supernatural Agent. If "the infinite" cannot be thought as one object distinct from "the finite," else it would be limited by the finite and so itself finite, then *a forteriori* it cannot be thought of as one supernatural *person* distinct from other persons, who favors one *polity* distinct from other polities. This is the pathway (partially) taken by most of the leading theologians of the axial age religious traditions. Stan Grenz's robustly trinitarian theology is an excellent example; he clearly saw the logical problems with the idea of a single infinite subject.

A second mode is *pragmatic*; here, the intensification of *compassionate action* leads to liberating efforts on behalf of those oppressed or excluded by the dominating policies of the elite within a supernatural agent coalition. Many evangelicals, especially those interested in challenging the racism, sexism, and classism within their coalitions have proactively developed new ways of inscribing the socius and questioned the extent to which particular interpretations of supernatural agents (and their role in personal and social transformation) are necessary conditions for fellowship. It is easy to see why such "emergent" movements are so vigorously and violently vilified by the (white, male, upper middle class) evangelical power elite. In fact, reaching out with authentic openness to members of out-groups (and defectors, like ex-evangelicals) is indeed dangerous for such religious in-groups, whose cohesion depends on clear statements about the boundaries of faith.

Third, there is a *mystical* mode in which the intensification of *contemplative awareness* leads to experiences that alleviate anxiety about being-limited by an infinite person-like supernatural agent and the need to protect the power of a particular group. One can find examples of this mode within all of the religious traditions that trace their roots to the axial age, and often evangelicals who begin to follow this trajectory explore meditative practices that evolved within other traditions (e.g., Buddhism, Daoism, Sufism, etc.). In Christianity, this mode is often linked to the apophatic way, in which the power of human language to comprehend the divine is rigorously denied. Evangelicals are permitted to express their ignorance about the essence of a transcendent supernatural agent, as long as they *also* express their confidence in kataphatic statements about the anthropomorphic attributes of that agent. In other words, the risky adventure of the iconoclastic trajectory must be domesticated by the sacerdotal.

For obvious reasons, most theologians have preferred the intellectual pathway. Not uncommonly, however, concerns about the plausibility of the idea of God (as an infinite Supernatural Agent who favors a human coalition) are driven by moral and aesthetic sensitivity as well as by conceptual reflection. It is important to ask why it is usually systematic theologians, rather than biblical scholars or historians (like Noll), who get drawn into evangelical heresy trials. Scholars of Scripture or Christian history are not required, and indeed sometimes actively discourage one another, from trying to provide a coherent account of the discrepancies within the texts and disparities across the eras they study. Systematic theologians, on the other hand, are pressured to follow out the logical implications of the assertions of their religious coalitions as far as possible – before appealing to mystery. If a theologian follows the intellectual mode too far, she gets in trouble. This is the real scandal of evangelical culture and its oppression of evangelical minds.

Noll decries the lack of integration between intellectual rigor and *piety* in evangelicalism. This is only a symptom, not the root, of the problem. Insofar as pious devotion imaginatively engages ritually-mediated, discarnate intentional forces concerned about "my" in-group, it *is* the hyper-active detection of coalitional gods – postulated as causal explanations for ambiguous natural phenomena. Insofar as pious activism is driven by an attempt to participate in and expand the kingdom of "our" God, it *is* the hyper-active protection of a supernatural coalition – interpreted as the best way to inscribe the global socius. In this sense, piety directly compromises inter-subjective discourse about natural phenomena and inter-communal discourse about social phenomena. It cannot be integrated with the intellectual rigor that is characteristic of the other sciences, which attempt to follow the trajectory in the upper right corner of *Figure 2* (Chapter 1, p. 64).

Sacerdotal appeals to "mystery" use the in-conceivability of infinity as a veil of ignorance – a learned ignorance that veils the hyperactivity of the religious family's shared imaginative engagement with their God. The inability of finite creatures to conceive the infinite (or even all finite things) suggests an (infinite) vacuum in human knowledge. Abhorred, the theogonic mechanisms quickly and easily fill it by detecting and protecting manifestations of a particular coalition's supernatural agent.

In contrast to the domesticating effect of the sacerdotal forces, the *iconoclastic* trajectory of theology de-personifies, de-politicizes and, in a certain sense, de-objectifies the existential conditions for axiological engagement. It is true that the "object" of theology is not like the objects of other disciplines; the relation between infinity and intentionality cannot be objectified like finite relations. That which conditions the existence of all finite valuations cannot itself be finite or even evaluated in the same sense. Rather than using this as an excuse for appealing to the mystery of a particular coalition's interpretation of a supernatural agent, however, the iconoclastic (a)theologian can explore other ways of making sense of this being-limited of thought (or being-thought of limitation) which can indeed be "objectified" (as the reader is currently doing).

The integration of anthropomorphic prudery and sociographic promiscuity is not merely *destructive* of religious images; as we can see from the other sciences, these theolytic forces also have a *creative* power. They can facilitate the construction of new hypotheses about the conditions for axiological engagement that avoid personifying or politicizing "infinity." I do not have the space to set out such a proposal here, so I devote the remainder of the chapter to a discussion of the possibility and promise of this iconoclastic trajectory.[8]

Theology after Pandora (and Eve)

Back to Pandora – but not the planet this time. The ancient Greeks were not the only ones to develop a myth in which the actions of the first woman are blamed for the evils in the world. Like Pandora, Eve's curiosity was supposed to have killed the race – or at least its chance for immortality. New and other gods would be invented; (mostly) male heroes whose supernatural powers would provide remedies for humanity.

---

8 For a detailed discussion of the iconoclastic trajectory in theology, see Shults, *Iconoclastic Theology*.

The stories of Pandora and Eve teach us that it is dangerous (especially for women) to question divine things, to look into the forbidden black boxes of divine intentionality. Inquiring too persistently into the mechanisms by which discarnate intentional forces punish and reward us threatens the shared imaginative engagement that holds the coalition together, and so it is taboo. As long as evangelicals, or members of any other religious in-group, protect the cohesion of their communities and institutions by encouraging detections of hidden supernatural powers only "we" know how to interpret by, for example, insisting that theologians sign coalitional statements of faith, they will not ever be able to engage in serious constructive scholarship in dialogue with other sciences.

I was an evangelical theologian long enough to know how my friends and former colleagues might respond to such claims. St. Paul claimed that God uses the "folly of what we preach to save those who believe," and that Christ crucified is "a stumbling block to Jews and folly to Gentiles" (1 Cor. 1:21–22). Blessed are those who are not scandalized, but signal their commitment to the coalition by faithfully adhering to apparent foolishness. Similar texts and similar strategies are present in other religions.

But why accept the Bible as the revelation of a Supernatural Agent in the first place? Why not accept the Qur'an, the Book of Mormon, or the Dhammapada instead? Human beings have evolved cognitive and coalitional mechanisms that short-circuit such questions. *We* know that our (interpretation of) shared imaginative engagement with *our* supernatural agents is true. Every other religious group (or denomination) says the same thing. At best, their leaders come to abstract "inter-religious" agreements that have little effect on everyday religious practice. At worst, they start "holy" (or "just") wars against one another. Is it any wonder that intellectuals in other fields hesitate to take (sacerdotal) theology seriously?

Several other objections will certainly arise from my religious family of origin. Wouldn't my proposal for taking more seriously the discoveries of the bio-cultural study of religion mean the dissolution of evangelicalism, indeed the destruction of Christianity itself? Doesn't the fact that Shults has clearly gone off the deep end prove that, in fact, engaging modern science and culture really is dangerous – too dangerous? Perhaps the conservative Christian political "right" is correct; sectarianism may be the only hope for protecting the purity of particular Christian coalitions.

I have several responses to these kinds of questions and concerns. First, we should begin by admitting that the dissolution of *other* supernatural coalitions is exactly the goal of most Christian evangelism and missions. If evangelicals want believers in other supernatural agents (whom they take to be "false" gods) to consider with all seriousness that they may be wrong, they should be willing to take their own medicine.

Second, do evangelicals want to believe what is true or do they want what they already believe to be true? Is being a Christian, or an evangelical, more important than being right – or even making sense? We have evolved to think that fitting into our coalition is indeed the most important thing in the world. We have learned to stifle our questions about the contents of the divine "black box" hidden in plain sight in the religious imagination and rituals of our in-groups. We are cognitively and coalitionally wired to ignore the psychological repression and political oppression caused by our own religious tribes.

Scholars, activists, and contemplatives are trained *not* to ignore them. My challenge to evangelicals who are also *iconoclasts* (in any of the three modes of intensification) is to take seriously the importance of the following questions: is contemporary shared imaginative engagement with supernatural agents the result of evolved hyper-active perceptive and cooperative strategies that helped our ancestors survive in small-scale societies? Are these strategies now obsolete in a complex, pluralistic social environment? If so, what can we do about it?

Third, iconoclastic theology does not necessarily lead to the destruction of social groups; the complete dissolution of evangelical coalitions is not the only option here. Like many other such religious in-groups, evangelicals have played an important role in developing strategies for caring for human persons and coalitions, including out-groups. The hard work ahead for the iconoclastic theologian (or activist, or contemplative) is to imagine and enact new and creative ways to live in community that do not rely on the mechanisms of the sacerdotal trajectory. This may very well, indeed we should expect that it would, include forms of axiological engagement that are inspired by exemplars like Jesus of Nazareth (among others).[9]

The best hope for theologians to join with other scholars of religion in serious inter-disciplinary conversation, and to participate with other groups in serious inter-cultural conversation, is to liberate the iconoclastic trajectory from the sacerdotal. Like their colleagues in other disciplines, theologians must learn to resist the evolutionary biases that make it so easy for people to keep engaging in religious sects. This does not at all mean giving up on the real *intensity* of the human experience of being-limited, the intense *reality* of being-conditioned in all of our axiological engagements. Reflecting on these really intense experiences of encountering infinite intensities remains an important task in human life.

---

9 Shults, "Ethics, Exemplarity and Atonement" in *Theology and the Science of Moral Action*, ed. van Slyke et al. (Routledge, 2012).

Even if we could reconstruct this discipline into a critique of axial age religious conceptions and new hypothesis-construction, would it be appropriate to call it "theology?" In fact this term has been used historically, from Aristotle to Zizek,[10] to refer to arguments about the existential conditions for human axiological engagement that do *not* appeal to human-like, coalition-favoring gods. In the long run, whether or not we keep the term "theology" is less important than undertaking the task of reconstructing this mode of inquiry so that it can fully enter into the important, ongoing academic dialogue on these issues. This can only happen if we honestly discuss how God is born(e) among us, however embarrassing the "facts of (religious) life" may be to evangelical (and other sacerdotal) theologians.

---

10   Aristotle, *Metaphysics* 1025a.19, 1064b.3; Slavoj Žižek, *The Parallax View* (Cambridge, MA: MIT Press, 2009).

CHAPTER 8

Wising Up: The Evolution of Natural Theology

The title of this chapter is a play on words, an adaptation and expansion of images and metaphors developed by Celia Deane-Drummond in her Boyle lecture and elsewhere, which I use as an entry point for reflecting on her theological proposals and their place within the broader context of the contemporary encounter between science and the Christian religion.[1] In more than one sense, her work illustrates the "wising up" of theology which, also in more than one sense, has been and must continue to "evolve" within its own complex niche of overlapping ecclesial, social, and academic environments.

My response to Deane-Drummond has two parts. First, I call attention to the adaptive value and significance of her proposal, which I call "the sophianic theodrama hypothesis," for the ongoing development of Christian theological responses to the empirical findings and theoretical formulations within sciences such as evolutionary biology and psychology. In the second part, I outline some challenges to this *way* of proposing, challenges which, in my view, must be taken yet more seriously even – and perhaps especially – by those in the vanguard of theological engagement with the natural sciences. What further adaptation, if any, will be necessary for "natural" theology to survive within the competitive intellectual environment of the contemporary academy? Can it find (or construct) its own niche, or will it be compelled to migrate or adapt in some other way?

Or, must we finally admit that natural theology is an increasingly engangered species whose natural habitat is the church, and that it can only survive in the academy when ecclesial, political, and/or social conventions construct environmental protection areas within departments or professional associations in which the cognitive and coalitional biases of anthropomorphic promiscuity and sociographic prudery are sheltered and nurtured?

The Sophianic Theo-drama Hypothesis as a Religious "Adaptation"

My use of the term "adaptation" is not intended negatively in any way. The transmission of any tradition from generation to generation requires a balance

---

1 This chapter is an adapted version of "Wising Up: The Evolution of Natural Theology," which was originally published in *Zygon: Journal of Religion & Science* 47, no. 3 (2012). That article was based on my response to Deane-Drummond's 2012 Boyle Lecture at St. Mary-le-Bow Church in London.

© F. LERON SHULTS, 2018 | DOI 10.1163/9789004360952_009

between maintaining the integrity and coherence of the system and developing new functionally adequate responses to environmental changes. This also applies to the tradition of Christian theology, and the sub-tradition of "natural theology" within it, which has indeed evolved since the first Boyle lectures, and now must continue to adapt. Metaphorically speaking, we can think of Christian theological hypotheses as complex functional strategies for nourishing and nurturing a particular set of religious communities within a late modern scientific and philosophical environment that sometimes feels very hostile indeed.

Some might find it tempting to repeat fossilized formulations without engaging any scientific challenges, others to concede to any and all scientific challenges without concern for communal integrity. One path leads to the petrification, the other to the dissolution of the Christian tradition. As clearly articulated in her Boyle lecture, and further elaborated elsewhere, especially in *Christ and Evolution: Wonder and Wisdom*,[2] Deane-Drummond takes the middle way between the twin temptations of ignoring and idolizing science. There is much wisdom in her approach – materially, as well as methodologically.

Deane-Drummond's material hypothesis is quite complex, but the central claim on which I will focus here can be summarized quite succinctly: Christians may interpret Jesus Christ as the *dramatic* expression of the *Wisdom* of God in a way that is compatible with contemporary *evolutionary* theory. The warrants and argumentation for this apparently simple claim are quite sophisticated. Her work is characterized by rigorous attempts to fulfill all four of what we might call the desiderata of constructive Christian theology: a faithful interpretation of the biblical witness, a critical appropriation of the theological tradition, a conceptual resolution of relevant philosophical issues, and a plausible elucidation of contemporary human experience.

Although it did not play a large role in her Boyle lecture, Deane-Drummond has argued elsewhere, in dialogue with current critical biblical scholarship, that early Christians interpreted Jesus in light of the Wisdom tradition of Hebrew literature. For example, in the Wisdom of Solomon, wisdom is portrayed as a feminine figure who fills all things and holds them together (1:6–7), who, more mobile than any motion, is creatively pervading and upholding all things (7:24–27). This language is applied to the risen Christ in the famous hymn of Colossians 1:15–20: "... for in him all things in heaven and on earth were created... all things have been created through him and for him. He himself is before all things, and in him all things hold together."

---

2   Deane-Drummond, *Christ and Evolution: Wonder and Wisdom* (Augsburg Fortress, 2009).

Deane-Drummond also appropriates a vast array of resources from different streams within the Christian theological tradition, relying most heavily on Eastern Orthodox, Roman Catholic, and Anglican theologians. She is committed to maintaining the intuitions behind the Chalcedonian creed while still squarely facing the shift in what "fully human" means today in light of evolutionary theory.

She also exhibits a commitment to the last two desiderata: doing theology in a way that is intellectually and existentially responsible. Deane-Drummond offers careful arguments for her position, engaging relevant philosophical debates on issues such as causality, and links them to real concerns facing humanity as a whole such as the environmental crisis. Her integration of the sophianic and the dramatic, especially as developed by Hans urs von Balthasar, provides a way of attending more carefully to the element of tragedy within the human longing for wisdom that characterizes *Homo sapiens*. Elsewhere I too once argued, although not nearly so extensively, that utilizing the dynamic and relational language of the sophianic tradition appears to be the wisest strategy to adopt if one's goal is reconstructing the classical doctrines of Christology in dialogue with contemporary science.[3]

There are certainly objections internal to the Christian tradition that could be and ought to be raised. Some might worry that her proposal is a form of adoptionism. Others might regret her lack of attention to resources within other Protestant traditions. Some would be concerned that her emphasis on contingency easily lends itself to a rejection of divine omnipotence, or inadequately protects the distinction between God and the world. What about the problem of evil? Her sophianic theo-drama hypothesis deals respectfully with the tragedy of creaturely suffering but does not ultimately explain why an omnibenevolent being allows it.

Of course such concerns are not unique to Deane-Drummond's proposal; these are the kinds of problems with which all Christian theologians must wrestle. In my judgment, the general adaptive strategy she has developed is one of the only options available for contemporary theologians who want to remain within the Christian tradition. Rather than focus on these internal questions, however, in the second part of my response, I want to look at the adaptive task with a wider lens. What is happening to the niche within which theology, especially "natural" theology, is attempting to adapt? Exactly why – and how – is it attempting to adapt within this niche?

---

3 Shults, *Christology and Science* (Wm. B. Eerdmans Publishing Co., 2008). That book was my last attempt to reformulate a Christian doctrine, before leaving behind what I now refer to as the sacerdotal trajectory of theology (see Chapters 6 and 7 above, and *Theology After the Birth of God*, Chapter 3).

Is the "Natural" Niche of Christian Theology Shrinking?

As I indicated in the first part of this chapter, theological hypotheses are a kind of *religious* adaptation. In other words, they are (whatever else they may be) strategies developed within religious coalitions to survive and thrive. The term "religious" is contentious in almost every environment, but I will continue to use it here in a way that is increasingly common among scientists in fields such as cognitive science, moral psychology, and cultural anthropology: shared imaginative engagement with axiologically relevant supernatural agents. As we have seen in earlier chapters, the disciplines that contribute to the bio-cultural study of religion offer compelling evidence that widespread ritual interaction with discarnate (or at least ontologically confused) intentional forces, which has been found in all known societies, is a result of evolved cognitive and coalitional biases.

In the sense we use the term today, "theology" emerged relatively late in human history, within complex literate states where *unity* of belief, ritual, and social identity was problematized by pluralistic encounters. During the axial age, the idea of *an* ultimate Supernatural Agency emerged in different ways across east, south, and west Asia. The *mono*-theistic construal of this Agency as *a* personal God is typical of the Abrahamic (west Asian) religions that trace their roots to this period. In Christian theology, affirming the transcendent intentionality of the *one* God is generally considered to be the basis for inclusion within (or exclusion from) *one* great Supernatural Coalition. In this sense, we could say that theology was an adaptive strategy that helped religious organizations *unify*, police, and transmit their preferred modes of imaginative supernatural engagement on a larger societal scale. Axial age religions provided the original and "natural" social niche within which theology evolved.

In another sense, however, theology – like science – is not "natural." Thinking scientifically – and theologically – is hard work, and requires extensive training; these intellectual engagement strategies must be cultivated. Thinking (as well as acting and feeling) religiously, however, *is* natural for many, if not most, people. Shared imaginative engagement with supernatural agents (or "gods") is common among human beings today because the cognitive and coalitional biases that promote such beliefs and behaviors are part of our shared phylogenetic inheritance.

As we have seen in earlier chapters, research in the disciplines that contribute to the bio-cultural study of religion suggests that gods are *born* in the human mind as a result of a wide array of cognitive mechanisms that engender belief in hidden agents when confronted with ambiguous or frightening phenomena. This first sort of mechanism helped (some of) our ancestors find (or avoid) important agents like predators, prey, protectors, or partners. However, the hypersensitivity of the cognitive tendency to detect intentionality also led to many false positives; faces are detected in clouds, ghosts in the shifting of shadows or smoke, divine blessing or punishment in unpredictable weather patterns, etc. But of course not all of these detected supernatural agents stick around.

Although gods may easily appear in the mental space of human life, it takes a village to nurture and care for them. In other words, supernatural agents must be *borne* in a special way within the social space of human life. The gods that stick around are typically those that are interpreted as having some social interest in and power over what happens within and to the in-group. Once detected, shared engagement with such gods – who may be watching in order to punish or reward – can lead to a decrease in cheating and defection to out-groups. This second sort of mechanism helps to explain *inter alia* the emergence of in-group altruistic behavior in a way that is consistent with natural selection. The cohesion of a group is protected when its members do not hurt one another, and are even willing to signal costly commitment to the coalition by hurting themselves (e.g., participating in painful rituals or other forms of self-sacrifice) or hurting members of out-groups (e.g., promoting exclusive or violent practices).

Empirical findings within the disciplines of the bio-cultural study of religion suggest that these detection/protection mechanisms come naturally to most people. So where does theology come in? Part of the "tragedy of the theologian" (to use Pascal Boyer's phrase) is that the vast majority of regular religious believers do not really *need* abstract doctrinal arguments about the incarnation of ultimate Supernatural Agents, for example, to hold together their everyday mental and social lives. Even if they can articulate the orthodox doctrine of an infinite and eternal God authorized by the church universal at Chalcedon, psychological studies (and a moment's reflection on our own experience as – or of – religious believers) show that under stress religious people's actual interpretation of events quickly collapse back into the detection of finite and temporally engaged supernatural agents (such as angels, saints or even the risen Jesus) who are interested in the protection of their own kith and kin. Those of us who have labored long in both academic and ecclesial environments know how difficult it is to get many believers to understand, or even to see the importance of, complicated doctrines like the incarnation.

The Evolution of Natural(ist) Theology

What does any of this have to do with the evolution of natural theology? *Theology* in general may have emerged in the axial age, but the environmental niche in which *natural* theology evolved was the competition of ideas within early modern science and philosophy, in which only the empirically sustainable and explanatorily powerful survived. Natural theology has traditionally been distinguished from *revealed* or confessional theology, which appeals explicitly to the detection of divine intentions (e.g., in a holy text), codifying and to some extent managing the coalition's shared engagement with its Supernatural Agent. This latter kind of theology serves an adaptive purpose, holding together the coalition in a more or less hostile social environment.

Now Deane-Drummond's project seems to blur the lines between revealed and natural theology; in my view, this distinction itself is a remnant of other ancient and modernist dualisms. However, we can still ask the question: in what niche and for what purpose does her sophianic theo-drama hypothesis operate? Her description of the task she has selected makes clear that her proposal is meant to function as a way of protecting the cohesion of (some parts of) the Christian tradition as it adapts to a changing conceptual environment. This frank acknowledgement that her natural theological efforts are intended to serve the church is refreshing.

But we might wonder about the viability of that other task, namely, the development of theological hypotheses that *could* function in the broader context of the academy or the public sphere as "defenses of Christianity in the wake of pressures from natural science." It seems to me that the latter would require argumentation that does *not* appeal directly or indirectly to controversial interpretations of the revelation of – or shared engagement with – the supernatural agents of one's own religious coalition. Otherwise, natural theology would still involve special pleading (as it always has). But are there *any* natural theological arguments that avoid such appeals? And if they could be developed, in what sense would they still be *theological*? Robert Boyle seems to have intuited these inherent tensions within natural theology when he indicated his desire that the original lectures series should not deal with controversies between Christians, i.e., with issues that might highlight – or even widen – fractures within the coalition.

Over the centuries, however, the conceptual environment within which such argumentation could be productive or even possible has been shrinking. Theologians who are concerned about the psychological and political health of Christian (or other) coalitions need to "wise up" to the fact that this academic niche is rapidly disappearing. Debates across the sciences and within the public sphere increasingly reject appeals to supernatural agency or coalitional authority in arguments about the causal nexus of the physical world or the normative organization of the social world. Many conservative Christian theologians and philosophers of religion have responded by forming their own enclaves within religious institutions.

Deane-Drummond takes the more courageous approach. She describes her task as demonstrating the possibility of a compatibility between a reconstructed articulation of the doctrine of the incarnation and a scientifically responsible acknowledgement of the explanatory power of cutting edge evolutionary theory. As she explicitly notes, however, science itself has no *need* for such demonstrations of compatibility. Neither the academy nor the public sphere need natural theology. What, then, is the environmental niche within which such proposals can serve a (re)productive function? Are they necessarily limited to the Church – or a church? Can they only survive within private or public institutions that provide ecological sanctuaries for guilds of confessional theologians and religious professionals?

Deane-Drummond's work has consistently called our attention to the ecological crises of our world, and urged theologians to contribute to their resolution.[4] As we have seen in earlier chapters, however, pouring fuel onto the firey imaginations of individuals and groups whose mental and social lives are saturated with anthropomorphically promiscuous and sociographically prudish intuitions can reinforce superstitious interpretations of the causes of (and solutions for) crises like global climate change. If theology is to contribute to real solutions to these sorts of problems, it will have to leave its sacerdotally sanctioned academic shelters and explore radically new options for a robustly *natural-ist* theology that follows what I referred to above as its *iconoclastic* trajectory: uncovering and resisting any and all sacred Images (icons) that pretend to mediate the supernatural agency or authority of a religious coalition.

---

4   See, e.g., Deane-Drummond, "Public Theology as Contested Ground: Arguments for Climate Justice," in *Religion and Ecology in the Public Sphere* (New York: T&T Clark, 2011), 189.

CHAPTER 9

What's the Use?

Robert Neville sticks to the iconoclastic trajectory of theology far more consistently than Celia Deane-Drummand; indeed, among contemporary scholars who self-identify as theologians, he is one of the most consistent followers of this trajectory. I will argue, however, that he is not consistent enough. For readers familiar with Neville's expansive literary corpus, one of the first questions that arises when confronted with Volume One of his trilogy on *Philosophical Theology* is: how many *magna opera* can one scholar produce?[1] For those of us interested in the details and the development of his metaphysical and epistemological hypotheses over the decades, a whole series of more serious intellectual questions also quickly emerge.

My primary interest here, however, is exploring an explicitly *pragmatic* question: what's the *use* of Neville's astonishingly consistent and carefully argued theoretical proposal in contemporary contexts shaped by radically pluralistic and globalizing forces? The answer, of course, is: "it depends." As Neville himself makes clear, whether or not a religious symbol carries over the value of ultimate reality in certain respects depends on a whole host of factors, including the purpose, maturity, and community of the interpreter.

The function of religious symbols within an interpretation also depends, however, on the extent to which individuals rely on or learn to contest evolved cognitive and coalitional biases that reinforce the tendency to detect human-like, coalition-favoring disembodied intentional forces. In the first section of this chapter I examine these dynamics in light of the same heuristic conceptual framework we have been exploring in earlier chapters. As one of the directors of the *Institute for the Bio-Cultural Study of Religion,* Neville is quite familiar with this literature and in the second section I point out some of the ways in which his theoretical project encourages the contestation of biases toward anthropomorphic symbols authorized by a particular in-group. My main concern, which comes to the forefront in the final section, is the extent to which Neville's "pastoral" practice of allowing (and even insisting upon) the continued *use* of such symbols for ultimacy can surreptitiously strengthen the superstitious and segregative tendencies he wants to enervate.

---

1 This chapter is an adapted version of "What's the Use? Pragmatic Reflections on Neville's Ultimates," which was originally published in the *American Journal of Theology & Philosophy* 36, no. 1 (2015).

Toward the end of *Philosophical Theology* Volume One, Neville points to
the intrinsic relation between what he calls systematic philosophical theology
and systematic practical theology. The latter has the task of determining the
truth of symbols of ultimacy in particular contexts. "How can we tell whether
popular religion carries over ultimate truth into its practitioners from context
to context?... That is, do the symbolic engagements in those contexts have in
their practical lives the truth of ultimate reality?" In order to answer such questions, practical theology has to understand not only the context but also the
functional network semiotic connections, the relevant iconic and indexical
referential dimensions and the extent to which "the individuals and groups
are ready or unready for accepting the symbols in a true way. But in order to do
any of this, systematic practical theology needs to hold on to the most sophisticated truth possible about what ultimate reality really is."[2]

One of the aspects of Neville's system that I most appreciate is his emphasis on the sense in which all religious symbols "break on the infinite."[3]
Insofar as they are intended to refer to finite/infinite contrasts within a sacred
canopy, such symbols inevitably break – determinate symbols cannot directly
refer to the indeterminate ontological act of creation. For the purposes of
this chapter, I am going to assume familiarity with these key aspects of Nevillian ontology and semiotics, referring readers to the relevant resources in the
footnotes.

I am also going to assume the reader's familiarity with theogonic reproduction theory, which we have been exploring throughout the earlier chapters.
However, it is important to emphasize once again that the mechanisms that
promote anthropomorphic promiscuity and sociographic prudery are *reciprocally reinforcing*. As we have seen, supernatural agents who are cared for and
ritually engaged within a coalition are easy imaginative targets for the hair-
triggered agency detection mechanisms of each new generation. Extensive
cross-cultural empirical research has demonstrated that activating people's
anxiety about the welfare of their kith and kin can increase their tendency to
interpret ambiguous phenomena as caused by potentially punitive disembodied agents. Conversely, priming individuals with thoughts about possible invisible watchers can reinforce a tendency to protect their in-group and become
antagonistic toward out-group members.

2   Neville, *Ultimates: Philosophical Theology, Volume One* (State University of New York Press,
    2014), p. 295. Unless otherwise noted, page numbers in the following paragraphs are from this
    book.
3   Neville, *The Truth of Broken Symbols* (State University of New York Press, 1996); Neville, *On
    the Scope and Truth of Theology: Theology as Symbolic Engagement* (T & T Clark, 2006).

This mutual intensification of superstitious interpretation and segregative inscription happens automatically and unconsciously, all too easily obscuring the powerful covert operations of theogonic reproduction. Understanding this is the key to understanding my concern with Neville's *use* of religious imagery in his "practical theology."

Neville's Anthropomorphic Prudery and Sociographic Promiscuity

Like most serious academics, Robert Neville resists explanations of the natural world that appeal to supernatural agents as causal forces. Like most public figures in pluralistic contexts, he also resists appeals to the supernatural authorities of particular in-groups. In fact, this dispositional tendency to contest the evolved cognitive and coalitional biases briefly outlined above is evident throughout Volume One of *Philosophical Theology*. Already in the Introduction Neville points toward the importance of the biological and social sciences for his project, and hints at themes that will pervade the book, including resistance to privileging "personal" ideas of God and "confessional" approaches to theology. Later in the book he sometimes explicitly incorporates insights from the bio-cultural sciences to support his arguments, as when he points to the role of the tendency to over-attribute agency to non-intentional things in fostering human-like interpretations of ultimate reality (254).

Neville is quite straightforward in his attitude toward anthropomorphic symbols for ultimacy: "We know from the concept of the ultimate as the ontological creative act that God does not have intentions. Metaphysics can tell us when a false inference is being drawn from an anthropomorphic symbol of divinity" (296). He insists that it follows from the concept of the ontological act of creation that it "cannot be internally intentional ... the personal connotations many people have with the term *God* should be carefully expunged from philosophical theology." However, this sentence is immediately followed by a qualification: "Of course, there might be situations in which highly personified symbols of ultimacy are well used for engaging ultimate reality" (280-281). It is this hasty "of course" that worries me, for pragmatic reasons to which I will return below.

Neville is also straightforward about the societal problems that arise from what I am calling sociographic prudery. For example, he insists that the in-group/out-group distinction is absurd if regarded as ultimately significant. It is also pernicious because "it leads people in the in-group to not pay attention to those in the out-groups, to not observe their diverse narratives and conditions" (158). Such distinctions can intensify anxious and violent reactions to cultural others. "Nevertheless," argues Neville, "the human need for intimate connections with the ultimate realities that might be depicted in sacred canopies means that we cannot do without ultimate narratives of some sort and some kinds of anthropomorphic symbols of ultimate realities" (158-159). It is the ease with which Neville asserts this "nevertheless" that worries me; is it really the case that we *cannot* do without "ultimate narratives" that include anthropomorphic symbols? It seems to me that in a growing number of contexts this is precisely what we *must* learn to do without.

For the most part, Neville's writings support the integration of anthropomorphic prudery and sociographic promiscuity depicted in *Figure 2* (see Chapter 1, p. 64). These forces are theolytic (god-dissolving) because they weaken the mechanisms of theogonic reproduction. Superstitious interpretations and segregative inscriptions are becoming more and more problematic in pluralistic, globalizing contexts. Increasingly, modern people are coming to believe that it is possible to make sense of the cosmos and act sensibly in society without appealing to supernatural agents as causal powers or moral regulators.

As we will discuss in more detail in Chapter 12, this shift is related to the spread of *naturalism* and *secularism*, both of which can be conceived as ways of adapting to the challenges of a new socio-ecological environment that is radically different from that of our upper Paleolithic ancestors. Both methodologically and metaphysically (in the sense introduced at the end of Chapter 1), Neville is a naturalist and a secularist.

He is also a theologian. Why is that relevant? Long before the rise of naturalism and secularism the intellectual elite of the large-scale religious traditions that emerged in the wake of the west Asian axial age had begun to think critically about anthropomorphic symbols for the divine. So did the "theologians" of those traditions that emerged out of the south and east Asian axial age traditions, as Neville notes at several places.[4] The idea of "God" as an infinite disembodied intentional Force was tentatively born(e) in the minds of monotheistic theologians who pressed the evolved biases toward anthropomorphic promiscuity and sociographic prudery to infinity.

This turned out to be too far. They realized that a truly infinite, absolutely transcendent reality could not be represented in the human mind. And so theologians worked hard to break idolatrous symbols, that is images (or icons, in the Platonic, not the Peircean sense) that pretended to represent the infinite. Ultimacy cannot be conceived (or perceived) as a Person (or anything else). *A forteriori* it really makes no sense to think of ultimate reality as preferring one

---

4 See, e.g., Neville ed., *Ultimate Realities* (State University of New York Press, 2001).

Polity over another. This "iconoclastic" trajectory in theology presses toward the integration of anthropomorphic prudery and sociographic promiscuity.

On the other hand, as active members of monotheistic coalitions, most theologians have also worked hard to defend hypotheses about the existential conditions for axiological engagement that *do* involve the interpretation of and ritual interaction with a supernatural Agent who cares for their Group. This latter "sacerdotal" trajectory has been the most dominant in theology by far. Reinforced by (hidden) cognitive and coalitional biases, it has not had much trouble domesticating the iconoclastic urges of even the most rigorous intellectuals, prophetic activists, and devoted contemplatives in those traditions. Moreover, regular believers have always found it relatively easy to ignore theological debates about the unknowable transcendence of God.

As cross-cultural psychological experiments have shown, people may memorize and repeat orthodox doctrinal formulations about God's transcendence (citing attributes like omniscience, omnipotence, and impassibility) but under time constraints or stressful conditions they automatically, and immediately fall back upon "theologically incorrect" interpretations guided by their evolved biases – imaginatively detecting finite gods who are interested in the practical lives, and have just enough power to punish or reward their kith and kin. In other words, even if a religious parishioner (or seminarian) was able to pass an exam on Nevillian metaphysics, as soon as she leaves the room and re-enters her everyday frame of reference she will continue reproducing anthropomorphic god-conceptions that hold her sacerdotal in-group together – unless and until she learns to contest the theogonic mechanisms.

Pragmatic Reflections on Religious Reproduction

Allowing the covert operations of these evolved biases to continue unchecked all too easily allows people to remain commited to their favored in-group superstitions and antagonistic out-group segregations. This is why I urge "practical theologians" to become more explicitly iconoclastic, and more intentional about pursuing a delicate conversation that is all too tempting to avoid. Postponing "the talk" about where babies come from and what it takes to care for them for too long can have devastating effects. Of course, it can be equally devastating if the conversation makes people feel attacked, afraid, or ashamed. The activities that lead to sexual and religious reproduction can feel terrific to our bodies, but baring our souls about them can feel terribly vulnerable. It is important to be sensitive when discussing these intimate issues but, as I have argued in earlier chapters, it is also important to be direct.

As I noted in Chapter 1, having "the talk" about *religious* reproduction should involve more than simply explaining how "it" works. It is equally important to work out the physical, emotional, and social consequences of "doing it." This is just as true for religious education as it is for sex education. We need a theological version of "the birds and the bees" that deals with the dynamics by which gods are reproduced in human minds, and the consequences of nurturing them in human groups. Part of the problem is that we are socialized not to ask where gods come from; we learn early that it is not polite to ask folks why they keep them around.

Until relatively recently, our understanding of the mechanisms that engender shared imaginative engagement with human-like disembodied agents associated with particular in-groups has been quite limited. The illuminative power of the disciplines that contribute to the bio-cultural study of religion challenges the plausibility of belief in ghosts, gods, and other culturally postulated disembodied intentional forces.

As a systematic *philosophical* theologian Neville seems to agree. In his efforts as a systematic *practical* theologian, however, he seems all too willing to allow and even encourage the use of personified religious symbols, even in rituals that have traditionally served to mark off the boundaries of an in-group (such as the Eucharist). In his discussion of worldviews in Chapter 4 of Philosophical Theology, Volume One, Neville suggests that "in most North Atlantic Christian congregations, few people would believe that they could manipulate God, shaman-wise, to get what they want in prayer" (89). Based on my own experience in literally hundreds of evangelical churches over the decades, I would argue the vast majority of religious people in such contexts believe *exactly* that. Neville is (understandably) dismissive of interpretations of 9/11 as God's punishment on America for the gays or the feminists (295), but this is precisely the sort of interpretation favored by some of the most tightly bound and fastest growing religious in-groups.

Continuing to foster symbolic engagements that utilize anthropomorphically promiscuous and sociographically prudish images all too easily reinforces the naturally evolved tendency to over-detect agents and over-protect groups. If the philosophical theologian does not explicitly challenge the validity ("truth," in Neville's sense) of such symbols in modern contexts, religious people on the "popular" side of the continuum, which is the vast majority of the population, will go on having unprotected imaginative intercourse within their own religious family of origin and reproducing "theologically incorrect" coalition-favoring supernatural progeny. The virtuoso speculative theologian may be an exception. She might theoretically be able to take such symbols in some respect that does not lead to their (mis)use or (ab)use in her social engagement with others.

But if she is *also* a practical theologian concerned about the real consequences of engaging such symbols within sacerdotal in-groups, why would she? Regardless of what she may say as a *philosophical* theologian, if she does not explicitly address the deleterious pragmatic effects of continuing to bear supernatural agents in supernatural coalitions in a violent-prone, ecologically fragile world, even those of us who can remember her sophisticated formulations will just smile and nod as we go on detecting the gods of our own group, thereby reinforcing the hidden mechanisms that activate cognitive and co-alitional biases that contribute to superstition and segregation. Why not just leave the gods out of (philosophical and practical) theology completely? Why not explicitly encourage people in late modern contexts to avoid *any use* of anthropomorphically promiscuous and sociographically prudish religious symbols?

In my view, not only is this sort of iconoclastic approach more likely to produce feasible pragmatic strategies for inscribing the socius in pluralistic contexts, it is also more consistent with Neville's own theoretical arguments. The basic thrust of his constructive work over the decades has challenged the personification of God and the authority of monotheistic Groups in thinking about *Ultimates*. Especially within socio-ecological niches in which survival no longer depends on the detection of gods that protect in-groups, symbolical engagements that incorporate iconic semiotic representations of the latter are (in the Nevillian sense of the terms) not only "broken" – they are "false" insofar as they promote inaccurate superstitious interpretations and aggressive segregative inscriptions that are in no sense "ultimate."

Scholars trained in the monotheistic (and other) traditions that emerged in the wake of the axial age have a unique role to play in the *practical* theological task to which Neville alludes in Volume One and addresses in more detail in Volume Two (and Three).[5] As the mechanisms that support the sacerdotal dominance of theology are increasingly unveiled, it will be easier to liberate the productive iconoclastic forces that have long been domesticated within these traditions. Our "pastoral" ministrations will either foster theogonic

---

5   Neville, *Existence: Philosophical Theology, Volume Two* (Albany: State University of New York Press, 2015); Neville, *Religion: Philosophical Theology, Volume Three* (Albany: State University of New York Press, 2016).

reproduction or promote the sort of theolytic retroduction that engenders naturalism and secularism. In the contexts in which most of us find ourselves, I argue that it makes good sense to become ever more explicit as we invite people to have "the talk" about the causes and consequences of shared symbolic engagement with the gods.

In his response to my original article in the *American Journal of Theology and Philosophy*, Neville focused mostly on the places where we agreed, and confirmed my assessment that he was a "seasoned practitioner" of a "vigorous iconoclasm" in his theology classes. However, he also wondered whether my proposals were insufficiently sensitive in practical theological contexts. For example, he noted that it would not be practical to refer "an unlettered midwestern Southern Baptist ... to the comfort of Abhinavagupta's ontological Shiva" when he is burying his children. The task at that point is to help "ground a grieving father in what is ultimately important in the face of death."[6] He also described how his own liberal Protestant father explained to him at a young age why the services at the Black Pentecostal church in St. Louis were so "emotional and nonsensical" and included "speaking in tongues"; this was "the only way they had," he said, "to make sense of their lives," and it helped them feel "freer and more purified."[7] Finally, Neville wonders whether I too quickly dismiss people's need to "engage what is ultimately important," and whether I think there is any place at all for the "ecstatic fulfillments" that religion can facilitate.

I will be the first to agree that the hospital room or funeral parlor is not the best time or place to challenge people's religious biases or press them toward a naturalist and secularist worldview. I also happily concede that there are individuals and groups in certain contexts for whom rigorous iconoclasm is not the (immediate) solution. I have made it clear in other writings that I value pastoral sensitivity as well as intensely ecstatic experiences, and that learning how to facilitate these without the need for supernatural beliefs and rituals is one of the most significant challenges (and opportunities) for the non-religious. However, I wonder whether discounting the capacity of emotionally needy, unlettered Baptists for ideological transformation, and condoning the nonsensical behaviors of Black Pentecostals who don't have any better way to make sense of their lives, is more patronizing and ultimately less pastorally sensitive

6   Neville, "Comments on F. LeRon Shults's 'What's the Use? Pragmatic Reflections on Neville's Ultimates'," *American Journal of Theology and Philosophy* 36, no. 1 (2015).

7   Ibid., 83–84.

than my suggestion that (outside of overwhelming psychologically distressful situations and alongside the effort to remedy politically oppressive situations) we have "the talk" with people as though they are adults who are capable of thinking critically and altering the ways in which they pursue the intensive ecstasies that life has to offer.

CHAPTER 10

Dis-integrating Psychology and Theology

I think that Neville would agree that doing the hard "pastoral" work of having "the talk" with people about where gods come from (and the costs of bearing them) can be intellectually, emotionally, and even existentially exhausting, even – or especially – when it occurs in an academic context. This chapter explores the *dis*-integrative dynamics within the ongoing process of relating the fields of psychology and theology, and argues that such dissolutive forces can play an important and valuable role in this interdisciplinary dialogue.[1] Healthy development requires that we sometimes let things fall apart. The main sections of the chapter point to the potentially generative power of disintegrating psychology, theology, selves, and gods. The conclusion addresses the existential fear and desire that often characterize human attempts to hold it all together, i.e., the tasks of "integration" in all its forms.

Integrating the disciplines of psychology and theology is a fascinating academic task, and one in which I was heavily involved for quite a long time. In this essay, however, I want to draw attention to the importance – and the value – of *dis*-integration. Although the generative forces of dis-integration are inexorably at work in all interdisciplinary engagement, they are too often inadequately emphasized and sometimes ignored or even suppressed. In my view, these ways of dealing with disintegrative dynamics are not good for the disciplines or for their disciples. Attending positively to the creative potential of dis-integrative negation can open up new possibilities for healthier ways of engaging within and across the fields of psychology and theology.

As I hope to make clear, my intention is not to dismiss the ideal of *integrity* – for scientists or for the sciences. However, I am resisting the idea that the *and* in "psychology and theology" can be reduced to a kind of linear function in which two *integers* are added together through a simple arithmetic conjunction. Instead we might imagine the relation of the disciplines within a nonlinear topological space that invites a more complicated infinitesimal calculus, in which finding *integrals* requires attention to differentials, to changes in the value of functions. The goal of "integration" should not be a final enumeration or sum(mary) of static ideas, but an ongoing generation of open systems of

---

1 This chapter is an adapted version of "Dis-Integrating Psychology and Theology," which was originally published as part of a special issue on interdisciplinary dialogue in the *Journal of Psychology & Theology* 40, no. 1 (2012).

© F. LERON SHULTS, 2018 | DOI 10.1163/9789004360952_011

dynamic inquiry in which the value of breaking things apart can be included in the equations.

Coming up with a provocative title is one thing – provoking in ways that move the discussion forward is quite another. In this context my limited goals are to point out some of the disruptive and dissolutive forces within the ongoing task of integration and to advocate an open, differentiated way of relating to them. This is consonant with the "relational" model of integration that psychologist Steven Sandage and I developed and defended in a variety of places.[2] We consistently emphasized the importance of facing the potentially transformative function of negation, of welcoming the "dark night" of the interdisciplinary soul as part of the ongoing quest for spiritual – and intellectual – enlightenment. In what follows, I press this point even further. In the conclusion, I will return to the necessary (and valuable) task of trying to hold it all together. First, however, let us acknowledge the value (and necessity) of letting go.

Letting It All Fall Apart

Good therapists know that the process of emotional healing usually (if not always) involves coming to a point where one is willing to let things fall apart. It is quite natural for human beings to try to hold things together; indeed, without integration of some kind we could not survive, much less thrive. When the integrator's hold on a particular integrative strategy becomes rigid and anxious, however, tightening one's grip on the "integrand" only makes the problem worse. Sometimes there must be a strategic dissolution before a new solution can be found. Healthy development in adapting to our natural and social environments occasionally calls for letting go of an integrand; the most radical transformations usually include intense moments of (at least partial) dis-integration.

Good theologians know this too. As James Loder argued, existential transformation involves a negation of negation, a facing of the Void in which one

---

2 Shults and Sandage, *The Faces of Forgiveness: Searching for Wholeness and Salvation* (Grand Rapids, MI: Baker Academic, 2003); Shults and Sandage, *Transforming Spirituality: Integrating Theology and Psychology* (Grand Rapids, MI: Baker Academic, 2006); Sandage and Shults, "Relational Spirituality and Transformation: A Relational Integration Model." *Journal of Psychology & Christianity* 26, no. 3 (2007). Although I was attempting to follow (what I now call) the iconoclastic trajectory of theology in the two books mentioned above, my contributions there were still held back by (what I can now see as) the theistic biases that motivate the sacerdotal trajectory of theology.

recognizes the incapacity of the self to hold together its lived world.[3] The great spiritual writers of the religious traditions testified to this in a variety of ways, pointing to the importance of acknowledging the inability of finite persons to secure themselves vis-à-vis the infinite ground (or abyss) of (non)existence. The vast majority of the world's population manages this anxiety through imaginative engagement with gods (or God), which is one of the main reasons why the dialogue between psychologists and theologians is important.

The human experience of the (dis)integrative processes of life is often characterized by fear and desire. We fear the pain of isolation and the threat of absorption; we desire to hold and be held in pleasurable communion with others. These powerful forces sometimes lead us to take extreme measures to maintain the "integrity" of the community and our place within it, even at the cost of putting up walls that inhibit (or even prohibit) authentic and lively communication, within and across these socially constructed boundaries. Letting things fall apart can be scary, but it can also be beautiful. Dis-integration happens whether we like it or not. The question is whether we can learn to open up our selves and our disciplines to the creative potential of these dis-solutive events.

Dis-integrating Psychology

When we approach the task of integrating "psychology" with theology (or any other discipline), it is important to acknowledge that we are not dealing with a fixed object. Psychology is not a substance that can be combined with others into new chemical compositions. Even if we stuck with the chemistry metaphor, dissolution of some kind is usually a precondition for a new solution. Simply perusing the abstracts at an APA conference or titles in the PsycINFO database is enough to bring the point home: psychology is not an integrated whole. Nor should it be. What makes the field interesting is the open and lively debate among (for example) developmental, moral, cognitive, evolutionary, social, and clinical psychologists. It is precisely this openness to the in-breaking of critical and dis-integrative voices (even from other disciplines) that keeps "psychology" alive.

Watching the carefully woven fabric of one's favorite theory being ripped apart can be an unpleasant experience, but sometimes this disassembly sparks insights that lead to new and more functional theoretical fashions that better

---

3 Loder, *The Transforming Moment*, 2 edition (Colorado Springs: Helmers & Howard Publishers, 1989).

fit the data. Dis-integration may be aggravating but it creates space and time for aggregating empirical data and concepts in novel ways. One of the reasons for the continuing explanatory power of the psychological disciplines is the way in which competing scientific paradigms (Kuhn) or research programs (Lakatos) continually dis-mantle one another, forcing the use of new conceptual threads for weaving together the patchwork of empirical findings.

These reflections also apply to "psychology and theology," which many consider to be its own (interdisciplinary) field or discipline. Here too we should be wary of the temptation to mark off intellectual boundaries too quickly. As I argued in chapter 2, the metaphor of a "field" should be construed not in geographical but in physical terms: a dynamic force field of interconnected and open explanatory events. Referring to this dialogue as a "discipline" should be understood not as an attempt to determine its departmental location but as a reminder to discipline our selves to remain interconnected and open during every event of explanation. Is this too idealistic? We might be optimistic about finding psychologists who, as scientists, would be willing to commit themselves to such explanatory openness. But can we really expect theologians to go along? This depends, of course, on what we mean by "theology."

Dis-integrating Theology

In many of the professional contexts in which readers of the *Journal of Psychology & Theology* operate, theology is explicitly tied to the beliefs, behaviors, and experiences of a particular religious coalition. The task of theologians working within and for the sake of a confessional community is often understood as protecting, articulating, and transmitting a coherent (integrated) set of doctrines. In this model part of the function of "theology" is holding together a coalition of believers, constituting and regulating its boundaries. This follows what I referred to above as the *sacerdotal* trajectory of theology, as opposed to the *iconoclastic* theological trajectory that breaks apart idolized conceptual schemes protected and policed by the religious elite. Theology in the service of the mono-theistic traditions has too often been characterized by a monopolistic obsession with unified propositional systems.

Some of the most significant developments in late modern theology, even within the sacerdotal trajectory, have been reactions against such rigid modes of "integration." In Christianity, for example, we can point to the more fluid and praxis-oriented tendencies of narrative, feminist, and liberation theologies. This is also evident in the renewal of interest in pneumatology among systematic theologians. In the west, Christian theology has most often privileged *logos* over *pneuma*, emphasizing the importance of the (logical) ordering of the cosmos and downplaying or even resisting the equally creative (pneumatic) forces that shake up our worlds, tossing them out of order. Both of these dynamics are part of life, and way in which we balance (or fail to balance) them "theologically" has powerful political and psychological ramifications.

The success of the integrative efforts promoted by the *Journal of Psychology & Theology* depends on good, healthy theological dis-integration. Resistance among (sacerdotal) theologians to interdisciplinary engagement can be a manifestation of an unhealthy over-attachment to a particular doctrinal integration, which is taken to be the indissoluble basis of the integrity of the coalition. This not only blocks open dialogue with and learning from other scientific disciplines (or religious coalitions), it also has the effect of petrifying (in both senses of the term) the theologian and the coalition.

The kind of "integration" that the journal aims to sponsor becomes truly possible when (iconoclastic) theologians are willing to engage in the vulnerable process of opening up their favorite formulations to radical critique as well as openly criticizing the fallacies and foibles they perceive in psychological theories and therapies. My point is not that theologians should give up on the ideal of systematicity but that the idealization of a finally closed systematic "integration" actually hinders the process of discovery and blocks the transformation of the disciplines and the disciplinarians within them.

Dis-integrating Selves

One of the reasons it is so difficult to let (even part of) our disciplines fall apart is that our own identities can so easily become fused with the disciplines (or coalitions) with which we identify. It is easy to understand why the petrification of psychological or theological formulations can seem attractive as a way of protecting against dis-integration. However, this strategy only represses and so intensifies the petrifying anxiety that can drive the self toward such strategies in the first place. "Integration" is accomplished (or attempted) by embodied *selves* in embedded social relations. Interdisciplinary engagement can only be truly transformative when it includes attention to the formal dynamics by which selves in relation "hold on" to the material issues under discussion.

Late modern thinkers have increasingly rejected early modern notions of selfhood that rely on categories such as essence and identity, preferring instead to emphasize the becoming and hybridity of "selves." One does not have to appeal to radical postmodernists to illustrate the point. In Robert Kegan's

theory of *The Evolving Self*,[4] for example, the developmental process is described as involving "evolutionary truces" in which the self learns new, more
complex ways of adapting to the tension between the longings for inclusion
and distinctness. Growth sometimes requires the "loss" of an old "self," a letting go and renegotiation of the way in which consciousness is ordered.[5] The
key point here is that healthy integration is an ongoing process that requires
an ongoing openness of selves to healthy dis-integration, which is sometimes
required in our adaptations to our natural and social worlds.

The literature of the major religious traditions is also attentive to the importance of dis-integrating selves. For example, the stories of Abraham,
Moses, David, Confucius, Zhuangzi, Arjuna, Buddha, Epictetus, Lucretius, Jesus, Paul, Mohammed, and Rumi include significant moments of "letting go"
and their teachings or reflections often encourage (albeit in different ways) a
humble openness to and acceptance of the role that dissolutive forces play in
transformation.

Of course, the preferred exemplars and doctrines of the world's religions
also sometimes manifest and even instill in their followers anxious ways of attempting to hold on to the self at all costs. Nevertheless, there are resources in
virtually all traditions for understanding and facilitating non-anxious ways of
differentiating selves within religious "families of origin." If healthy integration
requires an openness to some kind of deconstruction of *psyche*, where does
that leave *theos*?

Dis-integrating Gods

For many readers, this penultimate section will probably be the most dis-
concerting, and understandably so. However, take a moment to reflect on your
own journey in the integrative process. For most people I know, radical transformations in their self-understanding have included radical transformations
in their understanding of ultimate reality. Several studies have shown the powerful connection between people's working models of themselves (and others), which is shaped by their way of relating to a human "primary attachment
figure," and their working model of God.[6] Moreover, reports of longitudinal
therapeutic interventions suggest that people's God images are usually altered

---

4   Kegan, *The Evolving Self* (Harvard University Press, 1982).
5   See also Kegan, *In over Our Heads: The Mental Demands of Modern Life* (Harvard University
    Press, 1995).
6   See, e.g., Kirkpatrick, *Attachment, Evolution, and the Psychology of Religion*.

as they experience psychological and spiritual growth, for example from a judgmental, controlling Parent to less anthropomorphic conceptions of the divine.[7] This *psychological* process can be just as painful, and just as healing, as learning to let go of an unhealthy way of relating to an idealized image of another human being.

However, it is important to acknowledge that gods also play a powerful *political* role in the social worlds of human beings. Those of us raised in the Christian tradition are accustomed to speaking of "God," but monotheism is a recent development in human evolution, emerging alongside the development of complex literate states that required new forms of unified policing. The religious lives of most people (including monotheists) are typically characterized by shared imaginative engagement with a variety of supernatural agents (e.g., saints, angels, demons, jinn, ancestors, etc.). As long as the integration of "psychology and theology" is limited to attempts to hold together particular western notions of *psyche* and *theos*, the exercise will (at best) only be of value for provincial in-groups or (at worst) only reinforce alienating attitudes toward out-groups.

For this reason, it will become increasingly important to engage the empirical findings and theoretical insights of the bio-cultural study of religion, insofar as they shed light on the psychological and political significance of the way in which humans understand their shared engagement with "gods." In my view, this is one of the most significant tasks ahead for those interested in promoting integration within the disciplined conceptual fields of "psychology and theology." Overcoming the monopolizing sacerdotal forces that hinder real integration (*inter alia* by forbidding or punishing dis-integration) will require openness to the value of iconoclastic forces that can break open new possibilities for healthy reconstructive engagement across traditions and disciplines. The way in which we imagine our gods may help hold our in-groups together, but rigidly holding on to these images also holds us apart from out-groups and, indeed, crushes open discourse within our coalitions as well.

Holding It All Together

I anticipate that some readers will judge my comments here as insufficiently sensitive to the human need to hold things together. For the reasons outlined in the introduction and emphasized throughout this brief chapter, in this context I have stressed the importance of learning how to let things fall apart.

---

7   Wallin, *Attachment in Psychotherapy* (Guilford Press, 2007).

Nevertheless, life is not simply dis-integration, and dissolution is not always healthy. Both sides of this dialectic are necessary. For some of us, it is literally our "job" to work at integration. Doing this job well, however, will require ongoing attention to the generative and salutary potential of the forces of disruption, dissolution, and dis-integration.

The special issue of the *Journal of Psychology & Theology* in which an earlier version of this chapter first appeared, invited reflection on the past, present, and future of integration. I want to emphasize the pluperfect subjunctive and anterior future of *integrating*. The integrative process we are discussing is just that – a *process*. It is always and already occurring at the intersection, which is also the disjunction, of the passing of past "integratings" that may have not been but now once were and the arrival of new possibilities for "integratings" that may or may not be, but soon will or will not have been. "Integration" is what we do in the temporal space of the present that is constituted by this disjunctive intersection. Always dis-integrating and integrating, for better or for worse – as long as we all shall live. Our presence *is* our participation in the nexus of events within which we live and move and have our (dis)integrating. This applies to the way we relate psychology and theology as much as it does to anything else.

We cannot hold it all together. We cannot let it all fall apart. This tension is part of human life. Too often our efforts at integration are driven by an inordinate desire to hold onto old psychological and political patterns that have held us together – or by an inordinate fear of their dis-integration. But there is nothing to be afraid of, really. Can we learn to accept – and perhaps even enjoy – the dissolution of our being that makes possible our discovery of novel solutions? A final solution in which a static "integration" was totally secured against any future alteration is not a live option; indeed, it is death. The ongoing vivacity of attempts to integrate psychology and theology depends on the ability to develop healthy selves, healthy polities, and healthy disciplines. A hypochondriac obsession with maintaining the integrity of these systems by over-protecting them from contact with foreign elements does not promote health in any robust sense; vigorous integration only comes with and through energetic and risky engagement with the dis-integrative.

CHAPTER 11

Religion and the Prizing of Peace

Finding the balance between holding on too tightly and holding on too loosely is a challenge in politics as well as psychology. As we have seen, most people have held their "selves" together in human coalitions by appealing to the gods of their in-group. This can provide motivation for self-transformation as well as for altruistic behavior even toward out-group members. However, under stressful conditions, it can also intensify intergroup conflict and amplify superstitious beliefs that impede the discovery and implementation of feasible strategies for solving real-world problems. What about the very real problem of seeking "world peace?"

The awarding of the 2014 Peace Prize maintained the Norwegian Nobel Committee's relatively new tradition of highlighting non-traditional avenues for peace making.[1] During the last century the prize most often went to individuals who had contributed, in the words of Alfred Nobel's will, "to the abolition or reduction of standing armies and the formation and spreading of peace congresses." Recently, however, the award has sometimes gone to persons or groups who have promoted peace in other ways, such as raising awareness of climate change or fighting for human rights. In 2014 the Committee seemed to be sending the message: children are people too – and achieving and sustaining global peace will require renewed attention to the plight of young people, especially girls, in developing countries.

That year two well-known champions of children's rights, Kailash Satyarthi and Malala Yousafzai shared the Nobel Peace Prize "for their struggle against the suppression of children and young people and for the right of all children to education." Children, the Nobel Committee emphasized, "must go to school and not be financially exploited." In recent years, concern about these issues has increased in the international community, and a growing number of individuals and institutions have devoted themselves to remedying the situation. The good news is that today it is estimated that there are 78 million fewer child laborers in the world than there were in 2000. The bad news is that the same calculations indicate that at least 128 million children continue to suffer under exploitative labor conditions. The 2013/14 United Nations *Education for*

---

1 This chapter is an adapted version of "Reflections on the 2014 Nobel Peace Prize," which was based on a lecture to the Agder Academy of Sciences and Letters, and originally published in *Agder Vitenskapsakademi: Årbok 2014*, ed. Jahr et al., (Kristiansand: Portal, 2015).

© F. LERON SHULTS, 2018 | DOI 10.1163/9789004360952_012

All: Global Monitoring Report estimated that 250 million children and young people are not learning basic skills as a result of inadequate access to educational resources.

In its official announcement of the 2014 award winners, the Nobel Committee went out of its way to emphasize that it "regards it as an important point for a Hindu and a Muslim, an Indian and a Pakistani, to join in a common struggle for education and against extremism." The tensions between India and Pakistan are complex and long-standing, and the Committee concluded that the efforts of Satyarthi and Yousafzai may very well aid in "the realization of the fraternity of nations," another of the criterion for the Peace Prize set out by Alfred Nobel in his will.

This chapter is divided into three parts. The first two sections briefly describe the backgrounds and provide summaries of the accomplishments of the 2014 Peace Prize winners. One of my main areas of expertise is the scientific study of religion, and so in the final section I offer some reflections on the ambiguous role of religion in the prizing of peace in an increasingly pluralistic, interconnected, and ecologically fragile global context.

Kailash Satyarthi

Born on 11 January 1954 in the Vidisha district of India, Kailash Satyarthi was educated as an electrical engineer. At age 26 he gave up his plans for an engineering career and began fighting for the liberation of Indian children who had been sold or forced into slave labor. In the words of the Nobel Committee: "Showing great personal courage, Kailash Satyarthi, maintaining Gandhi's tradition, has headed various forms of protests and demonstrations, all peaceful, focusing on the grave exploitation of children for financial gain." Some of Satyarthi's strategies have been quite daring and dangerous, including mounting raids on factories – often protected by armed guards – where children were being held captive and forced to work. In fact, it has been estimated that his grassroots efforts have led to the rescue of over 83,500 child slaves.

Satyarthi realized early on, however, that freeing children from such situations was only half the battle. It was also necessary to provide them with resources and opportunities for making a new life. As the Committee pointed out, he "has also contributed to the development of important international conventions on children's rights." Satyarthi has campaigned worldwide against child labor, arguing that it is intrinsically linked to other social problems such as the perpetuation of illiteracy, poverty, and population growth.

He has worked with or served on the boards of numerous organizations committed to eradicating child labor and improving educational opportunities, including the Global March Against Child Labor, the South Asian Coalition on Child Servitude, and the International Center on Child Labor and Education.

In 1980, Satyarthi founded Bachpan Bachao Andolan (Save the Childhood Movement), which works with local villages and regional authorities to rehabilitate and educate children who are liberated from servitude. In 1994, he founded GoodWeave International (formerly known as Rugmark), a non-profit organization that monitors and audits companies that make hand-knotted carpets. Those companies that pass inspection are provided a GoodWeave logo that certifies that their product has been made without the use of child labor. Despite several attempts on his life over the decades, Satyarthi continues to work against child trafficking in India and worldwide, and for the rights of all children to attend school and to pursue a promising future.

Malala Yousafzai

Satyarthi's name may be new to some of you, but you would have to have been living in a cave the last three years not to have heard of Malala Yousafzai, the girl who – as the title of her autobiography puts it – "stood up for education and was shot by the Taliban." Born on 12 July 1997 in northwest Pakistan, Malala has often been in news headlines worldwide since October 2012, when a Taliban gunman attempted to assassinate her on a bus as she was returning home from an exam with her classmates. The bullet was not fatal, but she required multiple surgeries and was eventually treated in Great Britain, where she now attends school in Birmingham. Yousafzai was the first Pakistani and, at age 17, the youngest person ever to receive the Nobel Peace Prize.

In January 2009 she became a BBC blogger, anonymously providing a diary of her experiences in Swat Valley, and her reactions to the Taliban's destruction of schools and attempts to ban the education of girls. By late 2009 her identity had been revealed, and she became a public advocate for the rights of girls – and all children – to go to school. In the words of the Norwegian Nobel Committee: "Despite her youth, Malala Yousafzai has already fought for several years for the right of girls to education, and has shown by example that children and young people, too, can contribute to improving their own situations. This she has done under the most dangerous circumstances. Through her heroic struggle she has become a leading spokesperson for girls' rights to education."

Prior to winning the Nobel Peace Prize, Yousafzai had also received several other awards – including Pakistan's National Youth Peace Prize (2011), the Sakharov Prize (2013), the Simon de Beauvoir Prize (2013) and the Ambassador of Conscience Award (2013). She has spoken around the world on themes related to children's rights to education. One of her most well known speeches was given at the United Nations on her 16th birthday, which was declared "Malala Day." During that speech she declared: "I do not even hate the Talib who shot me. Even if there is a gun in my hand and he stands in front of me. I would not shoot him. This is the compassion that I have learnt from Muhammad-the prophet of mercy, Jesus Christ, and Lord Buddha."

Children, Religion, and Violence

In its official announcement the Nobel Committee noted that "in the poor countries of the world, 60% of the present population is under 25 years of age," and insisted that "it is a prerequisite for peaceful global development that the rights of children and young people be respected. In conflict-ridden areas in particular, the violation of children leads to the continuation of violence from generation to generation." It is hard to argue with the logic of the Committee: a peaceful future depends on properly educating the next generation of human beings. It is also hard to imagine that anyone would challenge the passion that seems to have motivated the Committee's selection process this year: who could possibly be against helping children?

Who indeed? In fact, there are all too many industrialists and fundamentalists who seem to value their own acquisition of revenue or their own interpretation of religion above all else, leading to policies and practices that enslave or encumber children and women – and this is not limited to the developing world. My assignment for the lecture upon which this chapter was based was to introduce the 2014 winners and to offer some brief evaluative and contextualizing comments, not to analyze the deleterious effects of capitalist excess and religious fanaticism. However, I would like to complete my task by drawing attention to the ambiguity of the role of religion in the promotion of peace.

As I noted in my introductory comments, the Nobel Committee stressed the significance of the religious background of the Peace Prize winners – a Hindu and a Muslim – and praised their common struggle against extremism and for education. In fact, both Kailash Satyarthi and Malala Yousafzai often refer to their religious faith as grounding their compassionate and courageous efforts to facilitate peace in general and the welfare of children in particular. There is no doubt that there are powerful resources for motivating altruism in the

Hindu and Islamic traditions. Like the other major religions that emerged in the wake of the axial age, their holy texts often encourage or even command justice and mercy.

However, it is also important to acknowledge that the global fight *for* the welfare of children, especially girls, will require fighting *against* the tide of traditional interpretations of such texts, which have in fact been used to sanction slavery and the oppression of women. The preference for boys over girls in the Brihadaranyaka Upanishad and Manusmriti, for example, is based on the assumption that only those reincarnated as males can perform the appropriate meritorious acts for liberating ancestors from sin. The fourth Surah of the Qur'an allows men to have sex with their slave girls and instructs husbands to beat their wives if they are unrepentantly disobedient or suspected of disloyalty. Similar examples of the privileging of men and the condoning of slavery and violence could be drawn from the Hebrew Bible and the New Testament.

Religion has in fact traditionally played a role in reinforcing sexism, as well as racism and classism – all biases that helped human groups hold together during the shift from small-scale hunter-gatherers to sedentary communities and the eventual emergence of regional empires. Today, however, most of us live in large-scale, pluralistic, globally interconnected societies. As we begin to feel the effects of our combined actions on an increasingly fragile ecological environment, we can no longer afford to appeal to supernatural agents in our attempts to make sense of nature and act sensibly in society. We must face these problematic aspects of the world's religious traditions head on. We certainly have a great deal to learn from Hinduism, Islam, and other axial religions as we try to cultivate peace both locally and across the globe.

But we have a harder row to hoe: *unlearning* the evolved tendency to imagine secretive and punitive gods who are invested in the behavior and survival of particular in-groups. If we hope to decrease the growing tensions between out-groups in a world with shrinking natural resources and an expanding population, we must develop the capacity to contest the evolved biases that engender religious superstition and segregation. Satyarthi and Yousafzai have significantly contributed to this process by questioning the attitudes and resisting the actions of those who try to manipulate the caste system in India and enforce universal Sharia law in Pakistan. In our rapidly globalizing context, the unveiling and unraveling of *all* forms of the religious oppression of women, children, and the poor have indeed become central tasks within the overall goal of promoting the prizing of peace.

Am I suggesting that these Nobel Prize winners should halt their efforts to make the world a more peaceful place? Obviously not. My point is that peace-making attempts that do not unveil the role that theist biases play in engendering and amplifying segregative behaviors and superstitious beliefs are leaving the job half-done. No, worse. They are inadvertently *feeding* the evolved biases that will continue to fuel the spiraling of intergroup conflict. There is no doubt that peace-making attempts that explicitly rely on and promote religious motivations can sometimes contribute to a temporary reduction of violence. However, statistical analyses of religion and conflict datasets indicate that competition among religious elites is often a *precursor* to conflicts, predicting their intensity and duration, and that the formation of inter-religious dialogue networks are usually a *reaction* to earlier clashes between religions.[2]

In other words, such efforts may in fact be more a symptom of, rather than a cure for, religious conflict. Neither Satyarthi nor Yousafzai are religious elites and, like millions of other religious individuals, they exhibit laudable virtues that have been cultivated in the context of their participation in supernatural rituals and nurtured by their supernatural beliefs. But if this is the only way to cultivate such virtues, our species is in trouble. If we are serious about making peace in our current, globally interconnected environment, we will have to find the courage to speak clearly and straightforwardly about the consequences of religious sects.

---

2   See, e.g., Isaacs, "Faith in Contention: Explaining the Salience of Religion in Ethnic Conflict," *Comparative Political Studies* 32, no. 1 (2016); and Vüllers et al., "Measuring the Ambivalence of Religion: Introducing the Religion and Conflict in Developing Countries (RCDC) Dataset," *International Interactions* 41, no. 5 (2015).

CHAPTER 12

Practicing Safe Sects

> We are all atheists about most of the gods that humanity has ever believed
> in. Some of us just go one god further.
> RICHARD DAWKINS

∴

For most of human history, theism – like racism, classism, and sexism – played
an important role in the emergence of ever more complex and expansive
forms of societal organization. The evolution and transmission of the perceptive and affiliative biases that foster religious reproduction facilitated the
psychological internalization and political institutionalization of personal
and social categories even – or especially – when those categories had the
effect of repressing or oppressing the needs and drives of some individuals. Shared imaginative engagement with axiological relevant supernatural
agents effectively held human beings together in increasingly differentiated
cultural coalitions, strengthening their capacity for in-group cooperation and
coordination.

In this sense, not practicing safe sects – *bearing gods* in mind and culture –
"worked." As we have seen throughout this book, however, there is another
sense in which (re)producing supernatual conceptions is *unsafe*. The mutual
intensification of superstitious beliefs and segregative practices that strengthens relationships within a religious coalition also covertly cements prejudice
against and antagonism toward members of other religious (or non-religious)
coalitions. This is the political tragedy of overtly *religious* attempts to promote
peace: struggling against the effects of sociographic prudery (e.g., aggressive
behaviors toward those who practice different supernatural rituals) while simultaneously embracing and encouraging anthropomorphic promiscuity
(e.g., idiosyncratic beliefs about the role of in-group gods in shaping society)
usually only makes things worse because the latter surreptitiously reinforces
the former (and vice versa).

All of this is complicated by the psychological tragedy that the mental dissolution of the gods imaginatively engaged by the religious coalition with
which one identifies can initially have a dis-integrating effect on one's sense of

© F. LERON SHULTS, 2018 | DOI 10.1163/9789004360952_013

self.[1] This is why having "the talk" about religious reproduction requires sensitivity and patience. Shaming people for wanting to bear gods only makes them angry or more anxious, which further activates the defense mechanisms of theistic bias. I have not tried to hide my god-dissolving intentions in the contraceptive essays of this book. The goal of the central chapters was to demonstrate the anaphrodisiacal effect of scientific and philosophical perspectives on religious reproduction. It turned out that even theology – at least when it followed its iconoclastic trajectory – had a role to play in contesting the evolved biases that engender shared imaginative engagement with coalition-favoring disembodied intentional forces.

On the other hand, it is also important to be clear and straightforward when discussing the consequences of "doing it." Religion did indeed promote the kind of small-group cohesion *Homo sapiens* needed to survive and thrive as they hunted and gathered in the upper Paleolithic. Today, however, most human beings live in large-scale, literate states governed by complex legal and political structures.

> In this very different social context, the strong cohesion that religion promotes is of much less benefit to most of us than it was to the inhabitants of the pre-Neolithic world. But *the intolerance and hostility that religion promotes toward out-groups are harmful and threatening to us all....* The benefits of strong social cohesion that religion engenders may at one time have outweighed the costs entailed by out-group intolerance and conflict, but this is *likely no longer the case.*[2]

I have argued that it is definitely no longer the case that engaging in religious sects is a viable strategy for survival, at least if we are concerned about the well-being of the whole human race (and other species that share our ecologically fragile global habitat), and not merely with our own in-group.

Like unfettered population growth in the natural world, the continued expansion of the supernatural population in the human Imaginarium negatively impacts all of us. It may seem like the problem lies primarily in the astonishing fertility of god-bearing conservative groups, including my own religious family of origin (American evangelicalism). Liberal Christians sometimes roll their

---

1   For a fuller analysis of other senses in which interreligious dialogue is "tragic," see Chapter 5 of *Theology after the Birth of God.*

2   Clarke et al., "Religion, Intolerance, and Conflict: Practical Implications for Social Policy," in *Religion, Intolerance, and Conflict: A Scientific and Conceptual Investigation* ed. Clarke et al. (Oxford: Oxford University Press, 2013), 272. Emphasis added.

eyes at the way in which conservatives flip-flop between attributing events to God or Satan (depending on the nature of the outcome).[3] However, progressive believers are just as susceptible to dissonance-reducing biases that lead them to project their own moral values onto the most salient supernatural agent of their in-group: an imagined contemporary Jesus.[4]

In its own way, religious liberal permissivism is just as problematic as religious conservative fertility. Challenging literalistic and xenophobic interpretations of holy texts is a good thing, but continuing to encourage participation in religious sects and reflection on the rehabituation of supernatural symbols only fuels the very biases about which liberals are so perplexed. Insofar as they fail to contest or challenge *theistic* credulity and conformity biases, and go on promoting or protecting religiously sectarian divisions of humanity, progressives are undermining their own efforts to fight racism, classism, sexism and other forms of prejudice and oppression.[5] This is why I have spent so much energy in earlier chapters on unveiling the hidden reciprocity that reinforces the *theogonic* mechanisms of anthropomorphic promiscuity and sociographic prudery.

Along the way, however, we have occassionally had the opportunity to observe the inverse (and more overt) operation of the *theolytic* mechanisms of anthropomorphic prudery and sociographic promiscuity (see *Figure 2*, Chapter 1, p. 64), as well as the effect of these iconoclastic forces in science, philosophy, and (to some extent) theology. In this final chapter, I discuss more empirical evidence for – and offer more philosophical reflections on the implications of – the reciprocal reinforcement of these "naturalistic" and "secularistic" tendencies. The central sub-sections, which make up the bulk of the chapter, describe the way in which these mechanisms promote the practice of "safe sects." The concluding sub-section introduces a relatively new methodology that can facilitate our evaluation of hypotheses about – and policies for altering – the dynamics of complex adaptive social systems. First, however, it is important to acknowledge and respond to one of the most common questions raised in reaction to theogonic reproduction theory.

---

3   Ray et al., "Attributions to God and Satan About Life-Altering Events," *Psychology of Religion and Spirituality* 7, no. 1 (2015).

4   Ross et al., "How Christians Reconcile Their Personal Political Views and the Teachings of Their Faith: Projection as a Means of Dissonance Reduction," *Proceedings of the National Academy of Sciences* 109, no. 10 (2012).

5   See, e.g., the analysis of Sumerau, "'Some of Us Are Good, God-Fearing Folks: Justifying Religious Participation in an LGBT Christian Church,'" *Journal of Contemporary Ethnography* 46, no. 1 (2017).

But Isn't "Religion" Supposed to be Good for Us?

In the early part of the 20th century, much of the scholarship on religion in the social sciences in general, and perhaps psychology and sociology in particular, focused on its negative effects. By the 1960s, many (if not most) of the leading lights in the relevant disciplines more or less openly criticized religion and anticipated its dissolution. The "return of religion" in the 1970s and 1980s (most obvious in the growth of fundamentalism in the United States and the Middle East) led many to reconsider. Perhaps religion is here to stay. Perhaps it is even good for us. A surge of interest in "positive" psychology in the 1990s had a profound effect on the social scientific study of religion, producing a wealth of studies that emphasized the apparent health benefits of religion and spirituality.[6] If religion makes us feel good – and act well – then why keep criticizing it?

First, it is important to be clear about what exactly we think is (or is not) supposed to be good for us. The term "religion" is in scare quotes in the title of the sub-heading above as a reminder of the definitional problems discussed briefly in Chapter 1. Religion is sometimes fuzzily defined in relation to, or even conflated with, qualities or behaviors like searching for meaning, having a worldview, feeling culturally connected, attending to the beauty of the universe, reflecting on ultimate concerns, or acting kindly toward others. Such things may very well be good for us, but the failure to offer operationalizable definitions of religion, to distinguish it from other salutogenic traits, or to clarify which facets of "religiosity" are being measured, leads to muddled claims about its alleged benefits.[7] For the purposes of this final chapter, I will continue to focus on the set of statistically measurable features that have captured our interest throughout this book: those related to shared imaginative engagement with axiologically relevant supernatural agents.

Is "religion" – in this sense – "good for us?" The second point to make is that engaging in religious sects has indeed had the effect of making (some) people

---

6   For an introduction to this literature, see Joseph et al., "Positive Psychology, Religion, and Spirituality," *Mental Health, Religion & Culture* 9, no. 3 (2006). For a recent example, see Boden, "Supernatural Beliefs: Considered Adaptive and Associated with Psychological Benefits," *Personality and Individual Differences* 86 (2015): 227.

7   For additional analysis of the problems with conflating "religion" with other traits, see Schuurmans-Stekhoven, "Are We, like Sheep, Going Astray: Is Costly Signaling (or Any Other Mechanism) Necessary to Explain the Belief-as-Benefit Effect?" *Religion, Brain & Behavior* 7, no. 3 (2016), and Schuurmans-Stekhoven, "'As a Shepherd Divideth His Sheep from the Goats': Does the Daily Spiritual Experiences Scale Encapsulate Separable Theistic and Civility Components?" *Social Indicators Research* 110, no. 1 (2013). See also Lechner et al., "Exploring the Stress-Buffering Effects of Religiousness in Relation to Social and Economic Change: Evidence From Poland," *Psychology of Religion and Spirituality* 5, no. 3 (2013).

feel good and act well. If believing in and ritually interacting with the disembodied (or otherwise ontologically confused) intentional forces postulated by one's ingroup did not have some survival value, the cognitive and coalitional biases that engender such behaviors would not have been naturally selected and socially transmitted. Even if it was not always good for particular individuals in earlier ancestral environments, especially the victims of sexism, racism, and classism, religion (in my sense of the term) was good for the survival of the species. As we have seen in earlier chapters, archaeologists and anthropologists have argued that the placebo effects of ritual healing practices, as well as the anxiolytic effects of believing the same thing and behaving in the same way as in-group members, most likely did promote health in some early ancestral contexts. And so it should not be that surprising that such stress-reducing and prosociality-producing traits could still be health-enhancing for some people, or that disaffiliation from religion in some contexts could have health-injuring effects.[8]

Third, much of the research on religion shaped by "positive" psychology suffers from severe methodological problems. For example, many of these studies fail to account for the well-documented "positivity bias" of religious people, who tend to over-report their sense of life satisfaction,[9] as well as their actual church attendance – especially in the U.S. where the vast majority of such studies appear.[10] Moreover, research on the alleged health benefits of religion in contemporary societies has been heavily skewed by its focus on believers, and relative lack of attention to non-believers. Although research on atheists has been "arrantly absent" for decades, the study of the non-religious (or irreligious) has been growing dramatically in recent years.[11] As we will see in more detail below, the proportion of atheists in the population has been increasing for quite some time, a trend that is likely to continue. This growth has led to increased attention from psychologists, sociologists, and other social scientists interested in the role of (non)religion in human life.[12]

---

8   Brooks, "Don't Stop Believing: Rituals Improve Performance by Decreasing Anxiety," *Organizational Behavior & Human Decision Processes* 137 (2016); Fenelon and Danielsen, "Leaving My Religion: Understanding the Relationship between Religious Disaffiliation, Health, and Well-Being," *Social Science Research* 57 (2016).

9   Headey et al., "Does Religion Make You Healthier and Longer Lived? Evidence for Germany," *Social Indicators Research* 119, no. 3 (2014).

10   Brenner, "Exceptional Behavior or Exceptional Identity?" *Public Opinion Quarterly* 75, no. 1 (2011).

11   Brewster et al., "Arrantly Absent: Atheism in Psychological Science from 2001 to 2012," *The Counseling Psychologist* 42, no. 5 (2014).

12   See, e.g., Bullivant and Lee, "Interdisciplinary Studies of Non-Religion and Secularity: The State of the Union," *Journal of Contemporary Religion* 27, no. 1 (2012); Zuckerman, *Living*

As we might expect from our review of the research on the bio-cultural study of religion in earlier chapters, the extent to which shared imaginative engagement with axiologically relevant supernatural agents is correlated with well-being varies across individuals and contexts.[13] This brings us to our fourth point. Many of the studies of the alleged health benefits of religion fail to attend to personality or other individual-level variables that are mediating the relevant effects. For example, recent psychological experiments following up such studies have found that health or well-being are mediated by factors like low personal locus of control, high levels of mentalizing, positive emotions, sense of community, intrinsic epistemological worldview commitment, or other health protective attitudes or behaviors, none of which are necessarily or even indirectly related to "religion."[14] Moreover, recent critiques of research on the relationship between religion and well-being have demonstrated that the failure to utilize multivariate and multiple regression analyses in many of the studies in this field has led some researchers to miss the fact that they are not measuring religiosity but other individual factors like virtue, agreeableness, conscientiousness, or confidence.[15]

——————————

the Secular Life: New Answers to Old Questions (New York: Penguin Press, 2014), Bullivant and Ruse, eds., The Oxford Handbook of Atheism (Oxford, 2013), Zuckerman et al., The Nonreligious: Understanding Secular People and Societies, (New York: Oxford University Press, 2016), and Cipriani and Garelli, Sociology of Atheism (Leiden: Brill, 2016).

13   See, e.g., Yoon et al., "Religiousness, Spirituality, and Eudaimonic and Hedonic Well-Being," Counselling Psychology Quarterly 28, no. 2 (2015); Shiah et al., "Religion and Subjective Well-Being: Western and Eastern Religious Groups Achieved Subjective Well-Being in Different Ways," Journal of Religion and Health 55, no. 4 (2016).

14   See, e.g., Osborne et al., "Examining the Indirect Effects of Religious Orientations on Well-being through Personal Locus of Control," European Journal of Social Psychology 46, no. 4 (2016), van Cappellen et al., "Religion and Well-Being: The Mediating Role of Positive Emotions," Journal of Happiness Studies 17, no. 2 (2016), van Cappellen et al., "Religiosity and Prosocial Behavior Among Churchgoers: Exploring Underlying Mechanisms," The International Journal for the Psychology of Religion 26, no. 1 (2016), Routledge et al., "Further Exploring the Link Between Religion and Existential Health: The Effects of Religiosity and Trait Differences in Mentalizing on Indicators of Meaning in Life," Journal of Religion and Health 56, no. 2 (2017), Steffen et al., "What Mediates the Relationship Between Religious Service Attendance and Aspects of Well-Being?" Journal of Religion and Health 56, no. 1 (2017), Speed, "Unbelievable?! Theistic/Epistemological Viewpoint Affects Religion – Health Relationship," Journal of Religion and Health 56, no. 1 (2017), and Ng and Fisher, "Protestant Spirituality and Well-Being of People in Hong Kong: The Mediating Role of Sense of Community," Applied Research in Quality of Life 11, no. 4 (2016).

15   This critique is spelled out in detail in Schuurmans-Stekhoven, "Is It God or Just the Data That Moves in Mysterious Ways? How Well-Being Research May Be Mistaking Faith for Virtue," Social Indicators Research 100, no. 2 (2010), Galen and Kloet, "Mental Well-Being in the Religious and the Non-Religious: Evidence for a Curvilinear Relationship," Mental

Fifth, a similar mistake in many studies of the relationship between religion and health is the failure to account for group-level variables that are mediating (or causing) feelings of well-being or prosocial behaviors. Is "religion" really the cause – or the only cause – of prosociality? A recent study involving experimental manipulation (priming) concluded that "religion is not fundamental to moral priming, and it is likely the perceived benefits of *being in a group* that enhances prosociality."[16] Another critical review of recent research in this field found that "this literature has often conflated belief in God with group involvement and failed to control for demographic and social network effects."[17] Once these controls are in place, multiple regression analyses indicate that the benefits of prosociality are more related to group membership in general rather than specifically religious content. In other words, it seems that it is *social* affiliation, or a sense of "belonging" in general, rather than *religious* affiliation, or belief in particular, that engenders feeling good and acting well.[18]

This is supported by psychological research showing that there is no evidence of a connection between the *content* of religious belief and variables like health, well-being, life satisfaction, etc. Categories of belief (atheist, agnostic, theist, etc.) are generally unrelated to reported global health, which suggests that "belief in God is not inherently linked to better health ... [and that] nonbelief in God is not associated with any type of health penalty."[19] A similar study demonstrated that generally speaking "Christians were no more or less healthy than the Religiously Unaffiliated."[20] Another survey analysis found that atheists do not differ from Christians or Buddhists on measures such as well-being, sociality, joviality, emotional stability, happiness, compassion, and

_____

      Health, Religion & Culture 14, no. 7 (2011), and Schuurmans-Stekhoven, "Spirit or Fleeting Apparition? Why Spirituality's Link with Social Support Might Be Incrementally Invalid," *Journal of Religion and Health*, 56, no. 4 (2017).

16   Thomson, "Priming Social Affiliation Promotes Morality – Regardless of Religion," *Personality and Individual Differences* 75 (2015): 195–200. Emphases added.

17   Galen et al., "Nonreligious Group Factors Versus Religious Belief in the Prediction of Prosociality," *Social Indicators Research* 122, no. 2 (2015): 411.

18   See, e.g., ten Kate et al., "The Effect of Religiosity on Life Satisfaction in a Secularized Context: Assessing the Relevance of Believing and Belonging," *Review of Religious Research* 59, no. 2 (2017).

19   Speed and Fowler, "What's God Got to Do with It? How Religiosity Predicts Atheists' Health," *Journal of Religion and Health* 55, no. 1 (2016): 305.

20   Speed and Fowler, "Good for All? Hardly! Attending Church Does Not Benefit Religiously Unaffiliated," *Journal of Religion and Health* 56, no. 3 (2017).

empathic concern.[21] It may well be that level of confidence about beliefs (or the extent to which one values one's beliefs) plays a role in promoting well-being. When comparing those with confident *disbelief* in God to those with confident *belief* in God, however, studies show no difference in well-being.[22]

A sixth response to this question is to point out all of the ways that religion is "bad for us." Throughout this book, I have been exploring the unpleasant consequences of engaging in religious sects, especially the ways in which it can directly promote superstitious interpretations of nature and amplify segregative inscriptions of society, which exacerbate the global (and local) challenges we all face today. It is important not to ignore the explictly negative effects that religion can have on human health and well-being. For example, recent converts may initially "feel good" participating in new religious movements, but this bump in some aspects of their mental health is correlated with reduced autonomy and submissiveness to "unjustified and meaningless requests," lowering optimal development and well-being.[23]

The negative health effects of religion are particularly obvious in relation to anxiogenic religious beliefs such as those related to demons, hell, or other forms of supernatural malevolence.[24] However, the problem does not seem to be limited to fundamentalist individuals who interpret the holy texts of their in-groups literally. A country level analysis of the World Values Survey (including 59 countries) found that life satisfaction was negatively predicted by religious belief. That is to say, "more religious countries showed lower mean levels of life satisfaction than less religious countries."[25] Moreover, some recent studies suggest that in comparison to the non-religious, religious people have either the same *or worse* health outcomes or levels of well-being, especially when statistical analyses are in place to control for other variables.[26]

---

21   Caldwell-Harris et al., "Exploring the Atheist Personality: Well-Being, Awe, and Magical Thinking in Atheists, Buddhists, and Christians," *Mental Health, Religion & Culture* 14, no. 7 (2011).

22   Galen and Kloet, "Mental Well-Being in the Religious and the Non-Religious."

23   Buxant and Saroglou, "Feeling Good, but Lacking Autonomy: Closed-Mindedness on Social and Moral Issues in New Religious Movements," *Journal of Religion and Health* 47, no. 1 (2008): 27.

24   Shariff and Aknin, "The Emotional Toll of Hell: Cross-National and Experimental Evidence for the Negative Well-Being Effects of Hell Beliefs," *PLoS ONE* 9, no. 1 (2014).

25   Plouffe and Tremblay, "The Relationship between Income and Life Satisfaction: Does Religiosity Play a Role?" *Personality and Individual Differences* 109 (2017): 70.

26   Hayward et al., "Externalizing Religious Health Beliefs and Health and Well-Being Outcomes," *Journal of Behavioral Medicine* 39, no. 5 (2016); Hayward et al., "Health and Well-Being Among the Non-Religious: Atheists, Agnostics, and No Preference Compared with

Finally, these rather obvious and direct negative effects of religion are in some ways less difficult to deal with than the *indirect* consequences of the surreptitious reciprocal reinforcement of theogonic mechanisms we have been exploring throughout this book. Even in those cases where shared imaginative ritual engagement with supernatural agents has some health benefits for some individuals in some contexts, it is still bad for the rest of us. The mutual amplification of theistic credulity and theistic conformity biases is covertly at work even in relatively peaceful communities of relatively stable religious individuals. Under psychologically, sociologically, or ecologically stressful conditions, however, these evolved biases toward explaining the world and organizing the social field by appealing to the supernatural agents of one's own religious ingroup can all too easily promote extreme ideologies and intergroup violence. We need to get better at contesting these biases. Can naturalist reasoning and secularist socializing help?

Safe Sects and Analytic Thought

In the next three sub-sections, I outline some of the more recent evidence for the claim that atheism – the attempt to make sense of nature and act sensibly in society without appealing to supernatural agents and authorities – can help us practice safe(r) sects. One of the main reasons for hope, as we will see in more detail below, is that although setting these theolytic mechanisms in motion often requires a great deal of intellectual and social investment, they reciprocally reinforce one another once they get going. Promoting anthropomorphic prudery and sociographic promiscuity will not automatically solve all of our problems, but it can contribute to the dissolution of pernicious theistic biases that complicate human life by triggering cognitive mistakes and coalitional conflicts. It may also loosen the hold that other biases, like sexism, classism, and racism have on human minds and cultures.

Many people will intitially be quite suspicious of these claims. This is not surprising since one of the most significant and well-documented consequences of thestic bias is a strong prejudice against non-theists.[27] There are

---

Religious Group Members," *Journal of Religion and Health* 55, no. 3 (2016); Hwang, "Atheism, Health, and Well-Being," in *The Oxford Handbook of Atheism* ed. Bullivant and Ruse (Oxford, 2013).

27   Andersson, "Atheism and How It Is Perceived: Manipulation of, Bias against, and Ways to Reduce the Bias," *Nordic Psychology*, 68, no. 3 (2016); Gervais et al., "Global Evidence of Extreme Intuitive Moral Prejudice against Atheists" *Nature Human Behavior* 1 (2017): 1–5.

complex reasons for this prejudice, which varies among individuals and across contexts.[28] In light of the role that theogonic mechanisms have played in promoting cooperation, coordination, and competition in human coalitions, we can understand why some people might not trust non-believers to behave. If atheists don't believe that punitive gods are watching them, how can we be sure they will follow the rules? As we will see in the next section, this anti-atheism is really biased; that is, non-believers in general are (at least) as prosocial as believers. Like all unfair group biases, however, prejudice against nonbelievers can have a negative effect on the physical and psychological well-being of the target group,[29] which might help to explain why atheism has not been growing even more rapidly worldwide. Happily, a growing body of research is beginning to shed light on the conditions under which – and the mechanisms by which – this anti-atheist prejudice can be dissolved.[30]

This section focuses on the strong correlation (and plausible causal link) between analytic thought and atheism. Non-believers tend to be more critically reflective, and are over-represented in the academy in general and among elite scientists in particular.[31] However, it is important to emphasize that not all atheists are created equally (or, better, not all followed the same developmental route into atheism). As research on atheists continues to grow, scholars

28   Hughes et al., "Tolerating the 'Doubting Thomas': How Centrality of Religious Beliefs vs. Practices Influences Prejudice against atheists," *Frontiers in Psychology* 6 (2015). Cragun et al., "On the Receiving End: Discrimination toward the Non-Religious in the United States," *Journal of Contemporary Religion* 27, no. 1 (2012); Cragun et al., "Perceived Marginalization, Educational Contexts, and (Non)Religious Educational Experience," *Journal of College and Character* 17, no. 4 (2016); Edgell et al., "From Existential to Social Understandings of Risk: Examining Gender Differences in Nonreligion," *Social Currents* 4, no. 6 (2017); Clobert et al., "East Asian Religious Tolerance versus Western Monotheist Prejudice: The Role of (In)tolerance of Contradiction," *Group Processes & Intergroup Relations* 20, no. 2 (2017).

29   Doane and Elliott, "Perceptions of Discrimination Among Atheists: Consequences for Atheist Identification, Psychological and Physical Well-Being," *Psychology of Religion and Spirituality* 7, no. 2 (2015); Weber et al., "Psychological Distress Among Religious Nonbelievers: A Systematic Review," *Journal of Religion and Health* 51, no. 1 (2012).

30   See, e.g., Labouff and Ledoux, "Imagining Atheists: Reducing Fundamental Distrust in Atheist Intergroup Attitudes," *Psychology of Religion and Spirituality* 8, no. 4 (2016); Gervais, "Finding the Faithless: Perceived Atheist Prevalence Reduces Anti-Atheist Prejudice," *Personality and Social Psychology Bulletin* 37, no. 4 (2011); Simpson and Rios, "The Moral Contents of Anti-Atheist Prejudice (and Why Atheists Should Care about It)," *European Journal of Social Psychology* 47, no. 1 (2017).

31   Caldwell-Harris, "Understanding Atheism/Non-Belief as an Expected Individual-Differences Variable," *Religion, Brain & Behavior* 2, no. 1 (2012).

are identifying a wide variety of pathways to – and types of – disbelief and disaffiliation.[32] Any particular individual's atheism will be the result of some complex combination of cognitive, motivational, social, and other factors.[33] Within this broad field of "atheodiversity," my special interest in what follows is in the connection between non-belief in supernatural agents (anthropomorphic prudery) and individual level variables like analytic cognitive style, education, and intelligence.

Throughout this book, and especially in Chapter 1, we have pointed to a large number of empirical studies demonstrating that religious individuals are more likely to make errors in reasoning related to detecting randomness and purposiveness (compared to non-religious individuals). We can turn this around and say it the other way: atheists and skeptics are better in general at contesting their teleological biases and other hyper-active anthropomorphizing tendencies. In some cases, of course, this may not be a conscious, effortful contestation; sometimes personality variables or contextual conditions make it relatively easy for the non-religious to abstain from religious sects. For whatever complex set of reasons, research consistently shows a strong correlation between low levels of religiosity and high levels of resistance to the biases that promote anthropomorphic promiscuity.

For example, the results of several psychological studies indicate that skeptics are less prone to illusory agency detection than paranormal believers and that non-believers are less likely to categorize ambiguous stimuli as face-like.[34] Other experiments have shown that analytically-thinking skeptics are better able to resist the biases that predispose humans toward religious and paranormal beliefs with *supernatural* content, which "is the only thing that joins religious and paranormal beliefs and ... the only thing that distinguishes

32   See, e.g., Silver et al., "The Six Types of Nonbelief: A Qualitative and Quantitative Study of Type and Narrative," *Mental Health, Religion & Culture* 17, no. 10 (2014), Schnell, "Dimensions of Secularity (DoS): An Open Inventory to Measure Facets of Secular," *The International Journal for the Psychology of Religion* 25, no. 4 (2015), and Stolz, "Institutional, Alternative, Distanced, and Secular," *Nordic Journal of Religion and Society* 30, no. 1 (2017).

33   See, e.g., Norenzayan, "Theodiversity," *Annual Review of Psychology* 67 (2016); Kalkman, "Three Cognitive Routes to Atheism: A Dual-Process Account," *Religion* 44, no. 1 (2014); Stewart, "The True (Non)believer? Atheists and the Atheistic in the United States," in *Sociology of Atheism* (Leiden: Brill Academic, 2016); Norenzayan and Gervais, "The Origins of Religious Disbelief", and Blanes and Oustinova-Stjepanovic, eds., *Being Godless: Ethnographies of Atheism and Non-Religion* (New York: Berghahn Books, 2017).

34   See, e.g., van Elk, "Paranormal Believers Are More Prone to Illusory Agency Detection than Skeptics," *Consciousness and Cognition* 22, no. 3 (2013), and van Elk, "Perceptual Biases in Relation to Paranormal and Conspiracy Beliefs," *PLoS ONE* 10, no. 6 (2015).

religiosity from non-religiosity."[35] Another experimental study comparing the relative tendencies toward false detection of anthropomorphic figures among religious believers and non-believers suggested that all humans may "be biased to perceive human characteristics where none exist, but religious and paranormal believers perceive them even more than do others."[36]

The point here is not that non-believers are completely free of such implicit biases, but that in general they are better able to resist them when trying to explain ambiguous phenomenon. They may exhibit such biases when under time pressure but, as the authors of one study on teleological reasoning put it, non-believers are able to inhibit such intuitions and "abandon them as guiding principles in reflective reasoning."[37] The authors of one experiment that found an association between low levels of religiosity and low levels of anthropomorphism concluded that the latter enables atheists "to interpret the non-animal world in terms of non-agentic forces and thus frees them from potential theistic conceptualization in their dealings with the world."[38] All of this makes sense in light of neuroscientific insights into the biological and cognitive underpinnings of religious misattributions. Such research suggests that supernatural beliefs and experiences are the result of "reduced error monitoring" in both interoceptive and exteroceptive inference processes, upon which religious prayer and hallucinations respectively rely.[39]

This mental advantage that atheists have over theists extends to intellectual reasoning in general, and even to the capacity for identifying valid logical deductions in particular. One study that assessed performance on logical (syllogistic) reasoning problems found that skeptics made fewer errors than believers. This evidence suggests that religious skeptics, compared to believers, are

---

35   Lindeman et al., "Skepticism: Genuine Unbelief or Implicit Beliefs in the Supernatural?" *Consciousness and Cognition* 42 (2016): 225.

36   Riekki et al., "Paranormal and Religious Believers Are More Prone to Illusory Face Perception than Skeptics and Non-believers," *Applied Cognitive Psychology* 27, no. 2 (2013).

37   Järnefelt et al., "The Divided Mind of a Disbeliever: Intuitive Beliefs about Nature as Purposefully Created among Different Groups of Non-Religious Adults," *Cognition* 140 (2015), 83; see also Heywood and Bering, "'Meant to Be': How Religious Beliefs and Cultural Religiosity Affect the Implicit Bias to Think Teleologically," *Religion, Brain & Behavior* 4, no. 3 (2014).

38   Talbot and Wastell, "Corrected by Reflection: The De-Anthropomorphized Mindset of Atheism," *Journal for the Cognitive Science of Religion* 3, no. 2 (2017): 121.

39   van Elk and Aleman, "Brain Mechanisms in Religion and Spirituality: An Integrative Predictive Processing Framework," *Neuroscience and Biobehavioral Reviews* 73 (2017). See also Zhong et al., "Biological and Cognitive Underpinnings of Religious Fundamentalism," *Neuropsychologia* 100 (2017).

both "more reflective and effective in logical reasoning tasks."[40] A recent meta-analysis reported that the vast majority of relevant research papers on this topic discovered a correlation between analytic thinking and religious disbelief. In the same article, the authors reported on four new empirical analyses that provided further confirmation of the impact of "the mere willingness to think analytically" on religious disbelief. All of this evidence suggests that statistically speaking "atheists and agnostics are more reflective than religious believers."[41]

But is this individual level difference between theists and atheists due primarily to cognitive *style* or cognitive *ability* (or both)? Several of the studies we have already cited in this section emphasize the role that the former plays in predicting religious (dis)belief.[42] There is also evidence that cognitive style has an impact on belief in God not only in the short term but over time as well. For example, one study showed that participants with a more intuitive – rather than reflective or analytic – cognitive style were more likely to report a stronger belief in God since childhood, regardless of familial religiosity during formative years.[43] Other studies have explored the differentiation between these variables. For example, one recent survey analysis suggested that religiosity is predicted by (lack of) analytic cognitive *style*, while social conservatism is predicted by (lower) cognitive *ability*.[44] Another survey analysis, explicitly designed to measure all three variables (religiosity, cognitive style and cognitive ability), found that "those with higher cognitive ability are less likely to accept religious doctrine or engage in religious behaviors and those with lower ability are more likely to accept religious doctrine and exhibit higher levels of fundamentalism."[45]

---

40   Pennycook et al., "Belief Bias during Reasoning among Religious Believers and Skeptics," *Psychonomic Bulletin & Review* 20, no. 4 (2013): 806. See also the classic study by Klaczynski and Gordon, "Self-Serving Influences on Adolescents' Evaluations of Belief-Relevant Evidence," *Journal of Experimental Child Psychology* 62, no. 3 (1997), which showed how the reasoning of religious adolescents was systematically biased toward protecting and promoting pre-existing religious beliefs.

41   Pennycook et al., "Atheists and Agnostics Are More Reflective than Religious Believers: Four Empirical Studies and a Meta-Analysis," *PLoS ONE* 11, no. 4 (2016): 1.

42   For other examples, see Pennycook et al., "Analytic Cognitive Style Predicts Religious and Paranormal Belief," *Cognition* 123, no. 3 (2012), and Pennycook, "Evidence That Analytic Cognitive Style Influences Religious Belief: Comment on Razmyar and Reeve," *Intelligence* 43 (2014).

43   Shenhav et al., "Divine Intuition: Cognitive Style Influences Belief in God," *Journal of Experimental Psychology: General* 141, no. 3 (2013).

44   Saribay and Yilmaz, "Analytic Cognitive Style and Cognitive Ability Differentially Predict Religiosity and Social Conservatism," *Personality and Individual Differences* 114 (2017).

45   Razmyar and Reeve, "Individual Differences in Religiosity as a Function of Cognitive Ability and Cognitive Style," *Intelligence* 41, no. 5 (2013): 667.

A high level of education is another variable often associated with low levels of religiosity. Although this connection is slightly more contentious, multiple studies have found that education levels are a strong predictor of both religious disbelief and religious disaffiliation.[46] It makes sense that being encouraged to think critically while learning about scientific explanations and humanistic interpretations would have a generally enervating effect on theogonic mechanisms. Statistical analyses suggest that, at the national level, student performance in science and mathematics is lower in countries with high levels of religiosity.[47] Not surprisingly, analytic thinking in an educational context predicts increased acceptance of evolution, which is generally considered a threat to religion.[48] All of this evidence suggests that "the conflicts between science and religion are not only the result of surface-level moral and epistemological conflicts, but are underpinned by divergent cognitive processes that promote religious belief and undermine scientific understanding."[49]

The negative relationship between religiosity and *intelligence* per se is even more well-documented and empirically validated. While intelligence has an impact on educational attainment, the latter is not necessarily what mediates the negative effect of the former on religious belief and behavior.[50] A recent meta-analysis of 63 studies documented the overwhelming consensus about the significant correlation between high intelligence and low religiosity across populations.[51] One study involving a sample of 137 nations found that the average intelligence of a population predicts the percentage of people who do not believe in God.[52] Another set of survey analyses found that the negative

---

46   See, e.g., Hungerman, "The Effect of Education on Religion: Evidence from Compulsory Schooling Laws," *Journal Of Economic Behavior & Organization* 104 (2014); Lewis, "Education, Irreligion, and Non-Religion: Evidence from Select Anglophone Census Data," *Journal of Contemporary Religion* 30, no. 2 (2015).

47   Stoet and Geary, "Students in Countries with Higher Levels of Religiosity Perform Lower in Science and Mathematics," *Intelligence* 62 (2017).

48   Gervais, "Override the Controversy: Analytic Thinking Predicts Endorsement of Evolution," *Cognition* 142 (2015).

49   McPhetres and Nguyen, "Using Findings from the Cognitive Science of Religion to Understand Current Conflicts between Religious and Scientific Ideologies," *Religion, Brain & Behavior* 7 (2017): 8.

50   Ganzach et al., "On Intelligence Education and Religious Beliefs," *Intelligence* 41, no. 2 (2013).

51   Zuckerman et al., "The Relation Between Intelligence and Religiosity: A Meta-Analysis and Some Proposed Explanations," *Personality and Social Psychology Review* 17, no. 4 (2013).

52   Lynn et al., "Average Intelligence Predicts Atheism Rates across 137 Nations," *Intelligence* 37, no. 1 (2009).

intelligence-religiosity link seems to be more robust across people than it is across countries, which highlights the importance of considering the role of other variables (such as education and quality of life) in moderating the link.[53] Is the negative correlation between intelligence and religiosity primarily a result of religious belief or religious affiliation? One study found that, at least in the case of older adults, religious belief seems to be the driver of this negative association.[54]

Research designs involving longitudinal within-families and cross-sectional analysis of changes in religious belief provide some warrant for inferring causality: over time the "more intelligent" tend to become "less religious."[55] Psychological experiments involving manipulation (priming), and guided by dual-process theories of cognition, also provide evidence of a causal relationship: triggering analytic processing *increases* religious disbelief in the laboratory.[56] Although there are many factors at play in the promotion of supernatural agent beliefs, in light of all the evidence discussed so far (here and in previous chapters) it makes sense to claim that they can be demoted by the activation of analytic thinking. Anticipating our discussion of the mutual intensification of theolytic mechanisms below, it is important to note that experimental studies demonstrate not only that analytic thinking promotes religious disbelief, but also that it *reduces prejudice*.[57]

As we have seen, both genetic and environmental factors contribute to individual levels of (non)religiosity. The same applies to analytic thinking style and intelligence. In fact, the heritability of these traits helps to explain the

---

53   Webster and Duffy, "Losing Faith in the Intelligence-Religiosity Link: New Evidence for a Decline Effect, Spatial Dependence, and Mediation by Education and Life Quality," *Intelligence* 55 (2016).

54   Ritchie et al., "Religiosity Is Negatively Associated with Later-Life Intelligence, but Not with Age-Related Cognitive Decline," *Intelligence* 46 (2014).

55   Ganzach and Gotlibovski, "Intelligence and Religiosity: Within Families and over Time," *Intelligence* 41, no. 5 (2013): 551.

56   Gervais and Norenzayan, "Analytic Thinking Promotes Religious Disbelief," *Science* 336, no. 6080 (2012). However, it is important to note that these results could not be replicated by Sanchez et al., "Direct Replication of Gervais & Norenzayan (2012): No Evidence That Analytic Thinking Decreases Religious Belief" *PLoS ONE* 12, no. 2 (2017). Another study found that *implicit* religiosity was not reduced when analytic thinking was primed: Yonker et al., "Primed Analytic Thought and Religiosity: The Importance of Individual Characteristics," *Psychology of Religion and Spirituality* 8, no. 4 (2016), suggesting the need to take individual characteristics into account in such priming studies.

57   Yilmaz et al., "Analytic Thinking, Religion, and Prejudice: An Experimental Test of the Dual-Process Model of Mind," *The International Journal for the Psychology of Religion* 26, no. 4 (2016).

negative correlation we have been discussing. In changing ecologies, problem solving abilities would be associated with the capacity and willingness to resist biases that evolved in earlier ecological contexts. This has led some scholars to conclude that intelligence was naturally selected as individuals who were "intellectuallly curious and thus open to non-instinctive possibilities" resisted religious (and other) biases and gained the survival advantage.[58] For most people, throughout most of human history, the bio-cultural pressures exerted by theistic conformity biases have usually overpowered intellectual resistance to theistic credulity biases.

Today human fertility is positively correlated with religiosity and negatively correlated with intelligence. Unless naturalism and secularization forces continue to gain ground and override projections based primarily on fertility rates, this means that as the world becomes more religious "the genes promoting religiosity will spread and the genes for intelligence will diminish." To face global challenges related to extreme climate change, excessive consumer capitalism, and escalating cultural conflict, we need scientific and philosophical reasoning that can resist superstitious appeals to hidden supernatural agents. That kind of analytic thinking "requires genes for high intelligence and is facilitated by genes for low religosity."[59] We will return below to the prospects for an adaptive atheism.

But don't lots of intelligent and well-educated people believe in God (not to mention angels, demons, saints, genies, devas, and other supernatural agents)? Indeed they do. For thousands of years the majority of the intellectual (and priestly) elite in large-scale societies have participated in and promoted unsafe sects, engendering all sorts of superstitious supernatural conceptions. Although a few ancient and early modern philosophers and naturalists explicitly resisted the temptations of anthropomorphic promiscuity, atheism did not become widespread until after the natural (and later evolutionary) origins of religion were discovered.[60]

Many smart people today still argue for the existence of god(s) and participate in the rituals of their religious in-groups. As we noted in Chapter 5,

---

58   Dutton and van der Linden, "Why Is Intelligence Negatively Associated with Religiousness?" *Evolutionary Psychological Science* 3, no. 4 (2017): 401; See also Kandler and Riemann, "Genetic and Environmental Sources of Individual Religiousness: The Roles of Individual Personality Traits and Perceived Environmental Religiousness," *Behavior Genetics* 43, no. 4 (2013).

59   Ellis et al., "The Future of Secularism: A Biologically Informed Theory Supplemented with Cross-Cultural Evidence," *Evolutionary Psychological Science* 3, no. 3 (2017): 238.

60   Collier, "The Natural Foundations of Religion," *Philosophical Psychology* 27, no. 5 (2014).

religiously affiliated scholars often make the same sort of errors in reasoning as their less-educated peers.[61] The intellectual habits and methodological practices fostered by science and (non-religious) philosophy can help entrain resistance to theistic credulity biases, but the latter continue to surreptitiously shape the abductive inferences of scholars engaged in religious sects. This is one example of the way in which even intellectuals can be powerfully (and covertly) influenced by evolved biases like motivated reasoning (or high levels of schizotypy).[62]

Even if one grants that analytical reasoning and atheism are correlated and causally connected, there might still be reasons not to promote either. Nonbelievers who contest their evolved intuitions, it might be argued, are generally more angry and less happy than believers, and so we ought to let sleeping theist biases lie. Here too it turns out that such claims are not warranted by the data. In fact, these anti-atheist prejudices are explicilty contradicted by the empirical evidence. The myth of the "angry atheist" is precisely that – a myth. A recent set of studies found that neither belief nor non-belief in God is correlated with measures of trait anger.[63] Perhaps even more surprisingly to many, an analysis of World Values Survey data found that belief in scientific-technological progress (a trait associated with anthropomorphic prudery) is a stronger predictor of life-satisfaction than religious belief in 69 out of 72 countries.[64]

The fact that atheists in general have stronger analytic and higher intellectual capacities does not mean that they will tend to be cold and emotionless. In fact, a set of studies on the relation between atheism, religion, and emotion found that individuals who identify themselves as religious "reported greater logical difficulty with respect to differentiating their emotions compared to atheists." Or, to put it positively: "atheists reported greater general facility with

61   Tobia, "Does Religious Belief Infect Philosophical Analysis?" See also Draper and Nichols, "Diagnosing Bias in Philosophy of Religion."

62   MacPherson and Kelly, "Creativity and Positive Schizotypy Influence the Conflict between Science and Religion," *Personality and Individual Differences* 50, no. 4 (2011).

63   Meier et al., "The Myth of the Angry Atheist," *The Journal of Psychology* 149, no. 3 (2015). However, it is important to acknowledge that some atheists can be as "dogmatic" about their disbelief as theists are about their belief. Fundamentalist tendencies in either direction can predispose individuals toward out-group prejudice. See Kossowska et al., "Many Faces of Dogmatism: Prejudice as a Way of Protecting Certainty against Value Violators among Dogmatic Believers and Atheists," *British Journal of Psychology* 108, no. 1 (2017), and Brandt and van Tongeren, "People Both High and Low on Religious Fundamentalism Are Prejudiced Toward Dissimilar Groups," *Journal of Personality and Social Psychology* (2015).

64   Stavrova et al., "Belief in Scientific-Technological Progress and Life Satisfaction: The Role of Personal Control," *Personality and Individual Differences* 96 (2016).

respect to focusing on, identifying, and describing their own emotions."[65] Many atheists also have intense experiences of joy, awe, and self-transcendence that are neither explicitly nor implicitly "religious."[66] Like believers, there are many sub-groups and individual differences between non-believers, but recent empirical research on these variables suggests that both religious and non-religious people can have strong empathizing tendencies.[67] Atheists may on average be smarter, more reflective, better educated, and more in control of their emotional states – but they are less moral than religious people. Right?

Safe Sects and Altruistic Behavior

Here too it is important to begin by confronting the anti-atheist prejudice that will lead many readers to quickly dismiss the idea that sociographic promiscuity can promote the construction and maintenance of good societies. Social psychological experiments have shown that people will tend to give more negative moral appraisals of atheists (compared to Christian theists) even when they performed exactly the same moral or immoral actions.[68] This stereotype against non-believers seems to be mediated by a lack of moral trust. People appear to expect atheists to behave badly, and somewhat automatically set up moral boundaries to protect themselves.[69] However, this implicit prejudice is reduced when non-believers are perceived to be prevalent in a society. In fact, experimentally induced reminders of atheist prevalence appear to *cause* a decrease in explicit distrust of atheists.[70] The results of other experiments

---

65   Burris and Petrican, "Hearts Strangely Warmed (and Cooled): Emotional Experience in Religious and Atheistic Individuals," *International Journal for the Psychology of Religion* 21, no. 3 (2011): 193.

66   Coleman et al., "Focusing on Horizontal Transcendence: Much More than a 'Non-Belief,'" *Essays in the Philosophy of Humanism* 21, no. 2 (2014).

67   Lindeman and Lipsanen, "Diverse Cognitive Profiles of Religious Believers and Nonbelievers." *The International Journal for the Psychology of Religion* 26, no. 3 (2016).

68   Wright and Nichols, "The Social Cost of Atheism: How Perceived Religiosity Influences Moral Appraisal," *Journal of Cognition and Culture* 14, no. 1–2 (2014).

69   Gervais et al., "Do You Believe in Atheists? Distrust Is Central to Anti-Atheist Prejudice," *Journal of Personality and Social Psychology* 101, no. 6 (2011); Gervais and Norenzayan, "Religion and the Origins of Anti-Atheist Prejudice," in *Intolerance and Conflict: A Scientific and Conceptual Investigation*, ed. Clarke et al. (Oxford, 2013); Edgell et al., "Atheists and Other Cultural Outsiders: Moral Boundaries and the Non-religious in the United States," *Social Forces* 95, no. 2 (2017).

70   Gervais, "Finding the Faithless: Perceived Atheist Prevalence Reduces Anti-Atheist Prejudice," *Personality and Social Psychology Bulletin* 37, no. 4 (2011).

indicate that priming thoughts about secular authority or government can have a similar effect.[71]

As the percentage of non-religious people in the population continues to grow, so does the recognition that we can be good without God.[72] But don't we need religion to make us moral? It is not difficult to understand why this myth is so widespread. For most of the history of the human species, moral norms have been justified by appeals to supernatural agents and authorities. As we have seen in earlier chapters, over time and in a variety of places bigger gods and bigger societies co-evolved; large-scale cooperation appears to have been enhanced by shared imaginative engagement with larger (smarter, stronger, and more punitive) supernatural agents.[73] The priestly and intellectual elites of the west Asian monotheistic religions that emerged in the wake of the axial age, and helped to fuel the growth of massive empires into the modern period, eventually postulated the existence of an infinitely powerful and eternally punitive God.

However, we have good reasons to believe that none of this implies that religions are *necessary* for morality. From the evolutionary history of social insects it is clear that mechanisms contributing to group social control and to the development of altruistic behavior in individual organisms have been operative for millions of years. Long before the emergence of *Homo sapiens*, eusocial animals – including many other primate species – were coordinating and competing with other groups.[74] Altruistic care of genetic kin, as well as direct and indirect reciprocity with cooperating kith, were among the moral behaviors that helped individual members of such species survive and protect their offspring long enough to reproduce.[75] Another important part of this

---

71   Gervais and Norenzayan, "Reminders of Secular Authority Reduce Believers' Distrust of Atheists," *Psychological Science* 23, no. 5 (2012).

72   Epstein, *Good Without God: What a Billion Nonreligious People Do Believe* (William Morrow Paperbacks, 2010); Blackford and Schuklenk, *50 Great Myths About Atheism* (Hoboken: Wiley-Blackwell, 2013).

73   See also Norenzayan and Shariff, "The Origin and Evolution of Religious Prosociality," *Science* 322, no. 5898 (2008).

74   de Waal, ed., *Evolved Morality: The Biology and Philosophy of Human Conscience* (Leiden: Brill Academic Publishers, 2014); Boehm, *Moral Origins: The Evolution of Virtue, Altruism, and Shame* (New York: Basic Books, 2012); Sinnott-Armstrong and Miller, eds., *Moral Psychology, The Evolution of Morality: Adaptations and Innateness, Vol. 1* (Cambridge, Mass: The MIT Press, 2007).

75   Krebs and Denton, "The Evolution of Sociality, Helping, and Morality," in *The Oxford Handbook of Secularism*, ed. Zuckerman and Shook (New York: Oxford University Press, 2017).

process, as we have seen, was the natural selection of genetic tendencies that reinforced individual's willingess to punish, and reward other's punishment of, cheaters and free-loaders.

Today, however, we have significant evidence that secular societies are able to promote prosociality at least as well as religious societies. One type of evidence comes from priming studies. Experimentally induced reminders of secular moral authority have the same sort of effect on altruistic behavior as reminders of supernatural authority. One study tested the impact of implicit priming on behavior in an anonymous dictator game and found that "implicit activation of concepts related to secular moral institutions restrained selfishness as much as did religious suggestion."[76] A similar study in Japan found that there was no difference between the amount of money allocated to strangers among the three priming conditions (religious, secular justic, control), with the somewhat surprising exception that theists allocated more money than atheists under the secular justice condition.[77]

The results of other priming experiments indicate that it is triggering thoughts about *social* affiliation in general – not religious affiliation in particular – that promote morality.[78] In other words, the same prosociality enhancing mechanisms that operate within religious sects can also operate within more sociographically promiscuous societies. Secular institutions like relatively transparent democratic legislatures, contract-enforcing courts, and policing authorities are more recent than religious institutions, but they too can foster large-scale trust and cooperation. Both gods and governments can "function as social monitors to encourage cooperation among individuals."[79] It is not supernatural beliefs nor even participation in supernatural rituals per se that engender altruistic behaviors. It turns out that the apparent prosocial benefits of religious sects are actually the benefits of *sociality* itself.

---

76   Shariff and Norenzayan, "God Is Watching You: Priming God Concepts Increases Prosocial Behavior in an Anonymous Economic Game," *Psychological Science* 18, no. 9 (2007): 807. See also Norenzayan, "Does Religion Make People Moral?" *Behaviour* 151, no. 2–3 (2014): 380, and Yilmaz and Bahçekapili, "Supernatural and Secular Monitors Promote Human Cooperation Only If They Remind of Punishment," *Evolution and Human Behavior* 37, no. 1 (2016).

77   Miyatake and Higuchi, "Does Religious Priming Increase the Prosocial Behaviour of a Japanese Sample in an Anonymous Economic Game?" *Asian Journal of Social Psychology* 20, no. 1 (2017).

78   Thomson, "Priming Social Affiliation Promotes Morality – Regardless of Religion." *Personality and Religious Differences*, 75 (2015).

79   Gervais and Norenzayan, "Reminders of Secular Authority Reduce Believers' Distrust of Atheists," *Psychological Science* 23, no. 5 (2012): 489.

We also have evidence from survey analyses that contradicts the idea that being religious makes people more altruistic. A broad analysis of U.S. survey data suggested that any form of voluntary association tends to make people feel better, safer, and more able to contribute to just and stable societies. It seems that membership in a secular bowling league, for example, is just as likely to boost charitable giving as affiliation within a religious organization.[80] Although some earlier studies found at least a weak correlation between factors related to religiosity and prosociality, such as *reported* generosity and willingness to help strangers, these have more recently come under serious critique due to their lack of conceptual clarity, their confounding of factors, and the fact that they are powerfully contradicted by experiments that test *actual* economic and other behaviors.[81] The bulk of the evidence indicates an absence of – or even a negative – correlation between religiosity and actual prosocial behavior.[82]

Religious people do give more to their own in-groups (e.g., churches), especially when they think they are being watched, but they seem less willing to act selflessly when there are religious or non-religious others who might benefit.[83] A recent analysis of social attitudes in 33 countries found several interesting correlations among variables related to religiosity, nastiness (defined in relation to readiness to cause pain for the satisfaction of doing harm) and morality (defined in relation to *conformity* to conventional standards of moral conduct). Western European countries, which are by far the most secular, have the lowest scores on all three factors.[84] In other words, in these contexts populations are more likely to be characterized by less parochial prosociality, less malicious aggression, and less shared imaginative engagement with supernatural agents.

Religious institutions have been around for such a long time, and are so deeply entangled with technologies that prime prosociality, that many people have become convinced that only they can produce morality. When it comes to real world altruism, however, religiosity in general does not seem to have any positive impact. For example, a well-known study of those who helped hide

80   Putnam, *Bowling Alone: The Collapse and Revival of American Community* (New York: Simon & Schuster, 2001); Putnam and Campbell, *American Grace: How Religion Divides and Unites Us* (New York: Simon & Schuster, 2012).

81   Sablosky, "Does Religion Foster Generosity?" *Social Science Journal* 51, no. 4 (2014).

82   For an overview, see Mitkidis and Levy, "False Advertising: The Attractiveness of Religion as a Moral Brand," in *The Attraction of Religion: A New Evolutionary Psychology of Religion* ed. Slone and van Slyke (Bloomsbury Academic, 2015).

83   Andreoni et al., "Diversity and Donations: The Effect of Religious and Ethnic Diversity on Charitable Giving," *Journal of Economic Behavior and Organization* 128 (2016).

84   Stankov and Lee, "Nastiness, Morality and Religiosity in 33 Nations," *Personality and Individual Differences* 99 (2016).

or rescue Jews during the Holocaust found that the moderately religious were predominantly non-rescuers. Most rescuers were either highly religious or non-religious, leading researchers to conclude that other factors like nonconformity or social responsibility were better predictors of altruistic behavior.[85] However, another more recent study of rescuers, which dealt with more variables in a broader statistical analysis, found that "religiosity and altruism are *negatively* related; the less religious one is, the more likely she is to rescue."[86]

One review of the literature on this topic concluded that "there is surprisingly little evidence for a moral effect of specifically religious beliefs."[87] On the contrary, a growing body of evidence indicates that *not* believing in God, *while* maintaining social affiliation with others, is better for you and promotes a *broader* sort of prosociality.[88] As we have seen throughout this book, religion promotes *parochial* or assortative prosociality, the dark side of which is anxiety about and antagonism toward out-groups. Non-religious people are on average at least as altruistic as religious people and, more importantly, their arena of moral concern applies to a wider range of subjects. While statistical survey analysis shows a (weak) relation beween respondents' religiosity and "benevolence" (forgiveness, loyalty, etc.), this concern for the welfare of others does not extend to "universalism" (tolerance, protection, etc., for all people).[89]

Belief in God does predict in-group prosociality and parochialism, but it also has the simultaneous effect of *decreasing universal concern* for those outside one's group.[90] This group-focused morality prevalent among religious believers may be connected not only to anxiety about protecting their own coalition, but also to the detection errors discussed above and in Chapter 1. Neuroanatomical evidence from scanning experiments and analysis of moral foundations questionnaires found that "increased adherence to *group-focused* moral foundations was associated with reduced ACC [anterior cingulate

---

85   Oliner and Oliner, *The Altruistic Personality: Rescuers of Jews in Nazi Europe* (New York: Touchstone, 1992).

86   Varese and Yaish, "The Importance of Being Asked: The Rescue of Jews in Nazi Europe," *Rationality and Society* 12, no. 3 (2000): 320. See also Beit-Hallahmi, "Morality and Immorality among the Irreligious," in Zuckerman, ed., *Atheism and Secularity* 1 (New York: Praeger, 2010).

87   Bloom, "Religion, Morality, Evolution," *Annual Review of Psychology* 63 (2012): 179.

88   For a review of this literature, see Galen, "Atheism, Wellbeing, and the Wager: Why Not Believing in God (With Others) Is Good for You," *Science, Religion and Culture* 2, no. 3 (2015).

89   Saroglou et al., "Values and Religiosity: A Meta-Analysis of Studies Using Schwartz's Model," *Personality and Individual Differences* 37, no. 4 (2004).

90   Galen et al., "Nonreligious Group Factors Versus Religious Belief in the Prediction of Prosociality," *Social Indicators Research* 122, no. 2 (2015).

cortex] and lateral PFC [prefrontal cortex] gray matter ... [suggesting that] people who adhere more stongly to group-focused moral foundations may be less able to detect and resolve conflict."[91] This latter inference is based on the fact that reduced ACC volume renders individuals less able to *detect* conflicts and discrepancies, and reduced lateral PFC volume renders them less able to *resolve* unavoidable conflicts, such as those related to social change.

While religious believers are often motivated to act "morally" by visions of an eschatological future promised by a supernatural agent, non-believers tend to "focus their moral concerns on social justice and the here-and-now."[92] Although they are motivated by different social cues, and use different criteria for judging an action as moral, for our purposes here the most interesting difference between theists and non-theists is that the former "tend to direct their prosociality more *parochially* toward ingroup members," while the morality of the latter has a "*more universal* scope."[93] A recent cross-national study found that household religiousness was inversely predictive of children's altruism, challenging the view that religiosity facilitates prosocial behavior and providing evidence that "the secularization of moral discourse will not reduce human kindness – in fact, it will do just the opposite."[94] When it comes to resisting the parochial prosociality and out-group antagonism driven by religious credulity and conformity biases, atheists also seem to have the moral advantage.[95]

Another body of evidence that supports the claim that people do not need religion to be moral comes from contemporary sociological analysis. A review of the Global Peace Index shows that it is "the least God-fearing nations that

---

91   Nash, et al., "Group-Focused Morality Is Associated with Limited Conflict Detection and Resolution Capacity: Neuroanatomical Evidence," *Biological Psychology* 123 (2017): 237. Emphasis added.

92   Caldwell-Harris, "Understanding Atheism/non-Belief as an Expected Individual-Differences Variable," *Religion, Brain & Behavior* 2, no. 1 (2012): 4.

93   Shariff et al., "Morality and the Religious Mind: Why Theists and Nontheists Differ," *Trends in Cognitive Sciences* 18, no. 9 (2014): 439. Emphasis added.

94   Decety et al., "The Negative Association between Religiousness and Children's Altruism across the World," *Current Biology* 25, no. 22 (2015): 3. Even critics of this controversial study concede that its dataset does reveal that generosity is (weakly) negatively correlated to frequency of household religious attendance and intrinsic religiosity: Shariff et al., "What Is the Association between Religious Affiliation and Children's Altruism?" *Current Biology* 26, no. 15 (2016).

95   The difference in the ability of religious and non-religious individuals to contest implicit in-group biases has also been detected in neuroscientific experiments. See, e.g., Huang and Han, "Shared Beliefs Enhance Shared Feelings: Religious/irreligious Identifications Modulate Empathic Neural Responses," *Social Neuroscience* 9, no. 6 (2014).

enjoy the greatest levels of peace. ...among the top ten most peaceful nations on earth, all are among the least God-believing – in fact, eight of the ten are specifically among the least theistic nations on earth. Conversely, of the bottom ten – the least peaceful nations – most of them are extremely religious."[96] When it comes to factors like women's rights, gay rights, and protecting the environment, "secular people actually possess a *stronger* or *more ethical* sense of social justice than their religious peers ... atheists and secular people are also the least likely to harbor ethnocentric, racist, or nationalist attitudes."[97]

As we have seen, these other forms of prejudice are reinforced by theism, and so it ought not to surprise us that atheists in general are less prone to sexism, classism, and racism. But what about the death and destruction caused by non-religious totalitarian regimes in the 20th century? All non-democratic regimes – religious or irreligious – have led to less than ideal living conditions for the members of the societies they oppressed. We can all agree that totalitarianism is bad, regardless of how it is justified. When we remove this factor and focus on the relation between social well-being and (ir)religiosity, "the least religious democracies fare better on nearly all indicators of social well-being."[98] Moreover, a recent analysis of a data set of over 700 ethnic groups in over 130 states found that religious identification was likely to contribute to the onset of civil war between ethnic group dyads, while atheist identification was not.[99]

Like secular democracies, atheist communities are a relatively new human experiment. Non-believers are certainly not immune to sexist, classicist, and racist biases. As we have seen, however, in general they are better at learning how to contest them. Increasingly, atheists are organizing, awakening to these and other concerns, and struggling to build even more just societies.[100] As the

---

96   Zuckerman, *Living the Secular Life*, 2014, 47. See also Zuckerman, *Society without God: What the Least Religious Nations Can Tell Us About Contentment* (New York: NYU Press, 2010), 183, and Zuckerman, *Faith No More: Why People Reject Religion* (New York: Oxford University Press, 2011).

97   Zuckerman, "Atheism, Secularity, and Well-Being: How the Findings of Social Science Counter Negative Stereotypes and Assumptions," *Sociology Compass* 3, no. 6 (2009): 954.

98   Zuckerman et al., *The Nonreligious*, p. 86. For detailed analysis of these indicators see Paul, "The Evolution of Popular Religiosity and Secularism: How First World Statistics Reveal Why Religion Exists, Why It Has Been Popular, and Why the Most Successful Democracies Are the Most Secular," in *Atheism and Secularity*, *1*, ed. Zuckerman (New York: Praeger, 2010).

99   Bormann et al., "Language, Religion, and Ethnic Civil War," *Journal of Conflict Resolution* 61, no. 4 (2017).

100   Cimino et al., *Atheist Awakening: Secular Activism and Community in America* (New York: Oxford University Press, 2014); Cimino and Smith, "Atheism, Class, and the Ghost of Karl

non-religious "come out of the closet," the prejudice against them is beginning to diminish, albeit slowly. Around the globe, atheists are coming together, developing new ways to respond to the maladaptive effects of theistic credulity and conformity biases, and learning how to take care of each other as well as religious others.[101]

It is also important to emphasize the role of contextual variance in shaping the relationship between atheism and altruistic behavior. Atheist identities can be expressed quite differently in various social contexts around the globe.[102] The well-being and moral flourishing of non-believers depends in large part on the social situation in which they find themselves. When surrounded by an oppressive religious majority, it is hardly surprising that atheists are stressed by constant encounters with theists and so keep to themselves. On the other hand, in more open, benign societies, the pattern tends to be that the non-religious have more life satisfaction and general social-well being.[103] One of the most significant variables that predicts a strong relation between low religiosity and well-being in a population is the proportion of the gross national product that a government spends on social welfare.[104] In Scandinavian

Marx," in *Sociology of Atheism*, ed. Cipriani and Garelli, (Leiden: Brill, 2016); Ledrew, "Discovering Atheism: Heterogeneity in Trajectories to Atheist Identity and Activism," *Sociology of Religion* 74, no. 4 (2013); Kettell, "Divided We Stand: The Politics of the Atheist Movement in the United States," *Journal of Contemporary Religion* 29, no. 3 (2014); Kettell, "Faithless: The Politics of New Atheism," *Secularism and Nonreligion* 2 (2013).

101   Smith and Cimino, "Atheisms Unbound: The Role of the New Media in the Formation of a Secularist Identity," *Secularism and Nonreligion* 1 (2012); Thiessen and Wilkins-Laflamme, "Becoming a Religious None: Irreligious Socialization and Disaffiliation," *Journal for the Scientific Study of Religion* 56, no. 1 (2017); Smith, "Creating a Godless Community: The Collective Identity Work of Contemporary American Atheists," *Journal for the Scientific Study of Religion* 52, no. 1 (2013); Sumerau and Cragun, "'I Think Some People Need Religion': The Social Construction of Nonreligious Moral Identities," *Sociology of Religion* 77, no. 4 (2016); Sumerau et al., "An Interactionist Approach to the Social Construction of Deities," *Symbolic Interaction* 39, no. 4 (2016).

102   Beaman, *Atheist Identities – Spaces and Social Contexts*, (Springer International Publishing, 2014); Zuckerman, ed., *Atheism and Secularity: Global Expressions. Vol 2.* (New York: Praeger, 2010).

103   Diener et al., "The Religion Paradox: If Religion Makes People Happy, Why Are So Many Dropping Out?" *Journal of Personality and Social Psychology* 101, no. 6 (2011).

104   Scheve et al., "Religion and Preferences for Social Insurance," *Quarterly Journal of Political Science* 1, no. 3 (2006). For a discussion of the psychological implications of these correlations, see Granqvist, "Mental Health and Religion from an Attachment Viewpoint: Overview with Implications for Future Research," *Mental Health, Religion & Culture* 17, no. 8 (2014).

countries like Norway, it is often the irreligious who have more positive experiences of social support and caring interaction.[105]

None of this should be taken to imply that religious individuals are "bad," any more that the previous section should be taken to imply that they are "dumb." The empirical studies we have been exploring shed light on cognitive and coalitional mechanisms that are manifested in quite different ways at the individual and population levels. Any particular theist you meet on the street may be far more intelligent and altruistic than the atheist walking next to her. And if you were to transplant either of them to a new context, their individual beliefs and behaviors might change considerably. However, that does not challenge the basic claims that emerge from all of this research. Generally speaking, anthropomorphically promiscuous tendencies lead people to make more errors when trying to explain perceptually ambiguous natural phenomena. Generally speaking, sociographically prudish tendencies lead people to avoid risky altruism toward morally ambiguous out-group members.

The Reciprocal Reinforcement of Theolytic Mechanisms

Happily, efforts at contesting these evolved biases by promoting either naturalism or secularism benefit from a spiraling interaction that intensifies their god-dissolving effectiveness.[106] When integrated, these theolytic forces create a trajectory that is diametrically opposed to the one produced by the god-bearing mechanisms operative in religious sects (*Figure 3*; compare to *Figure 1*, p. 3). In other words, pursuing anthropomorphically prudish explanations of nature and endorsing sociographically promiscuous inscriptions of society can also be reciprocally reinforcing.

We have already seen a great deal of warrant for this claim. As a sort of inversion of the third main hypothesis of theogonic reproduction theory, it is supported by much of the same evidence provided for the latter. This is particularly true for many of the survey data studies we have reviewed. If statistical analysis reveals that higher religious belief is correlated with lower support for broad governmental social care, for example, one can just as easily say it the other way around: lower religious belief is correlated with higher support. In other words, if naturalism and secularism really reinforce one another, then

105　Kvande et al., "Religiousness and Social Support: A Study in Secular Norway," *Review of Religious Research* 57, no. 1 (2015).

106　It is important to keep in mind that in this context I am continuing to use the terms naturalism and secularism in the sense introduced at the end of Chapter 1.

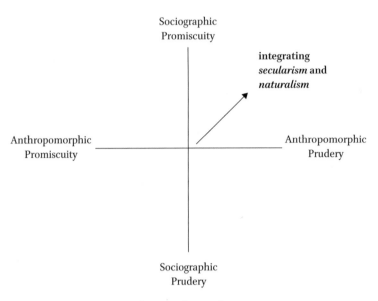

FIGURE 3   *Integrating secularism and naturalism*

we would expect the data to show exactly what we find: a strong *correlation* between variables related to disbelief in supernatural agents and variables related to dissaffection for supernatural authorities.

But do we also have evidence that there are *causal* relationships between some of the components of anthropomorphic prudery and sociographic promiscuity? Here too we can take the evidence from the priming and other experimental studies cited throughout this book as indirect support. In other words, if highly religious people are generally more likely to show aggression toward out-group members when mortality salience is triggered, for example, the converse also follows: less- or non-religious people are generally less likely to respond in that way. I leave it to the reader to review those studies again with the hypothesis about the mutual amplification of *theolytic* mechanisms in mind. However, we also have evidence that more directly supports the claim that these god-dissolving tendencies strengthen one another.

For example, one recent set of priming experiments tested the hypothesis that thinking about "science," which is at least methodologically naturalistic, and contains the idea of "broader moral vision of a society in which rationality is used for the mutual benefit of all," has a causal impact on moral judgments and behavior. It turns out that triggering thoughts about science can positively impact moral attitudes, such as responses to interpersonal violations and prosocial intentions. A post-experimental manipulation (playing an economics dictator game) revealed that priming naturalistic thoughts affected actual moral behaviors as well (leading to less economic exploitation).[107] Another study attempting to replicate these findings found that both priming words related to science in general and words related to science as a secular authority increased moral sensitivity.[108] Experiments involving young adults have found a somewhat automatic opposition between evaluations of explanations that use science and those that use God. These results indicate that "using scientific theories as ultimate explanations can serve as an automatic threat to religious beliefs, and vice versa."[109]

Analytic cognitive style predicts not only less belief in supernatural agents, as we saw above, but also less religious engagement (e.g., attendance or affiliation), an effect which one study found to be mediated through lower accepance of conventional religious beliefs.[110] Science, which is based on the analysis of naturalistic explanations of causal mechanisms, can provide people with a sense of order and predictability. Experimental studies suggest that it is at least as effective as religion in helping individuals regulate their responses to external threats, while simultaneously reducing their reliance on supernatural agents to make sense of and organize their social worlds.[111] Priming people with secular arguments that included claims that the world can be explained naturalistically through evolutionary theory lowered their explicit self-reports of religious belief – as well as measures of their *implicit* religiosity.[112] These sorts of empirical studies indicate that some of the components of sociographic promiscuity and anthropomorphic prudery have a direct and mutually intensifying relationship.

Another type of evidence for this reciprocity comes from social sciences like economics and sociology. We have already noted several cross-national studies that show religiosity tends to decline as the quality of life in a country

---

107   Ma-Kellams and Blascovich, "Does 'Science' Make You Moral? The Effects of Priming Science on Moral Judgments and Behavior," *PLoS ONE* 8, no. 3 (2013).

108   Yilmaz and Bahçekapili, "When Science Replaces Religion: Science as a Secular Authority Bolsters Moral Sensitivity," *PLoS ONE* 10, no. 9 (2015).

109   Preston and Epley, "Science and God: An Automatic Opposition between Ultimate Explanations," *Journal of Experimental Social Psychology* 45, no. 1 (2009): 240.

110   Pennycook et al., "Analytic Cognitive Style Predicts Religious and Paranormal Belief," *Cognition* 123, no. 3 (2012): 335.

111   Rutjens et al., "Step by Step: Finding Compensatory Order in Science," *Current Directions in Psychological Science* 22, no. 3 (2013), 250, 253. See also Rutjens et al., "Deus or Darwin: Randomness and Belief in Theories about the Origin of Life," *Journal of Experimental Social Psychology* 46, no. 6 (2010): 1080.

112   Shariff et al., "The Devil's Advocate: Secular Arguments Diminish Both Implicit and Explicit Religious Belief," *Journal of Cognition and Culture* 8, no. 3–4 (2008).

increases. The results of a recent study of the relationship between religiosity and economic development within each of the 50 U.S. states were consistent with these findings: state religiosity declines as the quality of life improves. The authors conclude that "religion may thus function as an adaptive emotional response to difficult living conditions."[113] These findings are also consistent with economic theories about the mechanisms of secularization. For example, one recent mathematical model predicts that if the marginal utility of secular leisure increases with more income availability, wealthier individuals who have higher cognitive abilities will be more likely to abandon "intuitive-believing" for "reflective-analytical" cognitive styles, which has a long-term positive effect on secularization.[114]

Even the presence of diverse marketing brands and their availability as tools for self-expression seems to have a negative impact on individual's self-reports (and demonstrations) of religious commitment.[115] As religion loses its authority in large secularized societies, and becomes marginalized as a matter of "personal preference over which the community and the state have no right of judgment," it can easily become something of a "consumer good like cars and toothpaste."[116] Secularization, in the sense in which I am using the term, tends to be found in contexts that are characterized not simply by economic growth, but also by lower income disparity (and other forms of psycho-sociological dysfunction). The fact that every first world country that has universal health coverage also has low religiosity is no coincidence. "[T]here is a direct *cause and effect* mechanism in which pragmatic secular progressive socioeconomic polices that use government assistance to modulate capitalism suppress mass faith by suppressing the economic and societal disparity and insecurity that mass religion depends on."[117]

---

113   Barber, "Why Is Mississippi More Religious Than New Hampshire? Material Security and Ethnicity as Factors," *Cross-Cultural Research* 49, no. 3 (2015): 323.

114   Strulik, "An Economic Theory of Religious Belief," *Journal of Economic Behavior and Organization* 128 (2016).

115   Cutright et al., "Finding Brands and Losing Your Religion?" *Journal of Experimental Psychology: General* 143, no. 6 (2014).

116   Bruce, "Authority and Freedom: Economics and Secularization," in *Religions as Brands: New Perspectives on the Marketization of Religion and Spirituality*, ed. Usunier and Stolz (Surrey, UK: Ashgate, 2014): 203.

117   Paul, "The Evolution of Popular Religiosity and Secularism: How First World Statistics Reveal Why Religion Exists, Why It Has Been Popular, and Why the Most Successful Democracies Are the Most Secular," 175–176, emphasis added. See also Paul, "The Chronic Dependence of Popular Religiosity upon Dysfunctional Psychosociological Conditions."

If some mechanisms that promote sociographic promiscuity had a direct and positive causal impact on anthropomorphic prudery, we would also expect to find evidence that a high level of exposure to secularism during childhood or adolescence would predict higher levels of naturalism during adulthood. This is what we do find. Regression analyses of survey data show that individuals who had exposure to "credibility enhancing displays" (CREDs) in their families of origin were far more likely to have supernatural agent beliefs later in life. Conversely, individuals who were "exposed to especially low levels of religious CREDs were most likely to currently report a lack of belief in God with high certainty."[118] Another recent study using priming manipulation techniques and a dictator game reported similar results.[119] This indicates that a sociographically promiscuous social network can have a causal effect on individual anthropomorphic prudery. The causal connection between these theolytic forces is further confirmed by studies showing that young people with secular backgrounds are more likely to commit themselves to a (naturalistic) academic life.[120]

The impact of secular contexts on naturalistic tendencies is evident already in childhood. Experimental studies have shown, not too surprisingly, that children raised in religious backgrounds are far more likely to judge a protagonist in a religious story to be a real person, whereas secular children will tend to regard them as fictional. Perhaps more surprisingly, the inability to differentiate fact from fiction carries over into the interpretation of generally fantastical (not explicitly religious) narratives. Psychological experiments have also shown that children from non-religious backgrounds are much better at identifying fictional characters. From this, researchers have concluded that "it is more plausible that a religious upbringing overcomes children's pre-existing doubts about whether ordinarily impossible events can occur than that a secular upbringing suppresses children's natural inclination toward credulity."[121]

---

118   Lanman and Buhrmester, "Religious Actions Speak Louder than Words," *Religion, Brain & Behavior* 7, no.1 (2015): 10. See also Lanman, "The Importance of Religious Displays for Belief Acquisition and Secularization," *Journal of Contemporary Religion* 27, no. 1 (2012): 57.

119   Hitzeman and Wastell, "Are Atheists Implicit Theists?" *Journal of Cognition and Culture* 17, no. 1–2 (2017).

120   Beit-Hallahmi, "Explaining the Secularity of Academics: Historical Questions and Psychological Findings," *Science, Religion and Culture* 2, no. 3 (2015).

121   Corriveau et al., "Judgments About Fact and Fiction by Children From Religious and Nonreligious Backgrounds," *Cognitive Science* 39, no. 2 (March 1, 2015): 375. See also Davoodi

The results of another study focusing on five-year olds indicated that children are likely to consider religious stories about miracles or prophecies as "nothing more than fantastical events such as anthropomorphic animals that are featured in fairy tales" *unless* they are "inculcated with religious sentiments and information by adults."[122] Children have traditionally been considered naturally gullible, becoming suspicious through experience over time. When it comes to fantasy figures in religious stories, however, psychological evidence suggests that younger children are more skeptical but become credulous as they get older – an effect amplified by family religiosity.[123] In other words, children are normally born with extremely sensitive agency detection mechanisms, but they may only falsely detect *supernatural* agents (i.e. become anthropomorphically promicuous in the religious sense) if they are regularly primed by sociographically prudish parents and priests. Conversely, children raised by secular parents are more likely to be better at distinguishing between enjoyable fantasies and fiction, on the one hand, and naturalistic explanations of the world, on the other.

There also seems to be a "hydraulic relation" between people's beliefs in religious sources of control (e.g., supernatual agents) and their perception of governmental instability. Chronic or fluctuating levels of perceived lack of control can increase individual's levels of support for either governmental or religious systems, which indicates that both can serve as compensatory systems of control.[124] Evidence from China revealed a "substitution effect" between strong legal institutions and the prevalence of religion in a region; both variables predict reduced corruption.[125] Experimental manipulation studies have shown that triggering perceptions of the instability of government leads to increases in belief in God. When God is depicted as a source of order

et al., "Distinguishing between Realistic and Fantastical Figures in Iran" *Developmental Psychology* 52, no. 2 (2016).

122 Kotaman and Tekin, "The Impact of Religion on the Development of Young Children's Factuality Judgments," *North American Journal of Psychology* 17, no. 3 (2015): 8.

123 See, e.g., Woolley and Cox, "Development of Beliefs about Storybook Reality," *Developmental Science* 10, no. 5 (2007); Vaden and Woolley, "Does God Make It Real? Children's Belief in Religious Stories from the Judeo-Christian Tradition," *Child Development* 82, no. 4 (2011).

124 Kay et al., "God and the Government: Testing a Compensatory Control Mechanism for the Support of External Systems.," *Journal of Personality and Social Psychology* 95, no. 1 (2008): 18, and Kay et al., "Religious Belief as Compensatory Control," *Personality and Social Psychology Review* 14, no. 1 (2010).

125 Xu et al., "Does Religion Matter to Corruption? Evidence from China," *China Economic Review* 42 (2017).

and control, participants in a North American study were less likely to defend the legitimacy of their governments. Conversely, "increased perceptions of political stability led to weaker beliefs in a controlling God."[126] All of these findings support the idea that some of the component mechanisms that promote naturalism and secularism are causally linked in reciprocally reinforcing relationships.

As we have seen throughout this book, it is important to account for the way in which personal and situational variances mediate (or moderate) these kinds of relationships. For example, one study found that an individual's level of religious belief shapes the extent to which mortality salience leads to an increase in belief in "social-moral" progress. The researchers concluded that whereas more religious (especially Protestant) individuals might "focus on the promise of a better world in the form of a supernatural hereafter," when less religious individuals are prompted to search for existential meaning they can be triggered to focus on a better future world here and now.[127] Contextual differences matter too. Regression analyses of data from 137 countries showed that disbelief in God increased in contexts where the human population acquired greater existential security, measured by factors such as economic development, proportion of people enrolled in higher education, health security, and more equal distribution of income.[128]

Scandinavian countries are often held up as exemplars of societies that promote safe sects. These contexts are among the least religious on the planet, and yet their people

> enjoy high levels of existential security, strong and stable governments with social safety nets, and they no longer witness passionate displays of religiosity in the public sphere. These factors were likely *mutually reinforcing*: increases in existential security reduced motivations to attend religious services, in turn causing further declines of religious belief, leading to a *cascade of irreligion*. Furthermore, these societies have gradually and successfully replaced religion with effective *secular institutions* that encourage cooperation and enjoy very high levels of *science education*,

126   Kay et al., "For God (or) Country: The Hydraulic Relation Between Government Instability and Belief in Religious Sources of Control," *Journal of Personality and Social Psychology* 99, no. 5 (2010): 733–734.

127   Rutjens et al., "A March to a Better World? Religiosity and the Existential Function of Belief in Social-Moral Progress," *The International Journal for the Psychology of Religion* 26, no. 1 (2015).

128   Barber, "A Cross-National Test of the Uncertainty Hypothesis of Religious Belief," *Cross-Cultural Research* 45, no. 3 (2011).

which further encourages and reinforces analytic thinking that fosters religious skepticism.[129]

These countries are far from perfect. However, their inhabitants do experience higher levels of well-being and exhibit more universal prosocial behaviors than inhabitants of more religious countries. The safe(r) sects one finds in Scandinavian (and other secular) contexts is largely a consequence of high investments of energy into supporting educational policies that promote anthropomorphic prudery and social policies that promote sociographic promiscuity.

The reciprocally reinforcing quality of these efforts may help to explain why the "cascade of irreligion" appears to be spreading around the world. As we saw in Chapter 6, atheists have been around at least since the axial age. Until the last two or three centuries, however, they have always made up a very small percentage of the population. With the rise of modern science and the establishment of non-sectarian governments, their numbers have steadily increased. Historians are continuing to uncover the significant role that atheists played in the emergence and growth of more socially just organizations and in the production and acceptance of more materially adequate explanations.[130] These developments, in turn, have made it far easier for people in a wide variety of contexts to make sense of the world and act sensibly in society without appealing to supernatural agents and authorities.

In 2007 it was estimated that non-believers in God numbered as high as 749 million worldwide, making them the fourth largest group after Christianity (2 billion), Islam (1.2 billion) and Hinduism (900 million).[131] By 2014 the "Unaffiliated" had overtaken Hindus by climbing to over 1.1 billion, making up 16.45% of the world's population.[132] A recent synthesis of research on cross-national trends shows that rates of religious attendance are declining or have "bottomed out" in most countries in Europe, the Americas (including the u.s.), Australia, and New Zealand.[133] Measuring the rise of "non-religion" can be difficult, in

---

129   Norenzayan and Gervais, "The Origins of Religious Disbelief," 24. Emphases added.

130   Stephens, *Imagine There's No Heaven: How Atheism Helped Create the Modern World* (New York: Macmillan, 2014); Watson, *The Age of Atheists: How We Have Sought to Live Since the Death of God* (New York: Simon and Schuster, 2014).

131   Zuckerman, "Atheism: Contemporary Numbers and Patterns," in Martin, ed., *The Cambridge Companion to Atheism* (Cambridge University Press: 2007).

132   Pew Research Center, "The Future of World Religions: Population Growth Projections, 2010–2050" (Washington, DC: Pew Research Center, May 11, 2015). See also Win-Gallup, "Global Index of Religiosity and Atheism" (Win-Gallup, 2012).

133   Brenner, "Research Synthesis: Cross-National Trends in Religious Service Attendance," *Public Opinion Quarterly* 80, no. 2 (2016). For more specific analyses of the decline in the

part because of terminological ambiguities; atheist, unaffiliated, unbeliever, and non-attender do not all mean the same thing. One of the most important tasks in the growing field of research on non-religion is the clarification and operationalization of terms.[134] Moreover, in some countries saying you are a non-believer carries a death sentence, which can deter self-reports of atheism. Still, even in many places in the Arab world, for example, atheism is on the rise, even if less visibly.[135]

Segregative inscriptions based on superstitious interpretations of punitive (or rewarding) gods are becoming more and more problematic in our current pluralistic, globalizing environment. Younger people in particular seem far less interested in "doing it." Survey data on high school students indicates that the next generation of young adults will be even less approving of religious organizations and find religion even less important in their lives.[136] In every country included in the European Social Survey, each new generation cohort reported being less religious than the last.[137]

The United States is not the exception some people consider it to be. In fact, American religiosity has been "declining for decades, and ... that decline has been produced by the same generational patterns that lie behind religious decline elsewhere in the West: each successive cohort is less religious than the preceding one."[138] The American Religious Identification Survey showed

---

U.S. and Britain, see McCaffree, *The Secular Landscape: The Decline of Religion in America* (New York, Palgrave Macmillan, 2017), Bruce, "The Sociology of Late Secularization: Social Divisions and Religiosity," *British Journal of Sociology* 67, no. 4 (2016), and Bruce and Voas, "Do Social Crises Cause Religious Revivals? What British Church Adherence Rates Show," *Journal of Religion in Europe* 9, no. 1 (2016).

134   Lee, "Talking about a Revolution: Terminology for the New Field of Non-Religion Studies," *Journal of Contemporary Religion* 27, no. 1 (2012); Lee, "Non-Religion," in *The Oxford Handbook of the Study of Religion*, ed. Stausberg and Engler (New York: Oxford University Press, 2016); Bullivant, "Defining 'Atheism,'" in *The Oxford Handbook of Atheism*, ed. Bullivant and Ruse (New York: Oxford University Press, 2013); Cragun, "Defining That Which Is Other to Religion," in *Religion: Beyond Religion*, ed. Zuckerman (New York: Macmillan, 2016).

135   Benchemsi, "Invisible Atheists," *New Republic* 246, no. 4 (2015).

136   Twenge et al., "*Generational and Time Period Differences in American Adolescents' Religious Orientation, 1966–2014*," *PLoS ONE* 10, no. 5 (May 11, 2015).

137   Voas, "The Rise and Fall of Fuzzy Fidelity in Europe," *European Sociological Review* 25, no. 2 (2009): 167. See also Bruce, "Post-Secularity and Religion in Britain: An Empirical Assessment," *Journal of Contemporary Religion* 28, no. 3 (2013).

138   Voas and Chaves, "Is the United States a Counterexample to the Secularization Thesis?," *American Journal of Sociology* 121, no. 5 (2016): 1517.

that less than a third of college students in 2013 self-identified as religious.[139] Another recent study found that nearly 40% of young adults in America are religiously unaffiliated, which is three times the rate among seniors; this age gap has been widening for decades.[140] Even more recently, analysis of the 2016 American Values Atlas showed that the unaffiliated make up the largest "religious" group in the U.S. (at 24%, compared to the second largest group, white evangelical Protestants, at 17%).[141]

While non-believers in countries like the U.S. are not (usually) in danger of losing their lives for their lack of religiosity, the label "atheist" remains heavily stigmatized. In order to overcome the problem that some atheists might not be willing to disclose their disbelief in response to pollsters explicitly asking about "atheism," one recent study utilized Bayesian estimation analysis to generate an indirect estimate of the prevalence of atheism. The results indicated that as many as 26% of all Americans could be atheists, a percentage far higher than the 11% range typically reported by conventional polls.[142]

Is the spread of irreligiosity, fueled by the reciprocal reinforcement of anthropomorphic prudery and sociographic promiscuity, likely to continue? Whatever the answer may be, I hope that the scientific and philosophical perspectives I have summarized and presented in the preceding pages will convince many readers that the expansion of atheism is not as scary as our evolved cognitive and coalitional biases have led us to think. It turns out that most demographic models predict that the number of unbelievers (or unaffiliators) will continue to grow, even if birth rates (which strongly favor the religious) remain the same.[143] Mathematical forecasting models, which can more easily include other parameters and be calibrated more precisely

139   Kosmin and Keysar, "Religious, Spiritual and Secular: The Emergence of Three Distinct Worldviews among American College Students," *American Religious Identification Survey* (Hartford, CT: Trinity College, 2013).

140   Jones et al., "Exodus: Why Americans Are Leaving Religion – and Why They're Unlikely to Come Back" (Washington, D.C.: Public Religion Research Institute, September 22, 2016). See also Funk and Smith, "Nones on the Rise: One-in-Five Adults Have No Religious Affiliation," (Washington: Pew Research Center, 2012).

141   Jones and Cox, "America's Changing Religious Identity" (Public Religion Research Institute, September 6, 2017), https://www.prri.org/wp-content/uploads/2017/09/PRRI -Religion-Report.pdf.

142   Gervais and Najle, "How Many Atheists Are There?" *Social Psychological & Personality Science.*, 2017.

143   See, e.g., Smith and Baker, *American Secularism: Cultural Contours of Nonreligious Belief Systems* (New York: NYU Press, 2015), 87, and the other demographic reports mentioned above.

in relation to earlier datasets, tend to predict an even more rapid growth of non-religion.[144]

As we have seen throughout this book, however, the interactions among the many mechanisms at work in religious reproduction, and the variations manifested both at the micro-level of individuals and the macro-level of cultures, are exceedingly complicated. Understanding and explaining these interactions and variations requires insights and methodological tools from a multitude of academic disciplines. Managing all this complexity can be taxing on the human brain. The good news is that recent advances in computer modeling techniques are opening up new ways to explore complex adaptive social systems, thereby enhancing our predictive and adaptive capacities.

The (Methodological) Joy of (Simulating) Sects

Computational modeling and simulation techniques have been around for several decades, but their application within the social sciences really only began to take off in the 1980s and 1990s.[145] During the first decade and a half of this century, the use of these sorts of methodologies has spread rapidly in a wide variety of disciplines, including biology, archaeology, cognitive science, economics, and sociology.[146] The excitement around these developments is

---

144   See, e.g., Stinespring and Cragun, "Simple Markov Model for Estimating the Growth of Nonreligion in the United States," *Science, Religion and Culture* 2, no. 3 (2015), and Abrams et al., "Dynamics of Social Group Competition: Modeling the Decline of Religious Affiliation," *Physical Review Letters* 107, no. 8 (2011).

145   Axelrod, *The Evolution of Cooperation*, (New York: Basic Books, 1984); Axelrod, *The Complexity of Cooperation: Agent-Based Models of Competition and Collaboration* (Princeton University Press, 1997); Gilbert and Doran, *Simulating Societies. The Computer Simulation of Social Phenomena*, (London: UCL Press, 1994); Conte et al., "Introduction: Social Simulation – a New Disciplinary Synthesis," in *Simulating Social Phenomena* (Springer, 1997); Epstein and Axtell, *Growing Artificial Societies: Social Science from the Bottom Up* (New York: Brookings Institution Press, 1996); Kohler and Gumerman, *Dynamics in Human and Primate Societies: Agent-Based Modeling of Social and Spatial Processes* (Oxford University Press, 2000).

146   Wolfram, *A New Kind of Science* (Champaign, IL: Wolfram Media, 2002); Gilbert and Troitzsch, *Simulation for the Social Scientist* (Berkshire, UK: McGraw-Hill Education, 2005); Epstein, *Generative Social Science: Studies in Agent-Based Computational Modeling* (Princeton University Press, 2006); Sun, *Cognition and Multi-Agent Interaction: From Cognitive Modeling to Social Simulation* (Cambridge University Press, 2006); Squazzoni, *Agent-Based Computational Sociology* (New York: John Wiley & Sons, 2012); Hamill and Gilbert, *Agent-Based Modelling in Economics* (New York: John Wiley & Sons, 2015); Alvarez,

due not only to the explanatory and predictive power of computer modeling and simulation, but also to the way in which such efforts raise new questions about (and provide new answers to) issues that bear on classical philosophical debates about epistemology, intelligibility, and rationality, as well as metaphysics, emergence, and causality.[147] Its philosophical implications and scientific impact have even led some scholars to refer to computer modeling and simulation as the "third pillar" of science, alongside theory and experimentation.[148]

Like all innovative methodologies, the use of computational models and simulation experiments raises important ethical questions as well. Could this new technology be used in ways that could lead to more human suffering? Could a criminal mastermind (or mega-church pastor or totalitarian dictator) use simulations to manipulate large numbers of people? Modeling tools provide significantly more analytic and predictive power than the human race has had to deal with before and so these sorts of questions must be taken seriously. It is also important to acknowledge that people are *already* and *always* manipulating and being manipulated by simulations. This is how human thought and action work. Thinking about what to do (or what we would like others to do) necessarily involves the construction of mental models and simulations. This means that the latter are already at work in the engineering and policing of all human groups – including families, religious sects, and secular institutions.

One of the virtues of *computational* simulation methodologies, however, is that they only work well if one is exceptionally clear about the assumptions, inferences, mechanisms, implications, and *intentions* of the proposed model. As is the case with every other technological advancement, there is no guarantee that computer simulations will not be misused or abused, but at least the process by which successful models are constructed invites critical analysis and debate. Throughout this book we have explored a wide array of mechanisms (at multiple levels of analysis) that operate within complex adaptive theogonic systems. Given all of this multiplicity and complexity, computational

---

ed., *Computational Social Science: Discovery and Prediction* (Cambridge University Press, 2016).

147 See, e.g., Squazzoni, *Epistemological Aspects of Computer Simulation in the Social Sciences* (Springer, 2009); Winsberg, *Science in the Age of Computer Simulation* (University of Chicago Press, 2010); DeLanda, *Philosophy and Simulation: The Emergence of Synthetic Reason* (Bloomsbury Publishing, 2011); Tolk, ed. *Ontology, Epistemology, and Teleology for Modeling and Simulation* (Springer, 2013).

148 Yilmaz, ed., *Concepts and Methodologies for Modeling and Simulation* (New York: Springer, 2015); others refer to these developments as initiating a "fourth paradigm" for science; see Hey et al., *The Fourth Paradigm: Data-Intensive Scientific Discovery* (Redmond, Wash.: Microsoft Research, 2009).

methodologies may be one of our best options for fostering an open dialogue about the way in which god-bearing mechanisms have shaped human life in the past – and about the ways in which naturalism and secularism may shape human life in the future.

In addition to forcing more conceptual clarity and offering more computational processing power, modeling and simulation methodologies have several other virtues, including providing the capacity

· to develop causal architectures that incorporate the dynamics of both micro- and macro-level mechanisms,
· to construct and execute experiments in artificial societies that would not otherwise be feasible or ethical,
· to explain the emergence of a complex social phenomenon by "growing it" from the bottom-up,
· to integrate insights from qualitative and quantitative research within the same computational model,
· to shift the burden of proof in long-standing theoretical debates about the causal dynamics at work in historical events, and
· to explore the dynamic possibility space of a social system in order to determine the parametric and probabilistic conditions for specific configurations.

These are some of the reasons why computational modeling and simulation are so valuable when it comes to dealing with phenomena as complex and important as (non)religion.

The appropriation of these new methodologies to analyze and explore the mechanisms at work in the emergence, development, and dissolution of religious beliefs and behaviors is relatively recent. However, interest in their explanatory power and potential within the relevant fields is growing rapidly.[149] Several of the early applications of computer modeling to issues related to religion produced simulations of the dynamics involved in shifts in religious faith and prejudice in human populations.[150] Over the years a variety

---

149   See, e.g., Nielbo et al., "Computing Religion: A New Tool in the Multilevel Analysis of Religion," *Method and Theory in the Study of Religion* 24, no. 3 (2012): 271, and Lane, "Method, Theory, and Multi-Agent Artificial Intelligence: Creating Computer Models of Complex Social Interaction," *Journal for the Cognitive Science of Religion* 1, no. 2 (2014).

150   For a review, and several examples, see Bainbridge, *God from the Machine: Artificial Intelligence Models of Religious Cognition* (Lanham, MD: AltaMira Press, 2006), and Bainbridge, "Artificial Intelligence Models of Religious Evolution," in *Evolution, Religion and Cognitive Science*, ed. Watts (Oxford: Oxford University Press, 2014). See also Iannaccone and Makowsky, "Accidental Atheists? Agent-Based Explanations for the Persistence of

of computational models have explored other aspects of religion, including mechanisms that contribute to what I have been referring to as anthropomorphic promiscuity and sociographic prudery. Simulation methodologies have provided new insights into the operation and transmission of these evolved cognitive and coalitional tendencies, including the error and risk management strategies that characterize religious imagination, ritual practices, and social networks.[151] These insights in turn have contributed to the refinement of classical and contemporary theories of religion, as well as the generation of new hypotheses for investigation in future empirical research.

The international network of scholars interested in applying these methodologies within the scientific study of religion continues to grow. For example, several interrelated research projects utilizing computational modeling and simulation methods are currently underway as part of a collaboration between the Center for Modeling Social Systems in Kristiansand, Norway, the Center for Mind and Culture in Boston, Massachusetts, and the Virginia Modeling, Analysis, and Simulation Center in Suffolk, Virginia. Our team has already developed several system-dynamics models of the role of religion in major transformational periods in human civilization, including the Neolithic

---

Religious Regionalism," *Journal for the Scientific Study of Religion* 46, no. 1 (2007). For examples of the application of computer modeling to the study of phenomena more generally related to cooperation within and competition among religions, see Upal, "The Structure of False Social Beliefs," *Proceedings of the 2007 IEEE Symposiun on Artificial Life* (2007), Lindström and Olsson, "Mechanisms of Social Avoidance Learning Can Explain the Emergence of Adaptive and Arbitrary Behavioral Traditions in Humans," *Journal of Experimental Psychology: General* 144, no. 3 (2015), Sun and Fleischer, "A Cognitive Social Simulation of Tribal Survival Strategies: The Importance of Cognitive and Motivati onal Factors," *Journal of Cognition and Culture* 12, no. 3–4 (2012), Choi et al., "The Coevolution of Parochial Altruism and War," *Science* 318, no. 5850 (2007), and Chiang, "Good Samaritans in Networks: An Experiment on How Networks Influence Egalitarian Sharing and the Evolution of Inequality," *PLoS ONE* 10, no. 6 (2015).

151   Nielbo and Sørensen, "Attentional Resource Allocation and Cultural Modulation in a Computational Model of Ritualized Behavior," *Religion, Brain & Behavior* 6, no. 4 (2016); Whitehouse et al., "The Role for Simulations in Theory Construction for the Social Sciences: Case Studies Concerning Divergent Modes of Religiosity," *Religion, Brain & Behavior* 2, no.3 (2012); Dávid-Barrett and Carney, "The Deification of Historical Figures and the Emergence of Priesthoods as a Solution to a Network Coordination Problem," *Religion, Brain & Behavior*, 6, no. 4 (2016); Roitto, "Dangerous but Contagious Altruism: Recruitment of Group Members and Reform of Cooperation Style through Altruism in Two Modified Versions of Hammond and Axelrod's Simulation of Ethnocentric Cooperation," *Religion, Brain & Behavior* 6, no. 2 (2016); Matthews et al., "Cultural Inheritance or Cultural Diffusion of Religious Violence? A Quantitative Case Study of the Radical Reformation," *Religion, Brain & Behavior* 3, no. 1 (2013).

transition (to agriculture and sedentation), the axial age transition (to large-scale cultures involving priestly elites), and the modernity transition (to less reliance on supernatural beliefs).[152]

We are also working on a series of agent-based models that simulate and explore the mechanisms at work in religious belief and behavior. Our goal is to develop a standard computational model of religious cognition. This will involve the construction of several component models, each of which will simulate a major theory that deals with an important mechanism (or set of mechanisms) at work in religion. The first step was developing a computational model of terror management theory, in which we we were able to simulate the dynamic interaction between mortality salience and religiosity (discussed in Chapter 1 above). Our simulation experiments were able to "grow" the complex phenomena studied by that theory, to replicate the findings of other psychological experiments on terror management, and to identify new patterns in these dynamics that had previously not been detected.[153] The second step was a model that could simulate the dynamics of intergroup conflict, allowing us to shed light on some of the parametric conditions under which mutually escalating religious violence can emerge.[154]

We are currently developing other models that aim to simulate the cognitive and coalitional dynamics analyzed by several other influential theories in the scientific study of religion discussed in the preceeding chapters, including identity fusion theory, sacred values theory, costly signaling theory, ritual competence theory, and ritual modes theory. Some of our projects are explicitly oriented toward developing computational models that are relevant for public policy discussions about the role of religion and secularization in contemporary societies. Intergroup conflicts that are based on (or amplified by) differences among religious groups have contributed to mass migration, refugee crises, and challenges related to the integration of immigrants into pluralistic contexts. Our goal is to utilize simulation methodologies to shed light on the causes (and effects) of these cultural phenomena, and to provide new computational tools for evaluating hypotheses about – and policies for – societal change.

---

152   See Shults and Wildman, "Modeling Çatalhöyük: Simulating Religious Entanglement and Social Investment in the Neolithic," in *Religion, History and Place in the Origin of Settled Life*, ed. Hodder (Colorado Springs, CO: University of Colorado Press, in press). Publications based on the other systems-dynamics models are under review.

153   See Shults et al., "Modeling Terror Management Theory: Computer Simulations of the Impact of Mortality Salience on Religiosity" *Religion, Brain & Behavior* 8, no. 1 (2018).

154   Shults et al., "Mutually Escalating Religious Violence: A Generative and Predictive Computational Model," *Social Simulation Conference Proceedings*, 2017.

Another set of models are aimed more explicitly at simulating the conditions under which – and the mechanisms by which – variables related to naturalism and secularism can be promoted within human populations. The first of these involved the construction of a computational architecture with simulated agents whose variables include supernatural belief, religious practice, belief in God, level of education, and felt existential security. The initial distribution of these variables (at the beginning of each simulation) was based on exploratory factor analyses and structural equation models grounded in the International Social Survey Programme data set, and country level analyses of well-being and existential security from the Human Development Report. Validation of these experiments showed that our agent-based model was up to three times more accurate in predicting the real-world rise and fall of the relevant variables than its nearest competitor (generalized linear regression analysis).[155]

The capacity of our model to "grow" a macro-level shift toward non-religiosity in a population, a shift that is generated by (or emergent from) micro-level agent behaviors and social network interactions, strengthens the plausibility of the argument that education and existential security are mechanisms that promote atheism within a society. The next step is to expand this model to include variables related to cultural pluralism and freedom of expression, factors that have also been identified in the empirical literature as drivers of religious disbelief and disaffiliation. We are also in the process of constructing several other computer models of non-religion, which will try to shed light on the mechanisms that contribute to the emergence of higher proportions of analytic and altruistic atheists who are affiliated in healthy social networks within human populations, thereby promoting the sort of reflective naturalistic explanations and peaceful secular organizations necessary for addressing the serious adaptive challenges we face in our current global environment.

Can we learn to practice safe sects? Can we learn how to live together without bearing gods? For all of the reasons we have been exploring in this book, most people will strongly resist the dissolution of (their) religion. They will keep fighting (mentally or physically) to hold onto the supernatural agents ritually engaged by their religious in-groups. It is highly unlikely that an adaptive atheist strategy will be universally embraced any time soon. In fact, it is not at all clear whether there is a viable route to the widespread practice of safe sects. It is quite clear, however, that explicitly oppressing religion – or trying

---

155   For details, see Gore et al., "Forecasting Change in Religiosity and Existential Security with an Agent-Based Model," *Journal of Artificial Societies and Social Simulation* 21, no. 1 (2018), and Shults et al., "Why Do the Godless Prosper? Modeling the Cognitive and Coalitional Mechanisms That Promote Atheism," *Psychology of Religion and Spirituality*, forthcoming.

to force people to repress their desire to engage in religious sects – will only make things worse.[156] As we have seen, attacking religious beliefs or mocking religious behaviors can all too easily activate and amplify theistic credulity and conformity biases, which tragically lead to the further entrenchment of the worldview and in-group defense mechanisms that fuel resistance to naturalism and secularism.

What's an atheist to do? Our review of the empirical findings in the biocultural study of (non)religion suggests several promising strategies, such as providing existential security, better education, more encounters with ideological others, greater freedom of expression, and opportunities to reflect on the mechanisms by which gods are born(e) in minds and cultures. Like racism, classism, sexism, and other biases involving constructed realities, theism often begins to dissolve when people start to understand the covert cognitive and coalitional tendencies that drive its operation and see the deleterious effects it has on human life. For many people, however, debiasing can only begin to happen when they (and those whom they love) feel safe and are given enough space, time, and encouragement to think through their options. This is why I have been calling for more of us to have "the talk" about religious reproduction, its causes *and* its consequences. No doubt some people will get offended when we start discussing the explicit details. Others will feel liberated. But we can't keep putting it off. This may be one of the most important conversations of our generation.

---

156   Northmore-Ball and Evans, "Secularization versus Religious Revival in Eastern Europe: Church Institutional Resilience, State Repression and Divergent Paths," *Social Science Research* 57 (2016); Toft et al., *God's Century: Resurgent Religion and Global Politics* (Norton, 2011); Shah et al., *Rethinking Religion and World Affairs* (Oxford University Press, 2012); Grim and Finke, *The Price of Freedom Denied: Religious Persecution and Conflict in the Twenty-First Century* (Cambridge University Press, 2011).

Bibliography

Abbink, Klaus, Jordi Brandts, Benedikt Herrmann, and Henrik Orzen. "Parochial Altruism in Inter-Group Conflicts." *Economics Letters* 117, no. 1 (2012): 45–48.

Abrams, Daniel M., Haley A. Yaple, and Richard J. Wiener. "Dynamics of Social Group Competition: Modeling the Decline of Religious Affiliation." *Physical Review Letters* 107, no. 8 (2011): 1–4.

Ali, Saba Rasheed, and Owen J. Gaasedelen. "Religion, Social Class, and Counseling," in William Ming Liu, ed., *The Oxford Handbook of Social Class in Counseling*, Oxford University Press, 2013.

Alvarez, R. Michael, ed. *Computational Social Science: Discovery and Prediction.* Cambridge University Press, 2016.

Andersson, Gerhard. "Atheism and How It Is Perceived: Manipulation of, Bias against and Ways to Reduce the Bias." *Nordic Psychology* 68, no. 3 (2016): 194–203.

Andreoni, James, A. Abigail Payne, Justin Smith, and David Karp. "Diversity and Donations: The Effect of Religious and Ethnic Diversity on Charitable Giving." *Journal of Economic Behavior and Organization* 128 (2016): 47–58.

Aning, Kwesi, and Mustapha Abdallah. "Islamic Radicalisation and Violence in Ghana." *Conflict, Security & Development* 13, no. 2 (2013): 149–167.

Arnason, Johann P., S.N. Eisenstadt, and Bjorn Wittrock, eds. *Axial Civilizations And World History*. Leiden: Brill Academic Pub, 2004.

Arweck, Elisabeth, Stephen Bullivant, and Lois Lee. *Secularity and Non-Religion*. London: Routledge, 2014.

Astuti, Rita, and Paul L. Harris. "Understanding Mortality and the Life of the Ancestors in Rural Madagascar." *Cognitive Science* 32, no. 4 (2008): 713–740.

Atkinson, Quentin D. "Religion and Expanding the Cooperative Sphere in Kastom and Christian Villages on Tanna, Vanuatu." *Religion, Brain & Behavior* 7 (2017): 1–19.

Atkinson, Quentin D., and Pierrick Bourrat. "Beliefs about God, the Afterlife and Morality Support the Role of Supernatural Policing in Human Cooperation." *Evolution and Human Behavior* 32, no. 1 (2011): 41–49.

Atran, Scott. *Cognitive Foundations of Natural History: Towards an Anthropology of Science*. Cambridge: Cambridge University Press, 1993.

Atran, Scott. *In Gods We Trust: The Evolutionary Landscape of Religion*. Oxford: Oxford University Press, 2002.

Atran, Scott. *Talking to the Enemy: Faith, Brotherhood, and the (Un)Making of Terrorists*. New York: HarperCollins, 2010.

Atran, Scott. "The Devoted Actor: Unconditional Commitment and Intractable Conflict across Cultures." *Current Anthropology* 57, no. 13 (2016): 192–203.

Atran, Scott, and Joseph Henrich. "The Evolution of Religion: How Cognitive By-Products, Adaptive Learning Heuristics, Ritual Displays, and Group Competition Generate Deep Commitments to Prosocial Religion." *Biological Theory* 5 (2010): 18–30.

Avalos, Hector. "Religion and Scarcity: A New Theory for the Role of Religion in Violence." In *The Oxford Handbook of Religion and Violence*, edited by Mark Juergensmeyer, Margo Kitts, and Michael Jerryson, 554–570. Oxford: Oxford University Press, 2013.

Axelrod, Robert. *The Evolution of Cooperation*. New York: Basic Books, 1984.

Axelrod, Robert. *The Complexity of Cooperation: Agent-Based Models of Competition and Collaboration*, Princeton University Press, 1997.

Bahna, Vladimir. "Explaining Vampirism: Two Divergent Attractors of Dead Human Concepts." *Journal of Cognition and Culture* 15 (2015): 285–298.

Bainbridge, William. "Artificial Intelligence Models of Religious Evolution." In *Evolution, Religion and Cognitive Science*, edited by F. Watts, 219–237. Oxford: Oxford University Press, 2014.

Bainbridge, William. *God from the Machine: Artificial Intelligence Models of Religious Cognition*. Lanham, MD: AltaMira Press, 2006.

Banerjee, Konika, and Paul Bloom. "Why Did This Happen to Me? Religious Believers' and Non-Believers' Teleological Reasoning about Life Events." *Cognition* 133, no. 1 (2014): 277–303.

Banyasz, Alissa M., David M. Tokar, and Kevin P. Kaut. "Predicting Religious Ethnocentrism: Evidence for a Partial Mediation Model." *Psychology of Religion and Spirituality*, 8, no. 1 (2014): 25–34.

Barber, Justin. "Believing in a Purpose of Events: Cross-Cultural Evidence of Confusions in Core Knowledge." *Applied Cognitive Psychology* 28 (2014): 1–6.

Barber, N. "A Cross-National Test of the Uncertainty Hypothesis of Religious Belief." *Cross-Cultural Research* 45, no. 3 (2011): 318–333.

Barber, Nigel. "Why Is Mississippi More Religious Than New Hampshire? Material Security and Ethnicity as Factors." *Cross-Cultural Research* 49, no. 3 (2015): 315–325.

Barnes, Kirsten, and Nicholas J.S. Gibson. "Supernatural Agency: Individual Difference Predictors and Situational Correlates." *International Journal for the Psychology of Religion* 23, no. 1 (2013): 42–62.

Barrett, Justin. "Cognitive Science, Religion, and Theology." In *The Believing Primate: Scientific, Philosophical and Theological Essays on the Origin of Religion*, edited by Jeffrey Schloss and Michael Murray, 76–99. New York: Oxford University Press, 2009.

Barrett, Justin. "Cognitive Constraints on Hindu Concepts of the Divine," *Journal for the Scientific Study of Religion* 37, no. 4, (1998): 608–619.

Barrett, Justin. "Dumb Gods, Petitionary Prayer and the Cognitive Science of Religion." In *Current Approaches in the Cognitive Science of Religion*, edited by Ilkka Pyysiäinen and Veikko Anttonen, 93–109. New York: Continuum, 2002.

Barrett, Justin. *Why Would Anyone Believe in God?* Walnut Creek, CA: AltaMira Press, 2004.

Barrett, Justin L., and Frank C. Keil. "Conceptualizing a Nonnatural Entity: Anthropomorphism in God Concepts." *Cognitive Psychology* 31, no. 3 (1997): 219–247.

Barrett, Justin, and Roger Trigg. "Cognitive and Evolutionary Studies of Religion." In *The Roots of Religion: Exploring the Cognitive Science of Religion*, edited by Roger Trigg and Justin Barrett, 1–16. Surrey, UK: Ashgate, 2014.

Bartke, Stephan, and Reimund Schwarze. "Risk-Averse by Nation or by Religion? Some Insights on the Determinants of Individual Risk Attitudes." *SOEPpaper*, no. 131 (2008).

Basedau, Matthias, Birte Pfeiffer, and Johannes Vüllers. "Bad Religion? Religion, Collective Action, and the Onset of Armed Conflict in Developing Countries" Journal of Conflict Resolution 60, no. 2 (2016): 226–255.

Bastian, Brock, Paul Bain, Michael D. Buhrmester, Angel Gomez, Alexandra Vazquez, Clinton G. Knight, and William B. Swann. "Moral Vitalism: Seeing Good and Evil as Real, Agentic Forces" *Personality and Social Psychology Bulletin* 41, no. 8 (2015): 1069–1081.

Batara, Jame Bryan L., Pamela S. Franco, Mequia Angelo M. Quiachon, and Dianelle Rose M. Sembrero. "Effects of Religious Priming Concepts on Prosocial Behavior Towards Ingroup and Outgroup." *Europe's Journal of Psychology* 12, no. 4 (2016): 635–644.

Batson, Daniel. "Rational Processing or Rationalization? The Effect of Disconfirming Information on a Stated Religious Belief." *Journal of Personality and Social Psychology* 32, no. 1 (1975): 176–184.

Bechtel, William. "Looking down, around, and up: Mechanistic Explanation in Psychology." *Philosophical Psychology* 22, no. 5 (2009): 543–564.

Beit-Hallahmi, Benjamin. "Explaining the Secularity of Academics: Historical Questions and Psychological Findings." *Science, Religion and Culture* 2, no. 3 (2015): 104–119.

Beit-Hallahmi, Benjamin. "Morality and Immorality among the Irreligious." In *Atheism and Secularity*, vol. 1, edited by Phil Zuckerman, 113–148. New York: Praeger, 2009.

Bellah, Robert N. *Religion in Human Evolution: From the Paleolithic to the Axial Age.* Cambridge, MA: Harvard University Press, 2011.

Bénabou, Roland, Davide Ticchi, and Andrea Vindigni. "Religion and Innovation." *The American Economic Review* 105, no. 5 (2015): 346–351.

Benchemsi, A. "Invisible Atheists." *New Republic* 246, no. 4 (2015): 24–31.

Bergunder, Michael. "What Is Religion?" *Method & Theory in the Study of Religion* 26, no. 3 (2014): 246–286.

Blackford, Russell, and Udo Schuklenk. *50 Great Myths About Atheism.* Hoboken, NJ: Wiley-Blackwell, 2013.

Blanes, Ruy Llera, and Galina Oustinova-Stjepanovic, eds. *Being Godless: Ethnographies of Atheism and Non-Religion*. New York: Berghahn Books, 2017.

Blogowska, Joanna, Catherine Lambert, and Vassilis Saroglou. "Religious Prosociality and Aggression: It's Real." *Journal for the Scientific Study of Religion* 52, no. 3 (2013): 524–536.

Bloom, Paul. *Descartes' Baby: How the Science of Child Development Explains What Makes Us Human*. New York: Basic Books, 2005.

Bloom, Paul. "Religion, Morality, Evolution." *Annual Review of Psychology* 63 (2012): 179–199.

Bluemke, M., J. Jong, D. Grevenstein, I. Miklouší, and J. Halberstadt. "Measuring Cross-Cultural Supernatural Beliefs with Self- and Peer-Reports." *PLoS ONE* 11, no. 10 (2016): 1–31.

Boden, Matthew Tyler. "Supernatural Beliefs: Considered Adaptive and Associated with Psychological Benefits." *Personality and Individual Differences* 86 (2015): 227–231.

Boehm, Christopher. *Moral Origins: The Evolution of Virtue, Altruism, and Shame*. New York: Basic Books, 2012.

Boehm, Christopher. "The Moral Consequences of Social Selection." *Behaviour* 151, no. 2–3 (2014): 167–183.

Bohman, Andrea, and Mikael Hjerm. "How the Religious Context Affects the Relationship between Religiosity and Attitudes towards Immigration." *Ethnic and Racial Studies* 37, no. 6 (2014): 937–957.

Bormann, Nils-Christian, Lars-Erik Cederman, and Manuel Vogt. "Language, Religion, and Ethnic Civil War." *Journal of Conflict Resolution* 61, no. 4 (2017): 744–771.

Borum, Randy. "Radicalization into Violent Extremism I: A Review of Social Science Theories." *Journal of Strategic Security* 4, no. 4 (2011): 7–36.

Botero, Carlos A., Beth Gardner, Kathryn R. Kirby, Joseph Bulbulia, Michael C. Gavin, and Russell D. Gray. "The Ecology of Religious Beliefs." *Proceedings of the National Academy of Sciences* 111, no. 47 (2014): 16784–16789.

Boucher, Helen C., and Mary A. Millard. "Belief in Foreign Supernatural Agents as an Alternate Source of Control When Personal Control Is Threatened." *The International Journal for the Psychology of Religion* 26, no. 3 (2016): 193–211.

Boudry, Maarten, and Johan Braeckman. "How Convenient! The Epistemic Rationale of Self-Validating Belief Systems." *Philosophical Psychology* 25, no. 3 (2012): 341–364.

Boudry, Maarten, and Jerry Coyne. "Fakers, Fanatics, and False Dilemmas: Reply to Van Leeuwen." *Philosophical Psychology*, 29, no.4, (2016): 622–627.

Bourrat, Pierrick. "Origins and Evolution of Religion from a Darwinian Point of View: Synthesis of Different Theories." In *Handbook of Evolutionary Thinking in the Sciences*, 761–780. ed. T. Hearnes and P. Huneman. New York: Springer, 2015.

Bourrat, Pierrick, Quentin D. Atkinson, and Robin I.M. Dunbar. "Supernatural Punish-ment and Individual Social Compliance across Cultures." *Religion, Brain & Behavior* 1, no. 2 (2011): 119–134.

Bouvet, Romain, and Jean-François Bonnefon. "Non-Reflective Thinkers Are Predis-posed to Attribute Supernatural Causation to Uncanny Experiences." *Personality and Social Psychology Bulletin* 41, no. 7 (2015): 955–961.

Bowlby, John. *A Secure Base: Clinical Applications of Attachment Theory*. New York: Taylor & Francis, 2005.

Bowlby, John. *Attachment*. New York: Basic Books, 2008.

Boyer, Pascal. *Religion Explained: The Evolutionary Origins of Religious Thought*. Re-print edition. New York: Basic Books, 2002.

Boyer, Pascal. *The Fracture of An Illusion: Science And The Dissolution Of Religion. Frankfurt Templeton Lectures 2008*. Edited by Elisabeth Grab-Schmidt, Jan Parker, and Thomas M. Schmidt. Göttingen: Vandenhoeck & Ruprecht, 2010.

Boyer, Pascal. *The Naturalness of Religious Ideas: A Cognitive Theory of Religion*. Berke-ley: University of California Press, 1994.

Boyer, Pascal, Rengin Firat, and Florian Van Leeuwen. "Safety, Threat, and Stress in In-tergroup Relations" Perspectives on Psychological Science 10, no. 4 (2015): 434–450.

Brandt, Mark J., and Daryl R. Van Tongeren. "People Both High and Low on Religious Fundamentalism Are Prejudiced Toward Dissimilar Groups." *Journal of Personality and Social Psychology*, 112, no. 1 (2015): 76–97.

Braxton, Donald. "Policing Sex: Explaining Demons in the Cognitive Economies of Re-ligion." *Journal of Cognition and Culture* 8, no. 1–2 (2008): 117–134.

Brenner, P.S. "Exceptional Behavior or Exceptional Identity?" *Public Opinion Quarterly* 75, no. 1 (2011): 19–41.

Brenner, P.S. "Research Synthesis: Cross-National Trends in Religious Service Atten-dance." *Public Opinion Quarterly* 80, no. 2 (2016): 563–583.

Breslin, Michael J., and Christopher A. Lewis. "Schizotypy and Religiosity: The Magic of Prayer." *Archive for the Psychology of Religion* 37, no. 1 (2015): 84–97.

Brewster, Melanie E., Matthew A. Robinson, Riddhi Sandil, Jessica Esposito, and Eliza-beth Geiger. "Arrantly Absent: Atheism in Psychological Science from 2001 to 2012." *The Counseling Psychologist* 42, no. 5 (2014): 628–663.

Bronkhorst, Johannes. "Can Religion Be Explained? The Role of Absorption in Various Religious Phenomena." *Method & Theory In The Study Of Religion* 29 (2017): 1–30.

Brooks, Alison Wood, Juliann Schroeder, Jane L. Risen, Francesca Gino, Adam Galinsky, Michael Norton, and Maurice E. Schweiter. "Don't Stop Believing: Rituals Improve Performance by Decreasing Anxiety." *Organizational Behavior & Human Decision Processes* 137 (2016): 71–85

Brownlee, Matthew T.J., Robert B. Powell, and Jeffery C. Hallo. "A Review of the Foun-dational Processes That Influence Beliefs in Climate Change: Opportunities for

Environmental Education Research." *Environmental Education Research*, no. 1 (2013): 1–20.

Bruce, Steve. "Authority and Freedom: Economics and Secularization." In *Religions as Brands: New Perspectives on the Marketization of Religion and Spirituality*, edited by Jean-Claude Usunier and Jorg Stolz, 191–204. Surrey, UK: Ashgate, 2014.

Bruce, Steve. "Post-Secularity and Religion in Britain: An Empirical Assessment." *Journal of Contemporary Religion* 28, no. 3 (2013): 369–384.

Bruce, Steve. "The Sociology of Late Secularization: Social Divisions and Religiosity." *British Journal of Sociology* 67, no. 4 (2016): 613–631.

Bruce, Steve, and David Voas. "Do Social Crises Cause Religious Revivals? What British Church Adherence Rates Show." *Journal of Religion in Europe* 9, no. 1 (2016): 26–43.

Bulbulia, Joseph. "Nature's Medicine: Religiosity as an Adaptation for Health and Co-operation." In *Where God and Science Meet*, edited by Patrick C. McNamara and Wesley J. Wildman, 87–121. New York: Praeger, 2006.

Bulbulia, Joseph. "Religious Costs as Adaptations That Signal Altruistic Intention." *Evolution and Cognition* 10, no. 1 (2004): 19–38.

Bulbulia, Joseph. "Spreading Order: Religion, Cooperative Niche Construction, and Risky Coordination Problems." *Biology & Philosophy* 27, no. 1 (2012): 1–27.

Bulbulia, Joseph, and Uffe Schjoedt. "Religious Culture and Cooperative Prediction under Risk: Perspectives from Social Neuroscience." In *Religion, Economy, and Cooperation*, edited by Illka Pyysiainen, 35–59. New York : De Gruyter, 2010.

Bulbulia, Joseph, and Richard Sosis. "Signalling Theory and the Evolution of Religious Cooperation." *Religion* 41, no. 3 (2011): 363–388.

Bulbulia, Joseph, Armin Geertz, Quentin Atkinson, Emma Cohen, Nicholas Evans, Pieter Francois, and Herbert Gintis, "The Cultural Evolution of Religion." In *Cultural Evolution: Society, Technology, Language and Religion*, edited by Peter Richerson and Morton Christiansen, 381–404. Cambridge: MIT Press, 2013.

Bulbulia, Joseph, et al. *The Evolution of Religion: Studies, Theories, & Critiques*. Santa Margarita, Calif: Collins Foundation Press, 2008.

Bulkeley, Kelly. *Big Dreams: The Science of Dreaming and the Origins of Religion*. New York: Oxford University Press, 2016.

Bullivant, Stephen. "Defining 'Atheism.'" In *The Oxford Handbook of Atheism*, edited by Stephen Bullivant and Michael Ruse, 11–21. New York: Oxford University Press, 2013.

Bullivant, Stephen, and Lois Lee. "Interdisciplinary Studies of Non-Religion and Secularity: The State of the Union." *Journal of Contemporary Religion* 27, no. 1 (2012): 19–27.

Burris, Christopher T., and Raluca Petrican. "Hearts Strangely Warmed (and Cooled): Emotional Experience in Religious and Atheistic Individuals." *International Journal for the Psychology of Religion* 21, no. 3 (2011): 183–197.

Bushman, Brad J., Robert D. Ridge, Enny Das, Colin W. Key, and Gregory L. Busath. "When God Sanctions Killing: Effect of Scriptural Violence on Aggression." *Psychological Science* 18, no. 3 (2007): 204–207.

Buxant, Coralie, and Vassilis Saroglou. "Feeling Good, but Lacking Autonomy: Closed-Mindedness on Social and Moral Issues in New Religious Movements." *Journal of Religion and Health* 47, no. 1 (2008): 17–31.

Caldwell-Harris, Catherine L. "Understanding Atheism/Non-Belief as an Expected Individual-Differences Variable." *Religion, Brain & Behavior* 2, no. 1 (2012): 4–23.

Caldwell-Harris, Catherine L., Angela L. Wilson, Elizabeth Lotempio, and Benjamin Beit-Hallahmi. "Exploring the Atheist Personality: Well-Being, Awe, and Magical Thinking in Atheists, Buddhists, and Christians." *Mental Health, Religion & Culture* 14, no. 7 (2011): 659–672.

Campbell, Troy H., and Aaron C. Kay. "Solution Aversion: On the Relation Between Ideology and Motivated Disbelief." *Journal of Personality and Social Psychology* 107, no. 5 (2014): 809–824.

Campbell, Maggie, and Johanna Ray Vollhardt. "Fighting the Good Fight: The Relationship between Belief in Evil and Support for Violent Policies." *Personality and Social Psychology Bulletin* 40, no. 1 (2014): 16–33.

Carroll, Joseph, Mathias Clasen, Emelie Jonsson, Alexandra Regina Kratschmer, Luseadra Mckerracher, Felix Riede, Jens-Christian Svenning, and Peter C. Kjærgaard. "Biocultural Theory: The Current State of Knowledge." *Evolutionary Behavioral Sciences*, 11, no. 1 (2015): 1–15.

Cauvin, Jacques. *The Birth of the Gods and the Origins of Agriculture*. Translated by T. Watkins. Cambridge University Press, 2000.

Cavrak, Sarah E., and Heather M. Kleider-Offutt. "Pictures Are Worth a Thousand Words and a Moral Decision or Two: Religious Symbols Prime Moral Judgments." *The International Journal for the Psychology of Religion* 25, no. 3 (2015): 173–192.

Chambers, Carissa. "Religiosity and Modern Prejudice: Points of Convergence and Points of Departure." Dissertation. Columbia University, 2016.

Chen, Xiang. "Why Do People Misunderstand Climate Change? Heuristics, Mental Models and Ontological Assumptions." *Climatic Change* 108, no. 1 (2011): 31–46.

Chiang, Yen-Sheng. "Good Samaritans in Networks: An Experiment on How Networks Influence Egalitarian Sharing and the Evolution of Inequality." *PLoS ONE* 10, no. 6 (2015): 1–13.

Choi, Jung-Kyoo, Samuel Bowles, and Samuel Bowles. "The Coevolution of Parochial Altruism and War." *Science* 318, no. 5850 (2007): 636–640.

Choma, Becky L., Reeshma Haji, Gordon Hodson, and Mark Hoffarth. "Avoiding Cultural Contamination: Intergroup Disgust Sensitivity and Religious Identification as Predictors of Interfaith Threat, Faith-Based Policies, and Islamophobia." *Personality and Individual Differences* 95 (2016): 50–55.

Chon, Don Soo. "Religiosity and Regional Variation of Lethal Violence Integrated Model." *Homicide Studies* 20, no. 2 (2016): 129–149.

Christopher Smith, and Richard Cimino. "Atheisms Unbound: The Role of the New Media in the Formation of a Secularist Identity." *Secularism and Nonreligion* 1 (2012): 17–31.

Chuah, Swee Hoon, Simon Gächter, Robert Hoffmann, and Jonathan H.W. Tan. "Religion, Discrimination and Trust across Three Cultures." *European Economic Review* 90 (2016): 280–301.

Chuah, Swee-Hoon, Robert Hoffmann, Bala Ramasamy, and Jonathan H.W. Tan. "Religion, Ethnicity and Cooperation: An Experimental Study." *Journal of Economic Psychology* 45 (2014): 33–43.

Ciftci, Sabri, Muhammad Asif Nawaz, and Tareq Sydiq. "Globalization, Contact, and Religious Identity: A Cross-National Analysis of Interreligious Favorability." *Social Science Quarterly* 97, no. 2 (2016): 271–292.

Cimino, Richard, and Christopher Smith. "Atheism, Class, and the Ghost of Karl Marx." In *Sociology of Atheism*, edited by Roberto Cipriani and Franco Garelli, 36–49. Leiden: Brill, 2016.

Cimino, Richard, and Christopher Smith. *Atheist Awakening: Secular Activism and Community in America*. Oxford: Oxford University Press, 2014.

Cipriani, Roberto, and Franco Garelli. *Sociology of Atheism*. Annual Review of the Sociology of Religion. Leiden: Brill, 2016.

Claidiere, N., T.C. Scott-Phillips, and D. Sperber. "How Darwinian Is Cultural Evolution?" *Philosophical Transactions of the Royal Society B: Biological Sciences* 369, no. 1642 (2014): 1–8.

Clarke, Steve, Russell Powell, and Julian Savulescu. "Religion, Intolerance, and Conflict: Practical Implications for Social Policy." In *Religion, Intolerance, and Conflict: A Scientific and Conceptual Investigation*, edited by Steve Clarke, Russell Powell, and Julian Savulescu, 266–272. Oxford University Press, 2013.

Clobert, Magali, Vassilis Saroglou, and Kwang-Kuo Hwang. "East Asian Religious Tolerance versus Western Monotheist Prejudice: The Role of (in)Tolerance of Contradiction." *Group Processes & Intergroup Relations* 20, no. 2 (2017): 216–232.

Coccia, Mario. "Socio-Cultural Origins of the Patterns of Technological Innovation: What Is the Likely Interaction among Religious Culture, Religious Plurality and Innovation? Towards a Theory of Socio-Cultural Drivers of the Patterns of Technological Innovation." *Technology in Society* 36 (2014): 13.

Cohen, Emma, Roger Mundry, and Sebastian Kirschner. "Religion, Synchrony, and Cooperation." *Religion, Brain & Behavior* 4, no. 1 (2014): 20–30.

Coleman III, Thomas J., Christopher F. Silver, and Jenny Holcombe. "Focusing on Horizontal Transcendence: Much More than a 'Non-Belief.'" *Essays in the Philosophy of Humanism* 21, no. 2 (2014): 1–18.

Collier, Mark. "The Natural Foundations of Religion." *Philosophical Psychology* 27, no. 5 (2014): 665–680.

Colzato, Lorenza S., Ilja Van Beest, Wery P.M. Van Den Wildenberg, Claudia Scorolli, Shirley Dorchin, Nachshon Meiran, Anna M. Borghi, and Bernhard Hommel. "God: Do I Have Your Attention?" *Cognition* 117, no. 1 (2010): 87–94.

Colzato, Lorenza S., Wery P.M. Van Den Wildenberg, Bernhard Hommel, and Antonio Verdejo García. "Losing the Big Picture: How Religion May Control Visual Attention." *PLoS ONE* 3, no. 11 (2008): 87–94.

Conte, Rosaria, Rainer Hegselmann, and Pietro Terna. "Introduction: Social Simulation – a New Disciplinary Synthesis." In *Simulating Social Phenomena*, edited by Rosaria Conte, Rainer Hegselmann, and Pietro Terna, 1–17. New York: Springer, 1997.

Corcoran, Katie E., David Pettinicchio, and Blaine Robbins. "A Double-Edged Sword: The Countervailing Effects of Religion on Cross-National Violent Crime." *Social Science Quarterly* (2017): 1–13.

Corriveau, Kathleen H., Eva E. Chen, and Paul L. Harris. "Judgments About Fact and Fiction by Children From Religious and Nonreligious Backgrounds." *Cognitive Science* 39, no. 2 (2015a): 353–382.

Cragun, Ryan. "Defining That Which Is Other to Religion." In *Religion: Beyond Religion*, edited by Phil Zuckerman, 1–16. New York: Macmillan, 2016.

Cragun, Ryan, and J.E. Sumerau. "God May Save Your Life, but You Have to Find Your Own Keys: Religious Attributions, Secular Attributions, and Religious Priming." *Archive For The Psychology Of Religion-Archiv Fur Religionspsychologie* 37, no. 3 (2015): 321–342.

Cragun, Ryan T., Victoria L. Blyde, J.E. Sumerau, Marcus Mann, and Joseph H. Hammer. "Perceived Marginalization, Educational Contexts, and (Non)Religious Educational Experience." *Journal of College and Character* 17, no. 4 (2016): 241–254.

Cragun, Ryan T., Barry Kosmin, Ariela Keysar, Joseph H. Hammer, and Michael Nielsen. "On the Receiving End: Discrimination toward the Non-Religious in the United States." *Journal of Contemporary Religion* 27, no. 1 (2012): 105–127.

Crescentini, Cristiano, Salvatore M. Aglioti, Franco Fabbro, and Cosimo Urgesi. "Virtual Lesions of the Inferior Parietal Cortex Induce Fast Changes of Implicit Religiousness/Spirituality." *Cortex* 54 (2014): 1–15.

Crescentini, Cristiano, Marilena Di Bucchianico, Franco Fabbro, and Cosimo Urgesi. "Excitatory Stimulation of the Right Inferior Parietal Cortex Lessens Implicit Religiousness/Spirituality." *Neuropsychologia* 70 (2015): 71–79.

Crespi, Bernard, and Kyle Summers. "Inclusive Fitness Theory for the Evolution of Religion." *Animal Behaviour* 92 (2014): 313–323.

Cutright, Keisha M., Tülin Erdem, Gavan J. Fitzsimons, and Ron Shachar. "Finding Brands and Losing Your Religion?" *Journal of Experimental Psychology: General* 143, no. 6 (2014): 2209–2222.

Dagnall, Neil, Kenneth Drinkwater, Andrew Parker, and Kevin Rowley. "Misperception of Chance, Conjunction, Belief in the Paranormal and Reality Testing: A Reappraisal." *Applied Cognitive Psychology* 28, no. 5 (2014): 711–719.

Dávid-Barrett, Tamás, and James Carney. "The Deification of Historical Figures and the Emergence of Priesthoods as a Solution to a Network Coordination Problem." *Religion, Brain & Behavior* 6, no. 4 (2016): 307–317.

Davies, Martin F., Murray Griffin, and Sue Vice. "Affective Reactions to Auditory Hallucinations in Psychotic, Evangelical and Control Groups." *British Journal of Clinical Psychology* 40, no. 4 (2001): 361–370.

Davis, Taylor. "Group Selection in the Evolution of Religion: Genetic Evolution or Cultural Evolution?" *Journal of Cognition and Culture* 15 (2015): 235–253.

Davis, Taylor. "The Goldberg Exaptation Model: Integrating Adaptation and By-Product Theories of Religion." *Review of Philosophy and Psychology* 8, no.3 (2017): 687–708.

De Cruz, Helen, and Johan De Smedt. *A Natural History of Natural Theology: The Cognitive Science of Theology and Philosophy of Religion*. Cambridge, MA: The MIT Press, 2014.

De Cruz, Helen, and Johan De Smedt. "Evolved Cognitive Biases and the Epistemic Status of Scientific Beliefs." *Philosophical Studies* 157, no. 3 (2012): 411–429.

De Cruz, Helen, and Johan De Smedt. "Naturalizing Natural Theology." *Religion, Brain & Behavior* 6, no. 4 (2016): 20–26.

De Muckadell, Caroline Schaffalitzky. "On Essentialism and Real Definitions of Religion." *Journal of the American Academy of Religion* 82, no. 2 (2014): 495–520.

Deane-Drummond, Celia. *Christ and Evolution: Wonder and Wisdom*. Augsburg Fortress, 2009.

Deane-Drummond, Celia. "Public Theology as Contested Ground: Arguments for Climate Justice." In *Religion and Ecology in the Public Sphere*. New York: T&T Clark, 2011.

Decety, Jean, Jason M. Cowell, Kang Lee, Randa Mahasneh, Susan Malcolm-Smith, Bilge Selcuk, and Xinyue Zhou. "The Negative Association between Religiousness and Children's Altruism across the World." *Current Biology* 25, no. 22 (2015): 2951–2955.

Delamontagne, R. Georges. "High Religiosity and Societal Dysfunction in the United States during the First Decade of the Twenty-First Century." *Evolutionary Psychology* 8, no. 4 (2010): 617–657.

DeLanda, Manuel. *Philosophy and Simulation: The Emergence of Synthetic Reason*. Bloomsbury Publishing, 2011.

Deleuze, Gilles. *Difference and Repetition*. Translated by Paul Patton. Revised ed. New York: Columbia University Press, 1995.

Deleuze, Gilles. *Empiricism and Subjectivity*. Translated by Constantin V. Boundas. New York: Columbia University Press, 2001.

Deleuze, Gilles. *Essays Critical And Clinical*. Minneapolis: University of Minnesota Press, 1997.

Deleuze, Gilles. *Expressionism in Philosophy: Spinoza*. Translated by Martin Joughin. New York : Cambridge, Mass: Zone Books, 1992.

Deleuze, Gilles. *Kant's Critical Philosophy: The Doctrine of the Faculties*. Minneapolis: University of Minnesota Press, 1985.

Deleuze, Gilles. *Nietzsche and Philosophy*. Translated by Janis Tomlinson. New York: Columbia University Press, 1983.

Deleuze, Gilles. *The Logic of Sense*. New York: Continuum, 2004.

Deleuze, Gilles. *Two Regimes of Madness: Texts and Interviews 1975–1995*. Semiotext, 2007.

Deleuze, Gilles, and Félix Guattari. *A Thousand Plateaus: Capitalism and Schizophrenia*. New York: Continuum, 2004.

Deleuze, Gilles, and Félix Guattari. *Anti-Oedipus: Capitalism and Schizophrenia*. Minneapolis: University of Minnesota Press, 1983.

Deleuze, Gilles, and Félix Guattari. *What Is Philosophy?* Translated by Hugh Tomlinson and Graham Burchell. Columbia University Press, 1996.

Dengah, H.J. François. "Being Part of the Nação: Examining Costly Religious Rituals in a Brazilian Neo-Pentecostal Church." *Ethos* 45, no. 1 (2017): 48–74.

Devine-Wright, Patrick, Jennifer Price, and Zoe Leviston. "My Country or My Planet? Exploring the Influence of Multiple Place Attachments and Ideological Beliefs upon Climate Change Attitudes and Opinions." *Global Environmental Change* 30 (2015): 68–79.

de Waal, Frans, ed. *Evolved Morality: The Biology and Philosophy of Human Conscience*. Leiden: Brill Academic Publishers, 2014.

Diener, Ed, Louis Tay, and David G. Myers. "The Religion Paradox: If Religion Makes People Happy, Why Are So Many Dropping Out?" *Journal of Personality and Social Psychology* 101, no. 6 (2011): 1278–1290.

Doane, Michael J., and Marta Elliott. "Perceptions of Discrimination Among Atheists: Consequences for Atheist Identification, Psychological and Physical Well-Being." *Psychology of Religion and Spirituality* 7, no. 2 (2015): 130–141.

Doebler, Stefanie. "Love Thy Neighbor? Relationships between Religion and Racial Intolerance in Europe." Politics and Religion 8, no. 4 (2015): 745–771.

Doebler, Stefanie. "Relationships Between Religion and Intolerance Towards Muslims and Immigrants in Europe: A Multilevel Analysis." *Review of Religious Research* 56, no. 1 (2014): 61–86.

Draper, Scott. "Effervescence and Solidarity in Religious Organizations." *Journal for the Scientific Study of Religion* 53, no. 2 (2014): 229–248.

Draper, Paul, and Ryan Nichols. "Diagnosing Bias in Philosophy of Religion." *The Monist* 96, no. 3 (2013): 420–446.

Du, Hongfei, and Peilian Chi. "War, Worries, and Religiousness." *Social Psychological and Personality Science* 7, no. 5 (2016): 444–451.

Dunham, Yarrow, Mahesh Srinivasan, Ron Dotsch, and David Barner. "Religion Insulates Ingroup Evaluations: The Development of Intergroup Attitudes in India." *Developmental Science* 17, no. 2 (2014): 311–319.

Dunkel, Curtis S., and Edward Dutton. "Religiosity as a Predictor of In-Group Favoritism within and between Religious Groups." *Personality and Individual Differences* 98 (2016): 311–314.

Dutton, Edward, and Dimitri van der Linden. "Why Is Intelligence Negatively Associated with Religiousness?" *Evolutionary Psychological Science* 3, no. 4 (2017): 392–403.

Edgell, Penny, Jacqui Frost, and Evan Stewart. "From Existential to Social Understandings of Risk: Examining Gender Differences in Nonreligion." *Social Currents* 4, no. 6 (2017): 556–574.

Edgell, Penny, Douglas Hartmann, Evan Stewart, and Joseph Gerteis. "Atheists and Other Cultural Outsiders: Moral Boundaries and the Non-Religious in the United States." *Social Forces* 95, no. 2 (2017b): 607–638.

Eilam, David, Rony Izhar, and Joel Mort. "Threat Detection: Behavioral Practices in Animals and Humans." *Neuroscience & Biobehavioral Reviews* 35, no. 4 (2011): 999–1006.

Eisenstadt, S.N., ed. *The Origins and Diversity of Axial Age Civilizations.* State University of New York Press, 1986.

Ellis, Lee. "Religious Variations in Fundamentalism in Malaysia and the United States: Possible Relevance to Religiously Motivated Violence." *Personality and Individual Differences* 107 (2017): 23–27.

Ellis, Lee, Anthony W. Hoskin, Edward Dutton, and Helmuth Nyborg. "The Future of Secularism: A Biologically Informed Theory Supplemented with Cross-Cultural Evidence." *Evolutionary Psychological Science* 3, no.3 (2017): 224–242.

Epley, Nicholas, Scott Akalis, Adam Waytz, and John T. Cacioppo. "Creating Social Connection through Inferential Reproduction: Loneliness and Perceived Agency in Gadgets, Gods, and Greyhounds." *Psychological Science* 19, no. 2 (2008): 114–120.

Epley, Nicholas, Benjamin A. Converse, Alexa Delbosc, George A. Monteleone, and John T. Cacioppo. "Believers' Estimates of God's Beliefs Are More Egocentric than Estimates of Other People's Beliefs." *Proceedings of the National Academy of Sciences* 106, no. 51 (2009): 21533–21538.

Epstein, Greg. *Good Without God: What a Billion Nonreligious People Do Believe.* New York: William Morrow Paperbacks, 2010.

Epstein, Joshua M. *Generative Social Science: Studies in Agent-Based Computational Modeling.* Princeton University Press, 2006.

Epstein, Joshua M., and Robert Axtell. *Growing Artificial Societies: Social Science from the Bottom Up.* New York: Brookings Institution Press, 1996.

Fenelon, Andrew, and Sabrina Danielsen. "Leaving My Religion: Understanding the Relationship between Religious Disaffiliation, Health, and Well-Being." *Social Science Research* 57 (2016): 49–62.

Fergus, Thomas A., and Wade C. Rowatt. "Uncertainty, God, and Scrupulosity: Uncertainty Salience and Priming God Concepts Interact to Cause Greater Fears of Sin." *Journal of Behavior Therapy and Experimental Psychiatry* 46 (2015): 93–98.

Fessler, Daniel M.T., Anne C. Pisor, and Carlos David Navarrete. "Negatively-Biased Credulity and the Cultural Evolution of Beliefs.(Report)." *PLoS ONE* 9, no. 4 (2014): 1–8.

Fincher, Corey L., and Randy Thornhill. "Parasite-Stress Promotes in-Group Assortative Sociality: The Cases of Strong Family Ties and Heightened Religiosity." *Behavioral and Brain Sciences* 35, no. 2 (2012): 61–79.

Fischer, Ronald, Rohan Callander, Paul Reddish, and Joseph Bulbulia. "How Do Rituals Affect Cooperation?" *Human Nature* 24, no. 2 (2013): 115–125.

Friesen, Amanda, and Aleksander Ksiazkiewicz. "Do Political Attitudes and Religiosity Share a Genetic Path?" *Political Behavior* 37, no. 4 (2015): 791–818.

Friesen, Justin, Troy H. Campbell, and Aaron C. Kay. "The Psychological Advantage of Unfalsifiability: The Appeal of Untestable Religious and Political Ideologies." *Journal of Personality and Social Psychology* 108, no. 3 (2015): 515–529.

Fuller, R.C., and D.E. Montgomery. "Body Posture and Religious Attitudes." *Archive For The Psychology Of Religion-Archiv Fur Religionspsychologie* 37, no. 3 (2015): 227–239.

Funk, Cary, and Greg Smith. "Nones on the Rise: One-in-Five Adults Have No Religious Affiliation." *Washington: Pew Research Center*, 2012.

Fux, Michal. "Cultural Transmission of Precautionary Ideas: The Weighted Role of Implicit Motivation." *Journal of Cognition and Culture* 16, no. 5 (2016): 415–435.

Galen, Luke. "Atheism, Wellbeing, and the Wager: Why Not Believing in God (With Others) Is Good for You." *Science, Religion and Culture* 2, no. 3 (2015): 54–69.

Galen, Luke. "Overlapping Mental Magisteria: Implications of Experimental Psychology for a Theory of Religious Belief as Misattribution," *Method & Theory in the Study of Religion* 29, no.3 (2017): 221–267.

Galen, Luke, Michael Sharp, and Alison McNulty. "Nonreligious Group Factors Versus Religious Belief in the Prediction of Prosociality." *Social Indicators Research* 122, no. 2 (2015): 411–432.

Galen, Luke, and James D. Kloet. "Mental Well-Being in the Religious and the Non-Religious: Evidence for a Curvilinear Relationship." *Mental Health, Religion & Culture* 14, no. 7 (2011): 673–689.

Gantman, Ana P., and Jay J. Van Bavel. "The Moral Pop-out Effect: Enhanced Perceptual Awareness of Morally Relevant Stimuli." *Cognition* 132, no. 1 (2014): 22–29.

Ganzach, Yoav, Shmuel Ellis, and Chemi Gotlibovski. "On Intelligence Education and Religious Beliefs." *Intelligence* 41, no. 2 (2013): 121–128.

Ganzach, Yoav, and Chemi Gotlibovski. "Intelligence and Religiosity: Within Families and over Time." *Intelligence* 41, no. 5 (2013): 546–552.

Garcia, Hector A. *Alpha God: The Psychology of Religious Violence and Oppression*. Amherst, New York: Prometheus Books, 2015.

Gearing, Robin Edward, Dana Alonzo, Alex Smolak, Katie McHugh, Sherelle Harmon, and Susanna Baldwin. "Association of Religion with Delusions and Hallucinations in the Context of Schizophrenia: Implications for Engagement and Adherence." *Schizophrenia Research* 126, no. 1 (2011): 150–163.

Gebauer, Jochen E., and Gregory R. Maio. "The Need to Belong Can Motivate Belief in God." *Journal of Personality* 80, no. 2 (2012): 465–501.

Gebauer, Jochen E., Delroy Paulhus, and Wiebke Neberich. "Big Two Personality and Religiosity Across Cultures: Communals as Religious Conformists and Agentics as Religious Contrarians." *Social Psychological and Personality Science* 4, no. 1 (2013): 21–30.

Gebauer, Jochen E., Wiebke Bleidorn, Samuel D. Gosling, Peter J. Rentfrow, Michael E. Lamb, and Jeff Potter. "Cross-Cultural Variations in Big Five Relationships With Religiosity: A Sociocultural Motives Perspective." *Journal of Personality and Social Psychology* 107, no. 6 (2014): 1064–1091.

Geertz, Armin W. "Long-Lost Brothers: On the Co-Histories and Interactions Between the Comparative Science of Religion and the Anthropology of Religion1." *Numen* 61, no. 2–3 (2014): 255–280.

Gelman, Susan, and Cristine Legare. "South African Children's Understanding of AIDS and Flu: Investigating Conceptual Understanding of Cause, Treatment and Prevention." *Journal of Cognition and Culture* 9, no. 3–4 (2009): 333–346.

Gervais, Will M. "Finding the Faithless: Perceived Atheist Prevalence Reduces Anti-Atheist Prejudice." *Personality and Social Psychology Bulletin* 37, no. 4 (2011): 543–556.

Gervais, Will M. "Global Evidence of Extreme Intuitive Moral Prejudice against Atheists." *Nature Human Behavior*, 1 (2017): 1–5.

Gervais, Will M. "Override the Controversy: Analytic Thinking Predicts Endorsement of Evolution." *Cognition* 142 (2015): 312–321.

Gervais, Will M. "Perceiving Minds and Gods." *Perspectives on Psychological Science* 8, no. 4 (2013): 380–394.

Gervais, Will M., and Joseph Henrich. "The Zeus Problem: Why Representational Content Biases Cannot Explain Faith in Gods." *Journal of Cognition and Culture* 10, no. 3 (2010): 383–389.

Gervais, Will M., and Maxine B. Najle. "How Many Atheists Are There?" *Social Psychological & Personality Science*, (2017): 1–8.

Gervais, Will M., and Maxine B. Najle. "Learned Faith: The Influences of Evolved Cultural Learning Mechanisms on Belief in Gods." *Psychology of Religion and Spirituality* 7, no. 4 (2015): 327–335.

Gervais, Will M., and Ara Norenzayan. "Analytic Thinking Promotes Religious Disbelief." *Science* 336, no. 6080 (2012a): 493–496.

Gervais, Will M., and Ara Norenzayan. "Like a Camera in the Sky? Thinking about God
Increases Public Self-Awareness and Socially Desirable Responding." *Journal of Experimental Social Psychology* 48, no. 1 (2012b): 298–302.

Gervais, Will M., and Ara Norenzayan. "Religion and the Origins of Anti-Atheist
Prejudice." *Intolerance and Conflict: A Scientific and Conceptual Investigation*, 2013,
126–145.

Gervais, Will M., and Ara Norenzayan. "Reminders of Secular Authority Reduce Believers' Distrust of Atheists." *Psychological Science* 23, no. 5 (2012c): 483–491.

Gervais, Will M., Ara Norenzayan, and Azim F. Shariff. "Do You Believe in Atheists?
Distrust Is Central to Anti-Atheist Prejudice." *Journal of Personality and Social Psychology* 101, no. 6 (2011a): 1189–1206.

Gervais, Will M., Aiyana K. Willard, Ara Norenzayan, and Joseph Henrich. "The Cultural Transmission of Faith: Why Innate Intuitions Are Necessary, but Insufficient,
to Explain Religious Belief." *Religion* 41, no. 3 (2011b): 389–410.

Gilbert, Nigel, and Jim Doran. *Simulating Societies. The Computer Simulation of Social
Phenomena*, London: UCL Press, 1994.

Gilbert, Nigel, and Klaus Troitzsch. *Simulation for the Social Scientist*. New York:
McGraw-Hill Education (UK), 2005.

Ginges, Jeremy, and Scott Atran. "What Motivates Participation in Violent Political Action" *Annals of the New York Academy of Sciences* 1167 (2009): 115.

Ginges, Jeremy, Scott Atran, Sonya Sachdeva, and Douglas Medin. "Psychology Out of
the Laboratory." *American Psychologist* 66, no. 6 (2011): 507–519.

Ginges, Jeremy, Ian Hansen, and Ara Norenzayan. "Religion and Support for Suicide
Attacks." *Psychological Science* 20, no. 2 (2009): 224–230.

Ginges, Jeremy, Scott Atran, Sonya Sachdeva, and Douglas Medin. "Religious Belief,
Coalitional Commitment, and Support for Suicide Attacks: Response to Liddle,
J.R., Machluf, K., and Shackelford, T.K." *Evolutionary Psychology* 8, no. 3 (July 1, 2010):
346–349.

Ginges, Jeremy, Hammad Sheikh, Scott Atran, and Nichole Argo. "Thinking from God's
Perspective Decreases Biased Valuation of the Life of a Nonbeliever." *Proceedings
of the National Academy of Sciences of the United States of America* 113, no. 2 (2016):
316–319.

Girard, René. *The Scapegoat*. Translated by Y. Freccero. Baltimore: Johns Hopkins University Press, 1986.

Girard, Réne. *Violence and the Sacred*. Johns Hopkins University Press, 1977.

Girotto, Vittorio, Telmo Pievani, and Giorgio Vallortigara. "Supernatural Beliefs: Adaptations for Social Life or by-Products of Cognitive Adaptations?" *Behavior*, 151 (2014):
385–402.

Good, Marie, Michael Inzlicht, and Michael J. Larson. "God Will Forgive: Reflecting on Gods Love Decreases Neurophysiological Responses to Errors." *Social Cognitive and Affective Neuroscience* 10, no. 3 (2013): 357–363.

Goplen, Joanna, and E. Ashby Plant. "A Religious Worldview Protecting One's Meaning System Through Religious Prejudice." *Personality and Social Psychology Bulletin* 41, no. 11 (2015): 1474–1487.

Gore, Ross, Carlos Lemos, F. LeRon Shults, and Wesley J. Wildman. "Forecasting Change in Religiosity and Existential Security with an Agent-Based Model," Journal of Artificial Societies and Social Simulation 21, no.1 (2018): 1–31.

Graeupner, Damaris, and Alin Coman. "The Dark Side of Meaning-Making: How Social Exclusion Leads to Superstitious Thinking." *Journal of Experimental Social Psychology* 69 (2017): 218–222.

Granqvist, Pehr. "Mental Health and Religion from an Attachment Viewpoint: Overview with Implications for Future Research." *Mental Health, Religion & Culture* 17, no. 8 (2014): 777–793.

Granqvist, Pehr, and Frances Nkara. "Nature Meets Nurture in Religious and Spiritual Development." *British Journal of Developmental Psychology* 35, no. 1 (2017): 142–155.

Gray, Russell D., and Joseph Watts. "Cultural Macroevolution Matters." *Proceedings of the National Academy of Sciences* 114, no. 30 (2017): 7846–7852.

Gregory, J.P., and T.S. Greenway. "Is There a Window of Opportunity for Religiosity? Children and Adolescents Preferentially Recall Religious-Type Cultural Representations, but Older Adults Do Not." *Religion, Brain & Behavior* (2017): 1–19.

Grenz, Stanley J. *Renewing the Center: Evangelical Theology in a Post-Theological Era.* Grand Rapids, MI: Baker Academic, 2000.

Grenz, Stanley J. *Revisioning Evangelical Theology.* Downers Grove, IL: IVP Academic, 1993.

Griffiths, Paul E., and John S. Wilkins. "Crossing the Milvian Bridge: When Do Evolutionary Explanations of Belief Debunk Belief?" In *Darwin in the Twenty-First Century: Nature, Humanity, God,* edited by Philip R. Sloan, Gerald McKenny, and Kathleen Eggleson, 201–231. Notre Dame, IL: University of Notre Dame Press, 2015.

Grim, Brian J., and Roger Finke. *The Price of Freedom Denied: Religious Persecution and Conflict in the Twenty-First Century.* Cambridge University Press, 2011.

Guen, Olivier, Rumen Iliev, Ximena Lois, Scott Atran, and Douglas L. Medin. "A Garden Experiment Revisited: Inter-Generational Change in Environmental Perception and Management of the Maya Lowlands, Guatemala." *Journal of the Royal Anthropological Institute* 19, no. 4 (2013): 771–794.

Guthrie, Stewart. "Animal Animism: Evolutionary Roots of Religious Cognition." In *Current Approaches in the Cognitive Science of Religion,* edited by Ilkka Pyysiäinen and Veikko Anttonen 38–67. London: Continuum, 2002.

Guthrie, Stewart. *Faces in the Clouds: A New Theory of Religion*. Oxford University Press, 1993.

Guthrie, Stewart, Joseph Agassi, Karin R. Andriolo, David Buchdahl, H. Byron Earhart, Moshe Greenberg, Ian Jarvie, et al. "A Cognitive Theory of Religion." *Current Anthropology*, 1980, 181–203.

Halberstadt, Jamin, and Jonathan Jong. "Scaring the Bejesus into People: The Role of Religious Belief in Managing Implicit and Explicit Anxiety." In *Motivation and its Regulation: The Control Within*, edited by Forgas and Harmon-Jones, 331–350. New York: Psychology Press, 2014.

Hall, Deborah L., David C. Matz, and Wendy Wood. "Why Don't We Practice What We Preach? A Meta-Analytic Review of Religious Racism." *Personality and Social Psychology Review* 14, no. 1 (2010): 126–139.

Hamill, Lynne, and Nigel Gilbert. *Agent-Based Modelling in Economics*. John Wiley & Sons, 2015.

Hanegraaff, W.J. "Reconstructing 'Religion' from the Bottom Up." *Numen* 63, no. 5–6 (2016): 576–605.

Hansen, Ian Grant, and Andrew Ryder. "In Search of 'Religion Proper.'" *Journal of Cross-Cultural Psychology* 47, no. 6 (2016): 835–857.

Harris, Paul L., Telli Davoodi, and Kathleen H. Corriveau. "Distinguishing between Realistic and Fantastical Figures in Iran," *Developmental Psychology* 52, no. 2 (2016): 221–231.

Harvey, Annelie J., and Mitchell J. Callan. "The Role of Religiosity in Ultimate and Immanent Justice Reasoning." *Personality and Individual Differences* 56, no. 1 (2014): 193–196.

Haselton, Martie G., Gregory A. Bryant, Andreas Wilke, David A. Frederick, Andrew Galperin, Willem E. Frankenhuis, and Tyler Moore. "Adaptive Rationality: An Evolutionary Perspective on Cognitive Bias." *Social Cognition* 27, no. 5 (2009): 733–763.

Haselton, Martie G., Daniel Nettle, and Damian Murray. "The Evolution of Cognitive Bias." *The Handbook of Evolutionary Psychology*. Second edition, edited by David Buss, 968–987. New York: Wiley, 2016.

Haselton, Martie G., and Daniel Nettle. "The Paranoid Optimist: An Integrative Evolutionary Model of Cognitive Biases." *Personality and Social Psychology Review* 10, no. 1 (2006): 47–66.

Hauser, Marc. *Moral Minds: The Nature of Right and Wrong*. New York: Harper Perennial, 2007.

Hayward, R., Neal Krause, Gail Ironson, Peter Hill, and Robert Emmons. "Health and Well-Being Among the Non-Religious: Atheists, Agnostics, and No Preference Compared with Religious Group Members." *Journal of Religion and Health* 55, no. 3 (2016a): 1024–1037.

Hayward, R., Neal Krause, Gail Ironson, and Kenneth Pargament. "Externalizing Religious Health Beliefs and Health and Well-Being Outcomes." *Journal of Behavioral Medicine* 39, no. 5 (2016b): 887–895.

Headey, Bruce, Gerhard Hoehne, and Gert Wagner. "Does Religion Make You Healthier and Longer Lived? Evidence for Germany." *Social Indicators Research* 119, no. 3 (2014): 1335–1361.

Healy, Amy Erbe, and Michael Breen. "Religiosity in Times of Insecurity: An Analysis of Irish, Spanish and Portuguese European Social Survey Data, 2002–12." *Irish Journal of Sociology* 22, no. 2 (2014): 4–29.

Heinskou, Marie Bruvik, and Lasse Suonperä Liebst. "On the Elementary Neural Forms of Micro-Interactional Rituals: Integrating Autonomic Nervous System Functioning Into Interaction Ritual Theory." *Sociological Forum* 31, no. 2 (2016): 354–376.

Heiphetz, Larisa, Jonathan D. Lane, Adam Waytz, and Liane L. Young. "How Children and Adults Represent God's Mind." *Cognitive Science* 40, no.1 (2016): 121–144.

Heiphetz, Larisa Alexandra, Elizabeth S. Spelke, and Mahzarin R. Banaji. "Patterns of Implicit and Explicit Attitudes in Children and Adults: Tests in the Domain of Religion." *Journal of Experimental Psychology: General*, 142, no. 3 (2013): 864–879.

Heiphetz, Larisa, Elizabeth S. Spelke, and Liane L. Young. "In the Name of God: How Children and Adults Judge Agents Who Act for Religious versus Secular Reasons." *Cognition* 144 (2015b): 134–149.

Henrich, Joseph. "The Evolution of Costly Displays, Cooperation and Religion: Credibility Enhancing Displays and Their Implications for Cultural Evolution." *Evolution and Human Behavior* 30, no. 4 (2009): 244–260.

Henry, Erika A., Bruce D. Bartholow, and Jamie Arndt. "Death on the Brain: Effects of Mortality Salience on the Neural Correlates of Ingroup and Outgroup Categorization." *Social Cognitive and Affective Neuroscience* 5, no. 1 (2010): 77–87.

Herek, Gregory M., and Kevin A. McLemore. "Sexual Prejudice." *Annual Review of Psychology* 64 (2013): 309–333.

Hey, Tony, Stewart Tansley, and Kristin Tolle, eds. *The Fourth Paradigm: Data-Intensive Scientific Discovery*. Redmond, WA: Microsoft Research, 2009.

Heywood, Bethany T., and Jesse M. Bering. "'Meant to Be': How Religious Beliefs and Cultural Religiosity Affect the Implicit Bias to Think Teleologically." *Religion, Brain & Behavior* 4, no. 3 (2014): 183–201.

Hill, Jonathan P. "Rejecting Evolution: The Role of Religion, Education, and Social Networks." *Journal for the Scientific Study of Religion* 53, no. 3 (2014): 575–594.

Hinds, Andrea L., Erik Z. Woody, Ana Drandic, Louis Schmidt, Michael van Ameringen, Marie Coroneos, and Henry Szechtman. "The Psychology of Potential Threat: Properties of the Security Motivation System." *Biological Psychology* 85, no. 2 (2010): 331–337.

Hirsch-Hoefler, Sivan, Daphna Canetti, and Ehud Eiran. "Radicalizing Religion? Religious Identity and Settlers' Behavior." *Studies in Conflict & Terrorism* 39, no. 6 (2016): 500–518.

Hitzeman, C., and C. Wastell. "Are Atheists Implicit Theists?" *Journal of Cognition and Culture* 17, no. 1–2 (2017): 27–50.

Hobson, Nicholas M., Devin Bonk, Michael Inzlicht, and Tsung-Min Hung. "Rituals Decrease the Neural Response to Performance Failure." *PeerJ* 5 (2017a): 1–26.

Hobson, Nicholas M., Francesca Gino, Michael I. Norton, and Michael Inzlicht. "When Novel Rituals Lead to Intergroup Bias: Evidence From Economic Games and Neurophysiology." *Psychological Science* 28, no. 6 (2017b): 733–750.

Hobson, Nicholas M., and Michael Inzlicht. "Recognizing Religion's Dark Side: Religious Ritual Increases Antisociality and Hinders Self-Control" Behavioral and Brain Sciences 39 (2016): 30–31.

Hobson, Nicholas M., Michael I. Norton, Francesca Gino, and Michael Inzlicht. "Mock Ritual Leads to Intergroup Biases in Behavior and Neurophysiology." Presentation at the Annual Meeting of the Association for Psychological Science, New York, NY, 2015.

Hodder, Ian. *The Leopard's Tale – Revealing the Mysteries of Çatalhöyük*. New York: Thames & Hudson, 2006.

Hodder, Ian, and Lynn Meskell. "The Symbolism of Çatalhöyük in Its Regional Context." In *Religion in the Emergence of Civilization: Catalhoyuk as a Case Study*, edited by Ian Hodder. New York: Cambridge University Press, 2010.

Hodges, Sara D., Carissa A. Sharp, Nicholas J.S. Gibson, and Jessica M. Tipsord. "Nearer My God to Thee: Self–God Overlap and Believers' Relationships with God." *Self and Identity* 12, no. 3 (2013): 337–356.

Hogg, Michael A., Janice R. Adelman, and Robert D. Blagg. "Religion in the Face of Uncertainty: An Uncertainty-Identity Theory Account of Religiousness." *Personality and Social Psychology Review* 14, no. 1 (2010): 72–83.

Holbrook, Colin, Keise Izuma, Choi Deblieck, Daniel M.T. Fessler, and Marco Iacoboni. "Neuromodulation of Group Prejudice and Religious Belief," *Social Cognitive and Affective Neuroscience* 11, no. 3 (2016): 387–394.

Hood, Bruce M., Marjaana Lindeman, and Tapani Riekki. "Is Weaker Inhibition Associated with Supernatural Beliefs?" *Journal of Cognition and Culture* 11, no. 1–2 (2011): 231–239.

Hope, Aimie L.B., and Christopher R. Jones. "The Impact of Religious Faith on Attitudes to Environmental Issues and Carbon Capture and Storage (CCS) Technologies: A Mixed Methods Study." *Technology in Society* 38 (2014): 48–59.

Hornbeck, Ryan, and Justin Barrett. "Refining and Testing 'Counterintuitiveness' in Virtual Reality: Cross-Cultural Evidence for Recall of Counterintuitive Representations." *The International Journal for the Psychology of Religion* 23, no. 1 (2013): 15.

Huang, Siyuan, and Shihui Han. "Shared Beliefs Enhance Shared Feelings: Religious/ Irreligious Identifications Modulate Empathic Neural Responses." *Social Neuroscience* 9, no. 6 (2014): 639–649.

Hughes, Jeffrey, Igor Grossmann, and Adam B. Cohen. "Tolerating the 'Doubting Thomas': How Centrality of Religious Beliefs vs. Practices Influences Prejudice against Atheists," *Frontiers in Psychology* 6 (2015): 1–10.

Hungerman, D.M. "The Effect of Education on Religion: Evidence from Compulsory Schooling Laws." *Journal of Economic Behavior & Organization* 104 (2014): 52–63.

Hunsberger, Bruce, and Lynne M. Jackson. "Religion, Meaning, and Prejudice." *Journal of Social Issues* 61, no. 4 (2005): 807–826.

Hwang, Karen. "Atheism, Health, and Well-Being," In *The Oxford Handbook of Atheism*, edited by Stephen Bullivant and Michael Ruse, 523–536. Oxford University Press, 2013.

Iannaccone, L.R., and M.D. Makowsky. "Accidental Atheists? Agent-Based Explanations for the Persistence of Religious Regionalism." *Journal for the Scientific Study of Religion* 46, no. 1 (2007): 1–16.

Immerzeel, Tim, and Frank Van Tubergen. "Religion as Reassurance? Testing the Insecurity Theory in 26 European Countries." *European Sociological Review* 29, no. 2 (2013): 359–372.

Inglehart, Ronald, and Christian Welzel. *Modernization, Cultural Change, and Democracy: The Human Development Sequence.* Cambridge University Press, 2005.

Inzlicht, Michael, Alexa M. Tullett, and Marie Good. "The Need to Believe: A Neuroscience Account of Religion as a Motivated Process." *Religion, Brain & Behavior* 1, no. 3 (2011): 192–212.

Irwin, Kyle, and Brandon C. Martinez. "The Effects of Protestant Theological Conservatism and Trust on Environmental Cooperation." *Journal for the Scientific Study of Religion* 56, no. 1 (2017): 199–212.

Isaacs, M. "Faith in Contention: Explaining the Salience of Religion in Ethnic Conflict." *Comparative Political Studies*, (2016): 1–32.

Järnefelt, Elisa, Caitlin F. Canfield, and Deborah Kelemen. "The Divided Mind of a Disbeliever: Intuitive Beliefs about Nature as Purposefully Created among Different Groups of Non-Religious Adults." *Cognition* 140 (July 2015): 72–88.

Jensen, Gary F. "Religious Cosmologies and Homicide Rates among Nations: A Closer Look," *Journal of Religion & Society* 8 (2006): 1–14.

Jensen, Jeppe Sinding. "Normative Cognition in Culture and Religion." *Journal for the Cognitive Science of Religion* 1, no. 1 (2013): 47–70.

Johnson, Dominic. *God Is Watching You: How the Fear of God Makes Us Human.* New York: Oxford University Press, 2015.

Johnson, Dominic. "What Are Atheists for? Hypotheses on the Functions of Non-Belief in the Evolution of Religion." *Religion, Brain & Behavior* 2, no. 1 (2012): 48–70.

Johnson, Dominic. "God's Punishment and Public Goods: A Test of the Supernatural Punishment Hypothesis in 186 World Cultures." *Human Nature* 16, no. 4 (2005): 410–446.

Johnson, Dominic. "The Wrath of the Academics: Criticisms, Applications, and Extensions of the Supernatural Punishment Hypothesis." *Religion, Brain & Behavior* (2017), 1–31.

Johnson, Dominic, and Oliver Krüger. "The Good of Wrath: Supernatural Punishment and the Evolution of Cooperation." *Political Theology* 5, no. 2 (2004): 159–176.

Johnson, Dominic, Daniel T. Blumstein, James H. Fowler, and Martie G. Haselton. "The Evolution of Error: Error Management, Cognitive Constraints, and Adaptive Decision-Making Biases." *Trends in Ecology and Evolution* 28, no. 8 (2013): 474–481.

Johnson, Dominic, Hillary L. Lenfesty, and Jeffrey P. Schloss. "The Elephant in the Room: Do Evolutionary Accounts of Religion Entail the Falsity of Religious Belief?" *Philosophy, Theology and the Sciences* 1, no. 2 (2014): 200–231.

Johnson, Kathryn A., Adam B. Cohen, Rebecca Neel, Anna Berlin, and Donald Homa. "Fuzzy People: The Roles of Kinship, Essence, and Sociability in the Attribution of Personhood to Nonliving, Nonhuman Agents." *Psychology of Religion and Spirituality* 7, no. 4 (2015): 295–305.

Johnson, Kathryn A., Yexin Jessica Li, and Adam B. Cohen. "Fundamental Social Motives and the Varieties of Religious Experience." *Religion, Brain & Behavior* 5, no. 3 (2015): 197–231.

Johnson, Megan K., Wade C. Rowatt, and Jordan Labouff. "Priming Christian Religious Concepts Increases Racial Prejudice." *Social Psychological and Personality Science* 1, no. 2 (2010): 119–126.

Jones, Robert P., and Daniel Cox. "America's Changing Religious Identity." Washington, DC: Public Religion Research Institute, September 6, 2017.

Jones, Robert P., Daniel Cox, Betsy Cooper, and Rachel Lienesch. "Exodus: Why Americans Are Leaving Religion – and Why They're Unlikely to Come Back." Washington, D.C.: Public Religion Research Institute, September 22, 2016.

Joseph, Stephen P., Alex Linley, and John Maltby. "Positive Psychology, Religion, and Spirituality." *Mental Health, Religion & Culture* 9, no. 3 (2006): 209–212.

Kahan, Dan M. "Cultural Cognition as a Conception of the Cultural Theory of Risk." In *Handbook of Risk Theory*, edited by S. Roeser, et al., 725–760. New York: Springer, 2011.

Kahan, Dan M., and Donald Braman. "Cultural Cognition and Public Policy." *SSRN Scholarly Paper*. Rochester, NY: Social Science Research Network, August 2, 2005.

Kahan, Dan M., Hank Jenkins-Smith, and Donald Braman. "Cultural Cognition of Scientific Consensus." *Journal of Risk Research* 14, no. 2 (2011): 147–174.

Kalkman, David Peter. "Three Cognitive Routes to Atheism: A Dual-Process Account." *Religion* 44, no. 1 (2014): 72–83.

Kanas, Agnieszka, Peer Scheepers, and Carl Sterkens. "Religious Identification and Interreligious Contact in Indonesia and the Philippines: Testing the Mediating Roles of Perceived Group Threat and Social Dominance Orientation and the Moderating Role of Context." *European Journal of Social Psychology* 46, no. 6 (2016): 700–715.

Kandler, Christian, and Rainer Riemann. "Genetic and Environmental Sources of Individual Religiousness: The Roles of Individual Personality Traits and Perceived Environmental Religiousness." *Behavior Genetics* 43, no. 4 (2013): 297–313.

Kapitány, Rohan, and Mark Nielsen. "Adopting the Ritual Stance: The Role of Opacity and Context in Ritual and Everyday Actions." *Cognition* 145 (2015): 13–29.

Kapitány, Rohan, and Mark Nielsen. "The Ritual Stance and the Precaution System: The Role of Goal-Demotion and Opacity in Ritual and Everyday Actions." *Religion, Brain & Behavior* 6 (2016): 1–16.

Kay, Aaron C., Danielle Gaucher, Ian Mcgregor, and Kyle Nash. "Religious Belief as Compensatory Control." *Personality and Social Psychology Review* 14, no. 1 (2010a): 37–48.

Kay, Aaron C., Danielle Gaucher, Jamie L. Napier, Mitchell J. Callan, and Kristin Laurin. "God and the Government: Testing a Compensatory Control Mechanism for the Support of External Systems." *Journal of Personality and Social Psychology* 95, no. 1 (2008): 18.

Kay, Aaron C., David A. Moscovitch, and Kristin Laurin. "Randomness, Attributions of Arousal, and Belief in God." *Psychological Science* 21, no. 2 (2010b): 216–218.

Kay, Aaron C., Steven Shepherd, Craig W. Blatz, Sook Ning Chua, and Adam D. Galinsky. "For God (or) Country: The Hydraulic Relation Between Government Instability and Belief in Religious Sources of Control." *Journal of Personality and Social Psychology* 99, no. 5 (2010c): 725–739.

Kazanas, Stephanie A., and Jeanette Altarriba. "Did Our Ancestors Fear the Unknown? The Role of Predation in the Survival Advantage." *Evolutionary Behavioral Sciences*, 11, no. 1 (2016): 83–91.

Kegan, Robert. *In over Our Heads: The Mental Demands of Modern Life*. Cambridge, MA: Harvard University Press, 1995.

Kegan, Robert. *The Evolving Self*. Cambridge, MA: Harvard University Press, 1982.

Kelemen, Deborah. "Why Are Rocks Pointy? Children's Preference for Teleological Explanations of the Natural World." *Developmental Psychology* 35, no. 6 (1999): 1440–1452.

Kelly-Hanku, Angela, Peter Aggleton, and Patti Shih. "'We Call It a Virus but I Want to Say It's the Devil Inside': Redemption, Moral Reform and Relationships with God among People Living with HIV in Papua New Guinea." *Social Science & Medicine* 119 (2014): 106–113.

Kettell, Steven. "Divided We Stand: The Politics of the Atheist Movement in the United States." *Journal of Contemporary Religion* 29, no. 3 (2014): 377–391.

Kettell, Steven. "Faithless: The Politics of New Atheism." *Secularism and Nonreligion* 2 (2013): 61–72.

Khenfer, Jamel, Elyette Roux, Eric Tafani, and Kristin Laurin. "When God's (Not) Needed: Spotlight on How Belief in Divine Control Influences Goal Commitment." *Journal of Experimental Social Psychology* 70 (2017): 117–123.

Kiessling, Florian, and Josef Perner. "God–Mother–Baby: What Children Think They Know." *Child Development* 85, no. 4 (2014): 1601–1616.

Kiper, Jordan, and Richard Sosis. "Shaking the Tyrant's Bloody Robe." *Politics and the Life Sciences* 35, no. 01 (2016): 27–47.

Kirkpatrick, Lee A. *Attachment, Evolution, and the Psychology of Religion.* New York: The Guilford Press, 2004.

Klaczynski, Paul A., and David H. Gordon. "Self-Serving Influences on Adolescents' Evaluations of Belief-Relevant Evidence." *Journal of Experimental Child Psychology* 62, no. 3 (1997): 317–339.

Klein, Naomi. *This Changes Everything: Capitalism vs. the Climate.* New York: Allen Lane, 2014.

Knight, Nicola, Paulo Sousa, Justin L. Barrett, and Scott Atran. "Children's Attributions of Beliefs to Humans and God: Cross-Cultural Evidence." *Cognitive Science* 28, no. 1 (2004): 117–126.

Kohler, Timothy A., and George G. Gumerman. *Dynamics in Human and Primate Societies: Agent-Based Modeling of Social and Spatial Processes.* Oxford University Press, 2000.

Kolbert, Elizabeth. *The Sixth Extinction : An Unnatural History.* New York: Henry Holt and Co, 2014.

Kollmuss, Anja, and Julian Agyeman. "Mind the Gap: Why Do People Act Environmentally and What Are the Barriers to Pro-Environmental Behavior?" *Environmental Education Research* 8, no. 3 (August 1, 2002): 239–260.

Kosmin, Barry A., and Ariela Keysar. "Religious, Spiritual and Secular: The Emergence of Three Distinct Worldviews among American College Students." American Religious Identification Survey. Hartford, CT: Trinity College, 2013.

Kossowska, Małgorzata, and Maciej Sekerdej. "Searching for Certainty: Religious Beliefs and Intolerance toward Value-Violating Groups." *Personality and Individual Differences* 83 (2015): 72–76.

Kossowska, Małgorzata, Aneta Czernatowicz-Kukuczka, and Maciej Sekerdej. "Many Faces of Dogmatism: Prejudice as a Way of Protecting Certainty against Value

Violators among Dogmatic Believers and Atheists." *British Journal of Psychology* 108, no. 1 (2017a): 127–147.

Kossowska, Małgorzata, Paulina Szwed, Eligiusz Wronka, Gabriela Czarnek, and Mirosław Wyczesany. "Anxiolytic Function of Fundamentalist Beliefs: Neurocognitive Evidence." *Personality and Individual Differences* 101 (2016): 390–395.

Kossowska, Małgorzata, Paulina Szwed, Aneta Czernatowicz-Kukuczka, Maciek Sekerdej, and Miroslaw Wyczesany. "From Threat to Relief: Expressing Prejudice toward Atheists as a Self-Regulatory Strategy Protecting the Religious Orthodox from Threat." *Frontiers in Psychology* 8 (2017b): 1–8.

Kotaman, Huseyin, and Ali Kemal Tekin. "The Impact of Religion on the Development of Young Children's Factuality Judgments." *North American Journal of Psychology* 17, no. 3 (2015): 525–540.

Krasnow, Max M., and Andrew W. Delton. "Are Humans Too Generous and Too Punitive? Using Psychological Principles to Further Debates about Human Social Evolution." *Frontiers in Psychology* 7 (2016): 1–5.

Krasnow, Max M., Andrew W. Delton, Leda Cosmides, and John Tooby. "Group Cooperation without Group Selection: Modest Punishment Can Recruit Much Cooperation." *PLoS ONE* 10, no. 4 (2015): 1–17.

Krebs, Dennis L., and Kaleda K. Denton. "The Evolution of Sociality, Helping, and Morality." In *The Oxford Handbook of Secularism*, edited by Phil Zuckerman and John Shook, 638–654. New York: Oxford University Press, 2017.

Kundt, Radek. *Contemporary Evolutionary Theories of Culture and the Study of Religion.* London: Bloomsbury Academic, 2015.

Kupor, Daniella M., Kristin Laurin, and Jonathan Levav. "Anticipating Divine Protection? Reminders of God Can Increase Nonmoral Risk Taking." *Psychological Science* 26, no. 4 (2015): 374–384.

Kurzban, Robert, Maxwell N. Burton-Chellew, and Stuart A. West. "The Evolution of Altruism in Humans." *Annual Review of Psychology* 66, no. 1 (2015): 575–599.

Kvande, Marianne, Randi Reidunsdatter, Audhild Løhre, Michael Nielsen, and Geir Espnes. "Religiousness and Social Support: A Study in Secular Norway." *Review of Religious Research* 57, no. 1 (2015): 87–109.

Labouff, J.P., and A.M. Ledoux. "Imagining Atheists: Reducing Fundamental Distrust in Atheist Intergroup Attitudes." *Psychology of Religion and Spirituality* 8, no.4 (2016): 330–340.

Labouff, J.P., W.C. Rowatt, M.K. Johnson, and C. Finkle. "Differences in Attitudes Toward Outgroups in Religious and Nonreligious Contexts in a Multinational Sample: A Situational Context Priming Study." *International Journal for the Psychology of Religion* 22, no. 1 (2012): 1–9.

Lane, Justin E. "Method, Theory, and Multi-Agent Artificial Intelligence: Creating Computer Models of Complex Social Interaction." *Journal for the Cognitive Science of Religion* 1, no. 2 (2014): 161–180.

Lane, Jonathan D., and Paul L. Harris. "Confronting, Representing, and Believing Counterintuitive Concepts: Navigating the Natural and the Supernatural." *Perspectives on
Psychological Science* 9, no. 2 (2014): 144–160.

Lang, Martin, Jan Krátký, John H. Shaver, Danijela Jerotijević, and Dimitris Xygalatas.
"Effects of Anxiety on Spontaneous Ritualized Behavior." *Current Biology* 25, no. 14
(2015): 1892–1897.

Lang, Martin, Panagiotis Mitkidis, Radek Kundt, Aaron Nichols, Lenka Krajčíková, and
Dimitris Xygalatas. "Music As a Sacred Cue? Effects of Religious Music on Moral
Behavior." *Frontiers in Psychology* 7 (2016): 1–13.

Lanman, Jonathan A. "The Importance of Religious Displays for Belief Acquisition and
Secularization." *Journal of Contemporary Religion* 27, no. 1 (2012): 49–65.

Lanman, Jonathan A., and Michael D. Buhrmester. "Religious Actions Speak Louder
than Words: Exposure to Credibility-Enhancing Displays Predicts Theism." *Religion,
Brain & Behavior* 7, no. 1 (2015), 3–16.

Laurin, Kristin. "Religion and Its Cultural Evolutionary By-Products." In *The Science
of Lay Theories*, edited by Claire Zedelius and Barbara Müller, 243–263. New York:
Springer, 2017.

Lawson, E. Thomas, and Robert N. McCauley. *Rethinking Religion: Connecting Cognition and Culture*. Cambridge University Press, 1993.

Lechner, Clemens M., Martin J. Tomasik, Rainer K. Silbereisen, and Jacek Wasilewski.
"Exploring the Stress-Buffering Effects of Religiousness in Relation to Social and
Economic Change: Evidence From Poland." *Psychology of Religion and Spirituality*
5, no. 3 (2013): 145–156.

Ledrew, Stephen. "Discovering Atheism: Heterogeneity in Trajectories to Atheist Identity and Activism." *Sociology of Religion* 74, no. 4 (2013): 431–453.

Lee, Lois. "Non-Religion." In *The Oxford Handbook of the Study of Religion*, edited by
Michael Stausberg and Steven Engler, 84–96. New York: Oxford University Press,
2016.

Lee, Lois. "Talking about a Revolution: Terminology for the New Field of Non-Religion
Studies." *Journal of Contemporary Religion* 27, no. 1 (2012): 129–139.

Leech, David, and Aku Visala. "The Cognitive Science of Religion: A Modified Theist
Response." *Religious Studies* 47, no. 3 (2011): 301–316.

Legare, Cristine H., and Susan A. Gelman. "Bewitchment, Biology, or Both: The Co-
Existence of Natural and Supernatural Explanatory Frameworks across Development." *Cognitive Science* 32, no. 4 (2008): 607–642.

Legare, Cristine H., and Andre L. Souza. "Evaluating Ritual Efficacy: Evidence from the
Supernatural." *Cognition* 124, no. 1 (2012): 1–15.

Legare, Cristine H., and André L. Souza. "Searching for Control: Priming Randomness
Increases the Evaluation of Ritual Efficacy." *Cognitive Science*, (2013): 152–161.

Lemos, Carlos, Ross Gore, and F. LeRon Shults. "Exploratory and Confirmatory Analyses of Religiosity: A Four-Factor Conceptual Model." (2017): *arXiv*: 1704.06112.

Lewis, James R. "Education, Irreligion, and Non-Religion: Evidence from Select Anglophone Census Data." *Journal of Contemporary Religion* 30, no. 2 (2015): 265–272.

Lewis-Williams, David. *Conceiving God : The Cognitive Origin and Evolution of Religion.* London: Thames & Hudson, 2010.

Lewis-Williams, David. *The Mind in the Cave: Consciousness and the Origins of Art.* London: Thames & Hudson, 2002.

Lewis-Williams, David, and David Pearce. *Inside the Neolithic Mind: Consciousness, Cosmos and the Realm of the Gods.* London: Thames & Hudson, 2009.

Liddle, James R., Karin Machluf, and Todd K. Shackelford. "Understanding Suicide Terrorism: Premature Dismissal of the Religious-Belief Hypothesis." *Evolutionary Psychology* 8, no. 3 (2010): 343–345.

Liénard, Pierre, and E. Thomas Lawson. "Evoked Culture, Ritualization and Religious Rituals." *Religion* 38, no. March (2008): 157–171.

Lifshin, Uri, Jeff Greenberg, David Weise, and Melissa Soenke. "It's the End of the World and I Feel Fine: Soul Belief and Perceptions of End-of-the-World Scenarios." *Personality and Social Psychology Bulletin* 42, no. 1 (2016): 104–117.

Lindeman, Marjaana, and Kia Aarnio. "Superstitious, Magical, and Paranormal Beliefs: An Integrative Model." *Journal of Research in Personality* 41, no. 4 (2007): 731–744.

Lindeman, Marjaana, and Jari Lipsanen. "Diverse Cognitive Profiles of Religious Believers and Nonbelievers." *The International Journal for the Psychology of Religion* 26, no. 3 (2016): 185–192.

Lindeman, Marjaana, and Annika M. Svedholm-Häkkinen. "Does Poor Understanding of Physical World Predict Religious and Paranormal Beliefs?" *Applied Cognitive Psychology* 30, no. 5 (2016): 736–742.

Lindeman, Marjaana, Sandra Blomqvist, and Mikito Takada. "Distinguishing Spirituality From Other Constructs: Not A Matter of Well-Being but of Belief in Supernatural Spirits." *The Journal of Nervous and Mental Disease* 200, no. 2 (2012): 167–173.

Lindeman, Marjaana, Annika M. Svedholm-Häkkinen, and Jari Lipsanen. "Ontological Confusions but Not Mentalizing Abilities Predict Religious Belief, Paranormal Belief, and Belief in Supernatural Purpose." *Cognition* 134 (2015): 63–76.

Lindeman, Marjaana, Annika M. Svedholm-Häkkinen, and Tapani Riekki. "Skepticism: Genuine Unbelief or Implicit Beliefs in the Supernatural?" *Consciousness and Cognition* 42 (2016): 216–228.

Lindström, Björn, and Andreas Olsson. "Mechanisms of Social Avoidance Learning Can Explain the Emergence of Adaptive and Arbitrary Behavioral Traditions in Humans." *Journal of Experimental Psychology: General* 144, no. 3 (2015): 688–703.

Lisdorf, Anders. "What's HIDD'n in the HADD?" *Journal of Cognition and Culture* 7, no. 3–4 (2007): 341–353.

Lloyd, Elisabeth A., and Vanessa J. Schweizer. "Objectivity and a Comparison of Methodological Scenario Approaches for Climate Change Research." *Synthese: An International Journal for Epistemology, Methodology and Philosophy of Science* 191, no. 10 (2014): 2049.

Lobato, Emilio, Valerie Sims, Matthew Chin, Jorge Mendoza, Valerie Sims, and Matthew Chin. "Examining the Relationship between Conspiracy Theories, Paranormal Beliefs, and Pseudoscience Acceptance among a University Population." *Applied Cognitive Psychology*, (2014): 617–625.

Loder, James E. *The Transforming Moment*. 2 edition. Colorado Springs, CO: Helmers & Howard Publishers, 1989.

Lori G. Beaman, ed. *Atheist Identities – Spaces and Social Contexts*. v.2. Boundaries of Religious Freedom: Regulating Religion in Diverse Societies. New York: Springer International Publishing, 2014.

Lynch, R., B.G. Palestis, and R.Trivers. "Religious Devotion and Extrinsic Religiosity Affect In-group Altruism and Out-group Hostility Oppositely in Rural Jamaica." *Evolutionary Psychological Science* 3, no. 4 (2017): 335–344.

Lynn, Richard, John Harvey, and Helmuth Nyborg. "Average Intelligence Predicts Atheism Rates across 137 Nations." *Intelligence* 37, no. 1 (2009): 11–15.

MacPherson, James S., and Steve W. Kelly. "Creativity and Positive Schizotypy Influence the Conflict between Science and Religion." *Personality and Individual Differences* 50, no. 4 (2011): 446–450.

Mahoney, Andrew. "The Evolutionary Psychology of Theology." In *The Attraction of Religion: A New Evolutionary Psychology of Religion*, edited by D. Jason Slone and James A. van Slyke, 189–210. London: Bloomsbury Academic, 2015.

Maij, David L.R., Hein T. Van Schie, and Michiel Van Elk. "The Boundary Conditions of the Hypersensitive Agency Detection Device: An Empirical Investigation of Agency Detection in Threatening Situations." *Religion, Brain & Behavior* 7 (2017) 1–29.

Ma-Kellams, C. "When Perceiving the Supernatural Changes the Natural: Religion and Agency Detection." *Journal of Cognition and Culture* 15, no. 3–4 (2015): 337–343.

Ma-Kellams, C., and J. Blascovich. "Does 'Science' Make You Moral? The Effects of Priming Science on Moral Judgments and Behavior." *PLoS ONE* 8, no. 3 (2013): 1–4.

Maltby, John, and Liza Day. "Religious Experience, Religious Orientation and Schizotypy." *Mental Health, Religion & Culture* 5, no. 2 (2002): 163–174.

Maltby, John, Iain Garner, Christopher Alan Lewis, and Liza Day. "Religious Orientation and Schizotypal Traits." *Personality and Individual Differences* 28, no. 1 (2000): 143–151.

Maltseva, K. "Prosocial Morality in Individual and Collective Cognition." *Journal of Cognition and Culture* 16, no. 1–2 (2016): 1–36.

Martin, Luther. *Deep History, Secular Theory: Historical and Scientific Studies of Religion.* Berlin: De Gruyter, 2014.

Martin, Luther H., and Donald Wiebe. "Pro-and Assortative-Sociality in the Formation and Maintenance of Religious Groups." *Journal for the Cognitive Science of Religion* 2, no. 1 (2014): 1–57.

Martinez, Brandon. "Is Evil Good for Religion? The Link between Supernatural Evil and Religious Commitment." *Review of Religious Research* 55, no. 2 (2013): 319–338.

Matthews, L.J., J. Edmonds, W.J. Wildman, and C.L. Nunn. "Cultural Inheritance or Cultural Diffusion of Religious Violence? A Quantitative Case Study of the Radical Reformation." *Religion, Brain & Behavior* 3, no. 1 (2013): 3–15.

Mauzay, D., A. Spradlin, and C. Cuttler. "Devils, Witches, and Psychics: The Role of Thought-Action Fusion in the Relationships between Obsessive-Compulsive Features, Religiosity, and Paranormal Beliefs." *Journal of Obsessive-Compulsive and Related Disorders* 11 (2016): 113–120.

Mavor, K.I., W.R. Louis, and B. Laythe. "Religion, Prejudice, and Authoritarianism: Is RWA a Boon or Bane to the Psychology of Religion?" *Journal for the Scientific Study of Religion* 50, no. 1 (2011): 22–43.

Maxwell-Smith, Matthew A., Clive Seligman, Paul Conway, Matthew A. Maxwell-Smith, and Irene Cheung. "Individual Differences in Commitment to Value-Based Beliefs and the Amplification of Perceived Belief Dissimilarity Effects." *Journal of Personality*, 83, no. 2 (2014): 127–141.

McCaffree, Kevin. *The Secular Landscape: The Decline of Religion in America.* New York, NY: Palgrave Macmillan, 2017.

McCauley, Clark, and Sophia Moskalenko. "Understanding Political Radicalization: The Two-Pyramids Model." *American Psychologist* 72, no. 3 (2017): 205–216.

McClure, Paul K. "Something besides Monotheism: Sociotheological Boundary Work among the Spiritual, but Not Religious." *Poetics* 62 (2017): 53–65.

McCorkle, Jr., William W., and Dimitris Xygalatas. "Social Minds, Mental Cultures – Weaving Together Cognition and Culture in the Study of Religion." In *Mental Cultures: Classical Social Theory and the Cognitive Science of Religion*, edited by Dimitris Xygalatas and William W. McCorkle, Jr., 1–10. Durham, UK: Acumen, 2013.

McGregor, Holly A., Joel D. Lieberman, Jeff Greenberg, Sheldon Solomon, Jamie Arndt, Linda Simon, and Tom Pyszczynski. "Terror Management and Aggression: Evidence That Mortality Salience Motivates Aggression Against Worldview-Threatening Others." *Journal of Personality and Social Psychology* 74, no. 3 (1998): 590–605.

McGregor, I., J. Hayes, and M. Prentice. "Motivation for Aggressive Religious Radicalization: Goal Regulation Theory and a Personality x Threat x Affordance Hypothesis."
Frontiers in Psychology 6 (2015): 1–18.

McGregor, Ian, Kyle Nash, and Mike Prentice. "Reactive Approach Motivation (RAM)
for Religion." *Journal of Personality and Social Psychology* 99, no. 1 (2010): 148–161.

McGregor, Ian, Mike Prentice, and Kyle Nash. "Approaching Relief: Compensatory Ideals Relieve Threat-Induced Anxiety by Promoting Approach-Motivated States." *Social Cognition* 30, no. 6 (2012): 689–714.

McKay, Ryan, and Harvey Whitehouse. "Religion and Morality." *Psychological Bulletin*
141, no. 2 (2015): 447–473.

McNamara, P., and K. Bulkeley. "Dreams as a Source of Supernatural Agent Concepts."
Frontiers in Psychology 6 (2015): 1–8.

McNamara, R.A., and J. Henrich. "Jesus vs. the Ancestors: How Specific Religious Beliefs
Shape Prosociality on Yasawa Island, Fiji." *Religion, Brain & Behavior* 7 (2017): 1–20.

McNamara, Rita Anne, Ara Norenzayan, and Joseph Henrich. "Supernatural Punishment, in-Group Biases, and Material Insecurity: Experiments and Ethnography
from Yasawa, Fiji." *Religion, Brain & Behavior* 6, no. 1 (2016): 34–55.

McPhetres, Jonathon, and Thuy-Vy T. Nguyen. "Using Findings from the Cognitive Science of Religion to Understand Current Conflicts between Religious and Scientific
Ideologies." *Religion, Brain & Behavior* 7 (2017): 1–12.

Meeusen, Cecil, Fiona Kate Barlow, and Chris G. Sibley. "Generalized and Specific
Components of Prejudice: The Decomposition of Intergroup Context Effects: In-
tergroup Context, Generalized and Specific Components of Prejudice." *European
Journal of Social Psychology* 47, no. 4 (2017): 443–456.

Meier, Brian P., Adam K. Fetterman, Michael D. Robinson, and Courtney M. Lap-
pas. "The Myth of the Angry Atheist." *The Journal of Psychology* 149, no. 3 (2015):
219–238.

Mikulincer, Mario. *Attachment in Adulthood: Structure, Dynamics, and Change*. Guilford Publications, 2007.

Miller, John H. and Scott E. Page. *Complex Adaptive Systems: An Introduction to Computational Models of Social Life*. Princeton University Press, 2007.

Minton, Elizabeth A., Lynn R. Kahle, and Chung-Hyun Kim. "Religion and Motives for
Sustainable Behaviors: A Cross-Cultural Comparison and Contrast." *Journal of Business Research*, (2015): 1937–1944.

Mitkidis, Panagiotis, and Gabriel Levy. "False Advertising: The Attractiveness of Religion as a Moral Brand." In *The Attraction of Religion: A New Evolutionary Psychology of Religion*, edited by D. Jason Slone and James A van Slyke, 159–168. London:
Bloomsbury, 2015.

Mitkidis, Panagiotis, Shahar Ayal, Shaul Shalvi, Katrin Heimann, Gabriel Levy, Miriam
Kyselo, Sebastian Wallot, Dan Ariely, and Andreas Roepstorff. "The Effects of

Extreme Rituals on Moral Behavior: The Performers-Observers Gap Hypothesis." *Journal of Economic Psychology* 59 (2017): 1–7.

Miyatake, Sanae, and Masataka Higuchi. "Does Religious Priming Increase the Prosocial Behaviour of a Japanese Sample in an Anonymous Economic Game?" *Asian Journal of Social Psychology* 20, no. 1 (2017): 54–59.

Miyazaki, Yuki. "Being Watched by Anthropomorphized Objects Affects Charitable Donation in Religious People," *Japanese Psychological Research* 59, no. 3 (2017): 221–229.

Mora, Louis Ernesto, Panayiotis Stavrinides, and Wilson McDermut. "Religious Fundamentalism and Religious Orientation Among the Greek Orthodox." *Journal of Religion and Health* 53, no. 5 (2014): 1498–1513.

Morgan, Thomas J.H. "Testing the Cognitive and Cultural Niche Theories of Human Evolution." *Current Anthropology* 57, no. 3 (2016): 370–377.

Mortreux, Colette, and Jon Barnett. "Climate Change, Migration and Adaptation in Funafuti, Tuvalu." *Global Environmental Change* 19, no. 1 (2009): 105–112.

Müller, Tim S., Nan Dirk De Graaf, and Peter Schmidt. "Which Societies Provide a Strong Religious Socialization Context? Explanations Beyond the Effects of National Religiosity." *Journal for the Scientific Study of Religion* 53, no. 4 (2014): 739–759.

Murray, Evan D., Miles G. Cunningham, and Bruce H. Price. "The Role of Psychotic Disorders in Religious History Considered." *The Journal of Neuropsychiatry and Clinical Neurosciences* 24, no. 4 (2012): 410–426.

Nash, Kyle, Thomas Baumgartner, and Daria Knoch. "Group-Focused Morality Is Associated with Limited Conflict Detection and Resolution Capacity: Neuroanatomical Evidence." *Biological Psychology* 123 (2017): 235–240.

Nelson-Pallmeyer, Jack. *Is Religion Killing Us? Violence in the Bible And the Quran*. New York: Continuum, 2005.

Neuberg, Steven L., Carolyn M. Warner, Stephen A. Mistler, Anna Berlin, Eric D. Hill, Jordan D. Johnson, Gabrielle Filip-Crawford, et al. "Religion and Intergroup Conflict." *Psychological Science* 25, no. 1 (2014): 198–206.

Neville, Robert C. "Comments on F. LeRon Shults's 'What's the Use? Pragmatic Reflections on Neville's Ultimates.'" *American Journal of Theology and Philosophy* 36, no. 1 (2015): 81–84.

Neville, Robert C. *Existence: Philosophical Theology, Volume Two*. Albany, NY: State University of New York Press, 2015.

Neville, Robert C. *On the Scope and Truth of Theology: Theology as Symbolic Engagement*. T & T Clark, 2006.

Neville, Robert C. *Religion: Philosophical Theology, Volume Three*. Albany, NY: State University of New York Press, 2016.

Neville, Robert C. *The Truth of Broken Symbols*. Albany, NY: State University of New York Press, 1996.

Neville, Robert C., ed., *Ultimate Realities*. State University of New York Press, 2001.

Neville, Robert C. *Ultimates: Philosophical Theology, Volume One*. State University of New York Press, 2014.

Newheiser, Anna-Kaisa, Miles Hewstone, Alberto Voci, Katharina Schmid, Andreas Zick, and Beate Küpper. "Social-Psychological Aspects of Religion and Prejudice: Evidence from Survey and Experimental Research." In *Religion, Intolerance, and Conflict: A Scientific and Conceptual Investigation*, edited by Steve Clarke, Russell Powell, and Julian Savulescu, 107–125. Oxford University Press, 2013.

Ng, Ben, and Will M. Gervais. "Religion and Prejudice." In *The Cambridge Handbook of the Psychology of Prejudice*, edited by Chris G. Sibley and Fiona Kate Barlow. Cambridge: Cambridge University Press, 2017.

Ng, Eddie, and Adrian Fisher. "Protestant Spirituality and Well-Being of People in Hong Kong: The Mediating Role of Sense of Community." *Applied Research in Quality of Life* 11, no. 4 (2016): 1253–1267.

Nielbo, Kristoffer L., and Jesper Sørensen. "Attentional Resource Allocation and Cultural Modulation in a Computational Model of Ritualized Behavior." *Religion, Brain & Behavior* 6, no. 4 (2015): 318–335.

Nielbo, Kristoffer L., Donald M. Braxton, and Afzal Upal. "Computing Religion: A New Tool in the Multilevel Analysis of Religion." *Method and Theory in the Study of Religion* 24, no. 3 (2012): 267–290.

Nielsen, Jytte Seested, Mickael Bech, Kaare Christensen, Astrid Kiil, and Niels Christian Hvidt. "Risk Aversion and Religious Behaviour: Analysis Using a Sample of Danish Twins." *Economics and Human Biology* 26 (2017): 21–29.

Nielsen, Mark, Rohan Kapitány, and Rosemary Elkins. "The Perpetuation of Ritualistic Actions as Revealed by Young Children's Transmission of Normative Behavior." *Evolution and Human Behavior* 36, no. 3 (2015): 191–198.

Niemyjska, Aleksandra, and Krystyna Drat-Ruszczak. "When There Is Nobody, Angels Begin to Fly: Supernatural Imagery Elicited by a Loss of Social Connection." *Social Cognition* 31, no. 1 (2013): 57–71.

Nola, Robert. "Do Naturalistic Explanations of Religious Beliefs Debunk Religion?" In *A New Science of Religion*, edited by Gregory W. Dawes and James Maclaurin, 162–188. London: Routledge, 2013.

Noll, Mark A. *Jesus Christ and the Life of the Mind*. Grand Rapids, MI: Eerdmans Publishing Co., 2013.

Noll, Mark A. *The Scandal of the Evangelical Mind*. Grand Rapids, MI: Eerdmans Publishing Co., 1995.

Nordin, Andreas. "Indirect Reciprocity and Reputation Management in Religious Morality Relating to Concepts of Supernatural Agents." *Journal for the Cognitive Science of Religion* 3, no. 2 (2016): 125–153.

Norenzayan, Ara. *Big Gods : How Religion Transformed Cooperation and Conflict*. Princeton University Press, 2013.

Norenzayan, Ara. "Does Religion Make People Moral?" *Behaviour* 151, no. 2–3 (2014): 365–384.

Norenzayan, Ara. "Theodiversity." *Annual Review of Psychology* 67 (2016): 465–488.

Norenzayan, Ara, and Will M. Gervais. "The Origins of Religious Disbelief." *Trends in Cognitive Sciences* 17, no. 1 (2013): 20–25.

Norenzayan, Ara, and Ian G. Hansen. "Belief in Supernatural Agents in the Face of Death." *Personality & Social Psychology Bulletin* 32, no. 2 (2006): 174–187.

Norenzayan, Ara, and Azim Shariff. "The Origin and Evolution of Religious Prosociality." *Science* 322, no. 5898 (2008): 58–62.

Norenzayan, Ara, Scott Atran, Jason Faulkner, and Mark Schaller. "Memory and Mystery: The Cultural Selection of Minimally Counterintuitive Narratives." *Cognitive Science* 30, no. 3 (2006): 531–553.

Norenzayan, Ara, Ilan Dar-nimrod, Ian G. Hansen, and Travis Proulx. "Mortality Salience and Religion: Divergent Effects on the Defense of Cultural Worldviews for the Religious and the Non-religious." *European Journal of Social Psychology* 39, no. 1 (2009): 101–113.

Norenzayan, Ara, Will M. Gervais, and Kali H. Trzesniewski. "Mentalizing Deficits Constrain Belief in a Personal God." *PLoS ONE* 7, no. 5 (2012): 1–8.

Norenzayan, Ara, Ian G. Hansen, Jasmine Cady, Ara Norenzayan, Ian G. Hansen, Jasmine Cady, and Ara Norenzayan. "An Angry Volcano? Reminders of Death and Anthropomorphizing Nature." *Social Cognition* 26, no. 2 (2008): 190–197.

Norenzayan, Ara, Azim F. Shariff, Will M. Gervais, Aiyana K. Willard, Rita A. McNamara, Edward Slingerland, and Joseph Henrich. "The Cultural Evolution of Prosocial Religions." *Behavioral and Brain Sciences* 39 (2016): 1–86.

Norman, Emma R., and Rafael Delfin. "Wizards under Uncertainty: Cognitive Biases, Threat Assessment, and Misjudgments in Policy Making." *Politics and Policy* 40, no. 3 (2012): 369–402.

Norris, Pippa, and Ronald Inglehart. "Are High Levels of Existential Security Conducive to Secularization? A Response to Our Critics," In *The Changing World Religion Map*, edited by S.D. Brunn. Berlin: Springer, 2015.

Norris, Pippa, and Ronald Inglehart. *Sacred and Secular: Religion and Politics Worldwide*. Cambridge University Press, 2011.

Northmore-Ball, Ksenia, and Geoffrey Evans. "Secularization versus Religious Revival in Eastern Europe: Church Institutional Resilience, State Repression and Divergent Paths." *Social Science Research* 57 (2016): 31–48.

Oliner, Samuel P., and Pearl M. Oliner. *The Altruistic Personality: Rescuers of Jews in Nazi Europe*. New York: Touchstone, 1992.

Osborne, Danny, Petar Milojev, and Chris G. Sibley. "Examining the Indirect Effects of Religious Orientations on Well-being through Personal Locus of Control." *European Journal of Social Psychology* 46, no. 4 (2016): 492–505.

Otsuki, Kei. "Ecological Rationality and Environmental Governance on the Agrarian Frontier: The Role of Religion in the Brazilian Amazon." *Journal of Rural Studies* 32 (2013): 411.

Palmer, Craig T., Ryan O. Begley, and Kathryn Coe. "Totemism and Long-Term Evolutionary Success." *Psychology of Religion and Spirituality* 7, no. 4 (2015): 286–294.

Paul, Gregory S. "The Chronic Dependence of Popular Religiosity upon Dysfunctional Psychosociological Conditions." *Evolutionary Psychology* 7, no. 3 (2009): 398–441.

Paul, Gregory S. "Cross-National Correlations of Quantifiable Societal Health with Popular Religiosity and Secularism in the Prosperous Democracies," *Journal of Religion & Society* 7 (2005): 1–17.

Paul, Gregory S. "The Evolution of Popular Religiosity and Secularism: How First World Statistics Reveal Why Religion Exists, Why It Has Been Popular, and Why the Most Successful Democracies Are the Most Secular." In *Atheism and Secularity, Volume 1: Issues, Concepts and Definitions*, edited by Phil Zuckerman, 149–208. New York: Praeger, 2010.

Pazhoohi, Farid, Martin Lang, Dimitris Xygalatas, and Karl Grammer. "Religious Veiling as a Mate-Guarding Strategy: Effects of Environmental Pressures on Cultural Practices." *Evolutionary Psychological Science* 3, no. 2 (2017): 118–124.

Pelletier-Baldelli, Andrea, Derek J. Dean, Jessica R. Lunsford-Avery, Ashley K. Smith Watts, Joseph M. Orr, Tina Gupta, Zachary B. Millman, and Vijay A. Mittal. "Orbitofrontal Cortex Volume and Intrinsic Religiosity in Non-Clinical Psychosis." *Psychiatry Research: Neuroimaging*, 222 (2014): 124–130.

Pennycook, Gordon. "Evidence That Analytic Cognitive Style Influences Religious Belief: Comment on Razmyar and Reeve (2013)." *Intelligence* 43 (2014): 21–26.

Pennycook, Gordon, James Cheyne, Derek Koehler, and Jonathan Fugelsang. "Belief Bias during Reasoning among Religious Believers and Skeptics." *Psychonomic Bulletin & Review* 20, no. 4 (2013): 806–811.

Pennycook, Gordon, Robert Ross, Derek Koehler, and Jonathan Fugelsang. "Atheists and Agnostics Are More Reflective than Religious Believers: Four Empirical Studies and a Meta-Analysis." *PLoS ONE* 11, no. 4 (2016): 1–18.

Pennycook, Gordon, James Cheyne, Nathaniel Barr, Derek Koehler, and Jonathan Fugelsang. "Cognitive Style and Religiosity: The Role of Conflict Detection." *Memory and Cognition* 42, no. 1 (2014): 1–10.

Pennycook, Gordon, James Cheyne, Nathaniel Barr, Derek Koehler, and Jonathan Fugelsang. "On the Reception and Detection of Pseudo-Profound Bullshit." *Judgment and Decision Making* 10, no. 6 (2015): 549–563.

Pennycook, Gordon, James Cheyne, Paul Seli, Derek Koehler, and Jonathan Fugelsang. "Analytic Cognitive Style Predicts Religious and Paranormal Belief." *Cognition* 123, no. 3 (2012): 335–346.

Peoples, H.C., and F.W. Marlowe. "Subsistence and the Evolution of Religion." *Human Nature* 23, no. 3 (2012): 253–269.

Peoples, H.C., P. Duda, and F.W. Marlowe. "Hunter-Gatherers and the Origins of Religion." *Human Nature*, (2016): 1–22.

Periss, V.A., and D.F. Bjorklund. "Playing for God's Team: The Influence of Belief in the Supernatural on Perceptions of Religious, Spiritual, and Natural Cues." *Journal of Cognition and Culture* 16, no. 3–4 (2016): 215–244.

Pew Research Center. "The Future of World Religions: Population Growth Projections, 2010–2050." Washington, DC: Pew Research Center, May 11, 2015.

Plantinga, Alvin. "Games Scientists Play." In *The Believing Primate: Scientific, Philosophical and Theological Reflections on the Origin of Religion*, edited by Jeffrey Schloss and Michael Murray, 139–167. Oxford: Oxford University Press, 2009.

Plouffe, Rachel A., and Paul F. Tremblay. "The Relationship between Income and Life Satisfaction: Does Religiosity Play a Role?" *Personality and Individual Differences* 109 (2017): 67–71.

Power, Eleanor A. "Discerning Devotion: Testing the Signaling Theory of Religion." *Evolution and Human Behavior*, 38, no. 1 (2017): 82–91.

Preston, Jesse, and Nicholas Epley. "Science and God: An Automatic Opposition between Ultimate Explanations." *Journal of Experimental Social Psychology* 45, no. 1 (2009): 238–241.

Preston, Jesse Lee, and Ryan S. Ritter. "Different Effects of Religion and God on Prosociality With the Ingroup and Outgroup." *Personality and Social Psychology Bulletin* 39, no. 11 (2013): 1471–1483.

Preston, Benjamin, Johanna Mustelin, and Megan Maloney. "Climate Adaptation Heuristics and the Science/Policy Divide." *Mitigation and Adaptation Strategies for Global Change* 20, no. 3 (2015): 467–497.

Previc, Fred H. "The Role of the Extrapersonal Brain Systems in Religious Activity." *Consciousness and Cognition* 15, no. 3 (2006): 500–539.

Purzycki, Benjamin G. "The Evolution of Gods' Minds in the Tyva Republic." *Current Anthropology* 57, no. S13 (2016): S88–S104.

Purzycki, Benjamin G., and Valerie Kulundary. "Buddhism, Identity, and Class: Fairness and Favoritism in the Tyva Republic." *Religion, Brain & Behavior*, 7 (2017): 1–22.

Purzycki, Benjamin G., and Rita Anne McNamara. "An Ecological Theory of Gods' Minds." In *Advances in Religion, Cognitive Science, and Experimental Philosophy*, edited by Helen De Cruz and Ryan Nichols, 143–167. London: Bloomsbury Academic, 2016.

Purzycki, Benjamin G., and Richard Sosis. "The Extended Religious Phenotype and the Adaptive Coupling of Ritual and Belief." *Israel Journal of Ecology & Evolution* 59, no. 2 (2013): 99–108.

Purzycki, Benjamin G., and Aiyana K. Willard. "MCI Theory: A Critical Discussion." *Religion, Brain & Behavior*, 6, no. 3 (2015): 207–248.

Purzycki, Benjamin G., Daniel N. Finkel, John Shaver, Nathan Wales, Adam B. Cohen, and Richard Sosis. "What Does God Know? Supernatural Agents' Access to Socially Strategic and Non-Strategic Information." *Cognitive Science* 36, no. 5 (2012): 846–869.

Purzycki, Benjamin G., Omar S. Haque, and Richard Sosis. "Extending Evolutionary Accounts of Religion beyond the Mind: Religions as Adaptive Systems." In *Evolution, Religion and Cognitive Science: Critical and Constructive Essays*, edited by F. Watts, 74–91. Oxford: Oxford University Press, 2014.

Purzycki, Benjamin G., Jordan Kiper, John Shaver, Daniel N. Finkel, and Richard Sosis. "Religion." In *Emerging Trends in the Social and Behavioral Sciences*, edited by Robert Scott and Stephan Kosslyn, 1–16. New York: John Wiley & Sons, 2015.

Purzycki, Benjamin G., Coren Apicella, Quentin D. Atkinson, Emma Cohen, Rita Anne Mcnamara, Aiyana K. Willard, Dimitris Xygalatas, Ara Norenzayan, and Joseph Henrich. "Moralistic Gods, Supernatural Punishment and the Expansion of Human Sociality." *Nature*, 530 (2016): 327–336.

Putnam, Robert D. *Bowling Alone: The Collapse and Revival of American Community*. New York: Touchstone Books, 2001.

Putnam, Robert D., and David E. Campbell. *American Grace: How Religion Divides and Unites Us*. New York: Simon & Schuster, 2012.

Puurtinen, Mikael, Stephen Heap, and Tapio Mappes. "The Joint Emergence of Group Competition and Within-Group Cooperation." *Evolution and Human Behavior* 36, no. 3 (2015): 211–217.

Pyysiäinen, Ilkka. *Supernatural Agents : Why We Believe in Souls, Gods and Buddhas*. Oxford University Press, 2009.

Pyysiäinen, Ilkka, and Marc Hauser. "The Origins of Religion: Evolved Adaptation or by-Product?" *Trends in Cognitive Sciences* 14, no. 3 (2010): 104–109.

Ramkissoon, Anmarie Kamanie, Casswina Donald, and Gerard Hutchinson. "Supernatural versus Medical: Responses to Mental Illness from Undergraduate University Students in Trinidad." *International Journal of Social Psychiatry* 63, no. 4 (2017): 330–338.

Ramsay, Jonathan E., Eddie M.W. Tong, Joyce S. Pang, and Avijit Chowdhury. "A Puzzle Unsolved: Failure to Observe Different Effects of God and Religion Primes on Intergroup Attitudes." *PLoS ONE* 11, no. 1 (2016): 1–21.

Ramsay, Jonathan, Joyce Pang, Megan Johnson Shen, and Wade Rowatt. "Rethinking Value Violation: Priming Religion Increases Prejudice in Singaporean Christians and Buddhists." *International Journal for the Psychology of Religion* 24, no. 1 (2014): 1–15.

Randles, Daniel, Michael Inzlicht, Travis Proulx, Alexa M. Tullett, and Steven J. Heine. "Is Dissonance Reduction a Special Case of Fluid Compensation? Evidence That

Dissonant Cognitions Cause Compensatory Affirmation and Abstraction." *Journal of Personality and Social Psychology* 108, no. 5 (2015): 697–710.

Ray, Shanna D., Jennifer D. Lockman, Emily J. Jones, and Melanie H. Kelly. "Attributions to God and Satan About Life-Altering Events." *Psychology of Religion and Spirituality* 7, no. 1 (2015): 60–69.

Razmyar, Soroush, and Charlie L. Reeve. "Individual Differences in Religiosity as a Function of Cognitive Ability and Cognitive Style." *Intelligence* 41, no. 5 (2013): 667–673.

Reddish, Paul, Ronald Fischer, and Joseph Bulbulia. "Let's Dance Together: Synchrony, Shared Intentionality and Cooperation." *PLoS ONE* 8, no. 8 (2013): 1–13.

Reed, Phil, and Natasha Clarke. "Effect of Religious Context on the Content of Visual Hallucinations in Individuals High in Religiosity." *Psychiatry Research* 215, no. 3 (2014): 594–598.

Reid, Vincent M., Kirsty Dunn, Robert J. Young, Johnson Amu, Tim Donovan, and Nadja Reissland. "The Human Fetus Preferentially Engages with Face-like Visual Stimuli." *Current Biology* 27, no. 12 (2017): 1825–1828.

Reser, Joseph P., and Janet K. Swim. "Adapting to and Coping With the Threat and Impacts of Climate Change." *American Psychologist* 66, no. 4 (2011): 277–289.

Rholes, W. Steven, and Jeffry A. Simpson, eds. *Adult Attachment: Theory, Research, and Clinical Implications*. New York: The Guilford Press, 2004.

Richerson, Peter, Lesley Newson, Cody Ross, Paul Smaldino, Timothy Waring, Matthew Zefferman, Ryan Baldini, et al. "Cultural Group Selection Plays an Essential Role in Explaining Human Cooperation: A Sketch of the Evidence." *Behavioral and Brain Sciences*, (2014): 1–68.

Richert, Rebekah A., Anondah R. Saide, Kirsten A. Lesage, and Nicholas J. Shaman. "The Role of Religious Context in Children's Differentiation between God's Mind and Human Minds." *British Journal of Developmental Psychology* 35, no. 1 (2017): 37–59.

Riekki, Tapani, Marjaana Lindeman, and Tuuka Raij. "Supernatural Believers Attribute More Intentions to Random Movement than Skeptics: An FMRI Study." *Social Neuroscience* 9, no. 4 (2014): 400–411.

Riekki, Tapani, Marjaana Lindeman, Marja Aleneff, Anni Halme, and Antti Nuortimo. "Paranormal and Religious Believers Are More Prone to Illusory Face Perception than Skeptics and Non-believers." *Applied Cognitive Psychology* 27, no. 2 (2013): 150–155.

Riggio, Heidi R., Joshua Uhalt, and Brigitte K. Matthies. "Unanswered Prayers: Religiosity and the God-Serving Bias." *The Journal of Social Psychology* 154, no. 6 (2014): 491–514.

Risen, Jane L. "Believing What We Do Not Believe: Acquiescence to Superstitious Beliefs and Other Powerful Intuitions." *Psychological Review*, 123, no. 2 (2015): 182–207.

Ritchie, Stuart J., Alan J. Gow, and Ian J. Deary. "Religiosity Is Negatively Associated with Later-Life Intelligence, but Not with Age-Related Cognitive Decline." *Intelligence* 46 (2014): 9–17.

Ritter, Ryan S., and Jesse Lee Preston. "Gross Gods and Icky Atheism: Disgust Responses to Rejected Religious Beliefs." *Journal of Experimental Social Psychology* 47, no. 6 (2011): 1225–1230.

Rock, Stella. "Introduction: Religion, Prejudice and Conflict in the Modern World." *Patterns of Prejudice* 38, no. 2 (2004): 101–108.

Roitto, Rikard. "Dangerous but Contagious Altruism: Recruitment of Group Members and Reform of Cooperation Style through Altruism in Two Modified Versions of Hammond and Axelrod's Simulation of Ethnocentric Cooperation." *Religion, Brain & Behavior* 6, no. 2 (2016): 154–168.

Rosenkranz, Patrick, and Bruce G. Charlton. "Individual Differences in Existential Orientation: Empathizing and Systemizing Explain the Sex Difference in Religious Orientation and Science Acceptance." *Archive for the Psychology of Religion* 35, no. 1 (2013): 119–146.

Ross, Lee D., Yphtach Lelkes, and Alexandra G. Russell. "How Christians Reconcile Their Personal Political Views and the Teachings of Their Faith: Projection as a Means of Dissonance Reduction." *Proceedings of the National Academy of Sciences* 109, no. 10 (2012): 3616.

Rossano, Matt. *Supernatural Selection: How Religion Evolved.* New York: Oxford University Press, 2010.

Rottman, Joshua, Liqi Zhu, Wen Wang, Rebecca Seston Schillaci, Kelly J. Clark, and Deborah Kelemen. "Cultural Influences on the Teleological Stance: Evidence from China." *Religion, Brain & Behavior* 7, no. 1 (2017): 17–26.

Routledge, Clay, Andrew Abeyta, and Christina Roylance. "An Existential Function of Evil: The Effects of Religiosity and Compromised Meaning on Belief in Magical Evil Forces." *Motivation and Emotion* 40, no. 5 (2016): 681–688.

Routledge, Clay, Jeff Greenberg, Melissa Soenke, and Tom Pyszczynski. "Death and End Times: The Effects of Religious Fundamentalism and Mortality Salience on Apocalyptic Beliefs." *Religion, Brain & Behavior* 6, no. 2 (2016).

Routledge, Clay, Christina Roylance, and Andrew Abeyta. "Further Exploring the Link Between Religion and Existential Health: The Effects of Religiosity and Trait Differences in Mentalizing on Indicators of Meaning in Life." *Journal of Religion and Health* 56, no. 2 (2017): 604–613.

Routledge, Clay, Christina Roylance, and Andrew Abeyta. "Miraculous Meaning: Threatened Meaning Increases Belief in Miracles." *Journal of Religion and Health* 56, no. 3 (2017): 776–783.

Rutjens, Bastiaan T., Joop van Der Pligt, and Frenk van Harreveld. "Deus or Darwin: Randomness and Belief in Theories about the Origin of Life." *Journal of Experimental Social Psychology* 46, no. 6 (2010): 1078–1080.

Rutjens, Bastiaan T., Frenk van Harreveld, and Joop van Der Pligt. "Step by Step: Finding Compensatory Order in Science." *Current Directions in Psychological Science* 22, no. 3 (2013): 250–255.

Rutjens, Bastiaan T., Frenk van Harreveld, Joop van Der Pligt, Michiel van Elk, and Tom Pyszczynski. "A March to a Better World? Religiosity and the Existential Function of Belief in Social-Moral Progress." *The International Journal for the Psychology of Religion*, 26, no. 1 (2015): 1–18.

Sääksvuori, Lauri, Lauri Sääksvuori, Tapio Mappes, and Mikael Puurtinen. "Costly Punishment Prevails in Inter Group Conflict." *Proceedings of the Royal Society B: Biological Sciences* 278, no. 1723 (2011): 3428–3436.

Saavedra, Javier. "Function and Meaning in Religious Delusions: A Theoretical Discussion from a Case Study." *Mental Health, Religion & Culture* 17, no. 1 (2014): 39–51.

Sablosky, R. "Does Religion Foster Generosity?" *Social Science Journal* 51, no. 4 (2014): 545–555.

Sadique, Kim, and Perry Stanislas, eds. *Religion, Faith and Crime: Theories, Identities and Issues.* Palgrave Macmillan, 2016.

Sagioglou, Christina, and Matthias Forstmann. "Activating Christian Religious Concepts Increases Intolerance of Ambiguity and Judgment Certainty." *Journal of Experimental Social Psychology* 49, no. 5 (2013): 933.

Salazar, Carles. "Religious Symbolism and the Human Mind: Rethinking Durkheim's Elementary Forms of Religious Life." *Method and Theory in the Study of Religion* 27 (2015): 82–96.

Saleam, James, and Ahmed A. Moustafa. "The Influence of Divine Rewards and Punishments on Religious Prosociality." *Frontiers in Psychology*, 2016.

Sanchez, Clinton, Brian Sundermeier, Kenneth Gray, and Robert J. Calin – Jageman. "Direct Replication of Gervais & Norenzayan (2012): No Evidence That Analytic Thinking Decreases Religious Belief." *PLoS ONE* 12, no. 2 (2017): 1–8.

Sandage, Steven J., and F. LeRon Shults. "Relational Spirituality and Transformation: A Relational Integration Model." *Journal of Psychology & Christianity* 26, no. 3 (2007): 261–269.

Saribay, S. Adil, and Onurcan Yilmaz. "Analytic Cognitive Style and Cognitive Ability Differentially Predict Religiosity and Social Conservatism." *Personality and Individual Differences* 114 (2017): 24–29.

Saroglou, Vassilis. "Believing, Bonding, Behaving, and Belonging." *Journal of Cross-Cultural Psychology* 42, no. 8 (2011): 1320–1340.

Saroglou, Vassilis. "Intergroup Conflict, Religious Fundamentalism, and Culture." *Journal of Cross-Cultural Psychology* 47, no. 1 (2016): 33–41.

Saroglou, Vassilis, Vanessa Delpierre, and Rebecca Dernelle. "Values and Religiosity: A Meta-Analysis of Studies Using Schwartz's Model." *Personality and Individual Differences* 37, no. 4 (2004): 721–734.

Schachner, Adena, Liqi Zhu, Jing Li, and Deborah Kelemen. "Is the Bias for Function-Based Explanations Culturally Universal? Children from China Endorse Teleological Explanations of Natural Phenomena." Journal of Experimental Child Psychology 157 (2017): 29–48.

Schaffer, Jonathan. "Cognitive Science and Metaphysics: Partners in Debunking." In Goldman and His Critics, edited by Korblith and McLaughlin, 337–368. New York: Wiley-Blackwell, 2016.

Schaller, Mark, Ara Norenzayan, Steven J. Heine, Toshio Yamagishi, and Tatsuya Kameda, eds. Evolution, Culture, and the Human Mind. New York: Psychology Press, 2009.

Schell, L.M., C.D. Lynn, J.J. Paris, and C.A. Frye. "Religious-Commitment Signaling and Impression Management amongst Pentecostals: Relationships to Salivary Cortisol and Alpha-Amylase." Journal of Cognition and Culture 15, no. 3–4 (2015): 299–319.

Scheve, Kenneth, David Stasavage, and others. "Religion and Preferences for Social Insurance." Quarterly Journal of Political Science 1, no. 3 (2006): 255–286.

Schilbrack, Kevin. "A Realist Social Ontology of Religion." Religion 47, no. 2 (2016): 161–178.

Schjoedt, Uffe, Jesper Sørensen, Kristoffer Laigaard Nielbo, Dimitris Xygalatas, Panagiotis Mitkidis, and Joseph Bulbulia. "Cognitive Resource Depletion in Religious Interactions." Religion, Brain & Behavior 3, no. 1 (2013): 39–55.

Schjoedt, Uffe, Hans Stødkilde-Jørgensen, Armin W. Geertz, Torben E. Lund, and Andreas Roepstorff. "The Power of Charisma – Perceived Charisma Inhibits the Frontal Executive Network of Believers in Intercessory Prayer." Social Cognitive and Affective Neuroscience 6, no. 1 (2011): 119–127.

Schmitt, David P., and Robert C. Fuller. "On the Varieties of Sexual Experience: Cross-Cultural Links Between Religiosity and Human Mating Strategies." Psychology of Religion and Spirituality 7, no. 4 (2015): 314–326.

Schnabel, Landon. "Religion and Gender Equality Worldwide: A Country-Level Analysis." Social Indicators Research 129, no. 2 (2016): 893–907.

Schnell, Tatjana. "Dimensions of Secularity (DoS): An Open Inventory to Measure Facets of Secular Identities." The International Journal for the Psychology of Religion 25, no. 4 (2015): 272–292.

Schofield, Malcolm B., Ian S. Baker, Paul Staples, and David Sheffield. "Mental Representations of the Supernatural: A Cluster Analysis of Religiosity, Spirituality and Paranormal Belief." Personality and Individual Differences 101 (2016): 419–424.

Schuurmans-Stekhoven, James Benjamin. "'As a Shepherd Divideth His Sheep from the Goats': Does the Daily Spiritual Experiences Scale Encapsulate Separable Theistic and Civility Components?" Social Indicators Research 110, no. 1 (2013): 131–146.

Schuurmans-Stekhoven, James Benjamin. "Is It God or Just the Data That Moves in Mysterious Ways? How Well-Being Research May Be Mistaking Faith for Virtue." Social Indicators Research 100, no. 2 (2010): 313–330.

Schuurmans-Stekhoven, James Benjamin. "Are We, like Sheep, Going Astray: Is Costly Signaling (or Any Other Mechanism) Necessary to Explain the Belief-as-Benefit Effect?" *Religion, Brain & Behavior* 7, no. 3 (2017): 258–262.

Schuurmans-Stekhoven, James Benjamin. "Measuring Spirituality as Personal Belief in Supernatural Forces: Is the Character Strength Inventory-Spirituality Subscale a Brief, Reliable and Valid Measure?" *Implicit Religion* 17, no. 2 (2014): 211–222.

Schuurmans-Stekhoven, James Benjamin. "Spirit or Fleeting Apparition? Why Spirituality's Link with Social Support Might Be Incrementally Invalid." *Journal of Religion and Health*, 56, no. 4 (2017): 1248–1267.

Schwadel, Philip, and Erik Johnson. "The Religious and Political Origins of Evangelical Protestants' Opposition to Environmental Spending." *Journal for the Scientific Study of Religion* 56, no. 1 (2017): 179–198.

Seguino, Stephanie. "Help or Hindrance? Religion's Impact on Gender Inequality in Attitudes and Outcomes." *World Development* 39, no. 8 (2011): 1308–1321.

Sela, Yael, Todd K. Shackelford, and James R. Liddle. "When Religion Makes It Worse: Religiously Motivated Violence as a Sexual Selection Weapon." In *The Attraction of Religion: A New Evolutionary Psychology of Religion*, edited by D. Jason Slone and James A. van Slyke, 111–131. London: Bloomsbury Academic, 2015.

Shah, Timothy Samuel, Alfred Stepan, and Monica Duffy Toft. *Rethinking Religion and World Affairs*. Oxford University Press, 2012.

Shariff, Azim, and Lara Aknin. "The Emotional Toll of Hell: Cross-National and Experimental Evidence for the Negative Well-Being Effects of Hell Beliefs." *PLoS ONE* 9, no. 1 (2014).

Shariff, Azim F., and Ara Norenzayan. "God Is Watching You Priming God Concepts Increases Prosocial Behavior in an Anonymous Economic Game." *Psychological Science* 18, no. 9 (2007): 803–809.

Shariff, Azim F., and Ara Norenzayan. "Mean Gods Make Good People: Different Views of God Predict Cheating Behavior." *International Journal for the Psychology of Religion* 21, no. 2 (2011): 85–96.

Shariff, Azim F., Stephanie R. Kramer, and Jared Piazza. "Morality and the Religious Mind: Why Theists and Nontheists Differ," *Trends in Cognitive Sciences* 18, no. 9 (2014): 439–441.

Shariff, Azim F., Ara Norenzayan, and Adam B. Cohen. "The Devil's Advocate: Secular Arguments Diminish Both Implicit and Explicit Religious Belief." *Journal of Cognition and Culture* 8, no. 3–4 (2008): 417–423.

Shariff, Azim F., Aiyana K. Willard, Teresa Andersen, and Ara Norenzayan. "Religious Priming" *Personality and Social Psychology Review* 20, no. 1 (2016): 27–48.

Shariff, Azim F., Aiyana K. Willard, Michael Muthukrishna, Stephanie R. Kramer, and Joseph Henrich. "What Is the Association between Religious Affiliation and Children's Altruism?" *Current Biology* 26, no. 15 (2016b): 699–700.

Shaver, John. H. "Why and How Do Religious Individuals, and Some Religious Groups, Achieve Higher Relative Fertility?" *Religion, Brain and Behavior* 7 (2017): 1–4.

Shaver, John H., Geoffrey Troughton, Chris G. Sibley, and Joseph A. Bulbulia. "Religion and the Unmaking of Prejudice toward Muslims: Evidence from a Large National Sample" *PLoS ONE* 11, no. 3 (2016): 1–25.

Sheikh, Hammad, Scott Atran, Jeremy Ginges, Lydia Wilson, Nadine Obeid, and Richard Davis. "The Devoted Actor as Parochial Altruist: Sectarian Morality, Identity Fusion, and Support for Costly Sacrifices," *Cliodynamics: The Journal of Quantitative History and Cultural Evolution* 5, no. 1 (2014): 23–40.

Sheikh, Hammad, Jeremy Ginges, and Scott Atran. "Sacred Values in the Israeli–Palestinian Conflict: Resistance to Social Influence, Temporal Discounting, and Exit Strategies." *Annals of the New York Academy of Sciences* 12991, no. 1 (2013): 11–24.

Sheikh, Hammad, Jeremy Ginges, Alin Coman, and Scott Atran. "Religion, Group Threat and Sacred Values." *Judgment and Decision Making* 7, no. 2 (2012): 110–118.

Shenberger, Jessica M., Brandt A. Smith, and Michael A. Zárate. "The Effect of Religious Imagery in a Risk-Taking Paradigm." *Peace and Conflict: Journal of Peace Psychology* 20, no. 2 (2014): 150–158.

Shenhav, Amitai, David G. Rand, and Joshua D. Greene. "Divine Intuition: Cognitive Style Influences Belief in God." *Journal of Experimental Psychology: General* 141, no. 3 (2013): 423–428.

Shiah, Yung-Jong, Frances Chang, Shih-Kuang Chiang, and Wai-Cheong Tam. "Religion and Subjective Well-Being: Western and Eastern Religious Groups Achieved Subjective Well-Being in Different Ways." *Journal of Religion and Health* 55, no. 4 (2016): 1263–1269.

Shtulman, Andrew. "Variation in the Anthropomorphization of Supernatural Beings and Its Implications for Cognitive Theories of Religion." *Journal of Experimental Psychology: Learning, Memory, and Cognition* 34, no. 5 (2008): 1123–1138.

Shults, F. LeRon. "Can Theism Be Defeated? CSR and the Debunking of Supernatural Agent Abductions." *Religion, Brain & Behavior* 6, no. 4 (2015): 349–355.

Shults, F. LeRon. "Can We Predict and Prevent Religious Radicalization?" In *Processes of Violent Extremism in the 21st Century: International and Interdisciplinary Perspectives*, edited by Gwenyth Øverland. Cambridge: Cambridge Scholars Press, forthcoming.

Shults, F. LeRon. *Christology and Science*. Grand Rapids, MI: Eerdmans Publishing Co., 2008.

Shults, F. LeRon. "De-Oedipalizing Theology: Desire, Difference and Deleuze." In Shults, F. LeRon and Jan-Olav Henriksen, eds., *Saving Desire: The Seduction of Christian Theology*. Grand Rapids. MI: Eerdmans Publishing Co., 2011.

Shults, F. LeRon. "Dis-Integrating Psychology and Theology." *Journal of Psychology and Theology* 40, no. 1 (2012): 21–25.

Shults, F. LeRon. "Ethics, Exemplarity and Atonement." In *Theology and the Science of Moral Action*, edited by James A. van Slyke, Gregory R. Peterson, Kevin S. Reimer, Michael L. Spezio, and Warren S. Brown, 164–178. New York: Routledge, 2012.

Shults, F. LeRon. "How to Survive the Anthropocene: Adaptive Atheism and the Evolution of Homo Deiparensis." *Religions* 6, no. 1 (2015): 1–18.

Shults, F. LeRon. *Iconoclastic Theology : Gilles Deleuze and the Secretion of Atheism.* Edinburgh University Press, 2014.

Shults, F. LeRon. "Reflections on the 2014 Nobel Peace Prize." In *Agder Vitenskapsakademi: Årbok 2014*, edited by Ernst Håkon Jahr, Rolf Tomas Nossum, and Leiv Storesletten, 232–236. Kristiansand: Portal, 2015.

Shults, F. LeRon. "Science and Religious Supremacy: Toward a Naturalist Theology of Religions." In *Science and the World's Religions, Volume III: Religions and Controversies*, edited by Wesley Wildman and Patrick McNamara, 73–100. New York: Praeger, 2012.

Shults, F. LeRon. "Spiritual Entanglement: Transforming Religious Symbols at Çatalhöyük." In *Religion in the Emergence of Civilization: Çatalhöyük as a Case Study*, edited by Ian Hodder, 73–98. Cambridge University Press, 2010.

Shults, F. LeRon. "The Problem of Good (and Evil): Arguing about Axiological Conditions in Science and Religion." In *Science and the World's Religions, Volume I: Origins and Destinies*, edited by Wesley J. Wildman and Patrick McNamara, 39–68. New York: Praeger, 2012.

Shults, F. LeRon. "Theology after the Birth of God: A Response to Commentators." *Syndicate: A New Forum for Theology* 2, no. 3 (2015): 91–108.

Shults, F. LeRon. *Theology after the Birth of God : Atheist Conceptions in Cognition and Culture*. New York: Palgrave Macmillan, 2014.

Shults, F. LeRon. "Transforming Religious Plurality." *Studies in Interreligious Dialogue* 20, no. 2 (2010): 148–170.

Shults, F. LeRon. "What's the Use? Pragmatic Reflections on Neville's Ultimates." *American Journal of Theology & Philosophy* 36, no. 1 (2015): 69–80.

Shults, F. LeRon. "Wising Up: The Evolution of Natural Theology." *Zygon* 47, no. 3 (2012): 542–548.

Shults, F. LeRon, and Steven J. Sandage. *The Faces of Forgiveness: Searching for Wholeness and Salvation*. Grand Rapids, MI: Baker Academic, 2003.

Shults, F. LeRon, and Steven J. Sandage. *Transforming Spirituality: Integrating Theology and Psychology*. Grand Rapids, MI: Baker Academic, 2006.

Shults, F. LeRon, and Wesley J. Wildman. "Modeling Çatalhöyük: Simulating Religious Entanglement and Social Investment in the Neolithic." In *Religion, History and Place in the Origin of Settled Life*, edited by Ian Hodder. Colorado Springs, CO: University of Colorado Press, forthcoming.

Shults, F. LeRon, Ross Gore, Carlos Lemos, and Wesley J. Wildman. "Why Do the God-less Prosper? Modeling the Cognitive and Coalitional Mechanisms That Promote Atheism." *Psychology of Religion and Spirituality*, forthcoming.

Shults, F. LeRon, Ross Gore, Wesley J. Wildman, Justin E. Lane, Chris Lynch, and Monica Toft. "Mutually Escalating Religious Violence: A Generative and Predictive Computational Model." *Social Simulation Conference Proceedings*, 2017.

Shults, F. LeRon, Justin E. Lane, Saikou Diallo, Christopher Lynch, Wesley J. Wildman, and Ross Gore. "Modeling Terror Management Theory: Computer Simulations of the Impact of Mortality Salience on Religiosity." *Religion, Brain & Behavior* 8, no. 1 (2018): 77–100.

Sibley, Chris G., and Joseph Bulbulia. "Faith after an Earthquake: A Longitudinal Study of Religion and Perceived Health before and after the 2011 Christchurch New Zealand Earthquake" *PLos ONE* 7, no. 12 (2012): 1–10.

Silver, Christopher F., Thomas J. Coleman, Ralph W. Hood, and Jenny M. Holcombe. "The Six Types of Nonbelief: A Qualitative and Quantitative Study of Type and Narrative." *Mental Health, Religion & Culture* 17, no. 10 (2014): 990–1001.

Silvia, Paul J., Emily C. Nusbaum, and Roger E. Beaty. "Blessed Are the Meek? Honesty–humility, Agreeableness, and the HEXACO Structure of Religious Beliefs, Motives, and Values." *Personality and Individual Differences* 66 (2014): 19–23.

Simpson, Ain, and Kimberly Rios. "The Moral Contents of Anti-Atheist Prejudice (and Why Atheists Should Care about It)." *European Journal of Social Psychology*, 2017,

Singh, Dhairyya, and Garga Chatterjee. "The Evolution of Religious Belief in Humans: A Brief Review with a Focus on Cognition." *Journal of Genetics* 96, no. 3 (2017): 517–524.

Singh, Manvir. "The Cultural Evolution of Shamanism." *Behavioral and Brain Sciences*, in press.

Sinnott-Armstrong, Walter, and Christian B. Miller, eds. *Moral Psychology-The Evolution of Morality: Adaptations and Innateness, Vol. 1.* Cambridge, Mass: The MIT Press, 2007.

Slone, D. Jason. *Theological Incorrectness : Why Religious People Believe What They Shouldn't.* Oxford University Press, 2007.

Smith, Aaron C.T. *Thinking about Religion: Extending the Cognitive Science of Religion.* New York, NY: Palgrave Macmillan, 2014.

Smith, Jesse M. "Creating a Godless Community: The Collective Identity Work of Contemporary American Atheists." *Journal for the Scientific Study of Religion* 52, no. 1 (2013): 80–99.

Smith, Brandt A., and Michael A. Zárate. "The Effects of Religious Priming and Persuasion Style on Decision-Making in a Resource Allocation Task." *Peace and Conflict: Journal of Peace Psychology*, 21, no. 4 (2015): 665–668.

Smith, Buster G., and Joseph O. Baker. *American Secularism: Cultural Contours of Non-religious Belief Systems.* New York: NYU Press, 2015.

Soler, M., and H.L. Lenfesty. "Coerced Coordination, Not Cooperation." *Behavioral and Brain Sciences* 39 (2014): 39.

Sosis, Richard. "Religious Behaviors, Badges, and Bans: Signaling Theory and the Evolution of Religion." In *Where God and Science Meet: How Brain and Evolutionary Studies Alter Our Understanding of Religion,* edited by Patrick McNamara and Wesley J. Wildman, 61–86. New York: Praeger, 2006.

Sosis, Richard, and Jordan Kiper. "Religion Is More Than Belief: What Evolutionary Theories of Religion Tell Us About Religious Commitment." In *Challenges to Religion and Morality: Disagreements and Evolution,* edited by Michael Bergmann and Patrick Kain, 256–276. Oxford University Press, 2014.

Speed, David "Unbelievable?! Theistic/Epistemological Viewpoint Affects Religion – Health Relationship." *Journal of Religion and Health* 56, no. 1 (2017): 238–257.

Speed, David, and Ken Fowler. "Good for All? Hardly! Attending Church Does Not Benefit Religiously Unaffiliated." *Journal of Religion and Health* 56, no. 3 (2017): 986–1002.

Speed, David, and Ken Fowler. "What's God Got to Do with It? How Religiosity Predicts Atheists' Health." *Journal of Religion and Health* 55, no. 1 (2016): 296–308.

Sperber, Dan. *Explaining Culture: A Naturalistic Approach.* Oxford, UK: Blackwell Publishers, 1996.

Sperber, Dan, and Deirdre Wilson. *Relevance: Communication and Cognition.* 2nd edition. Oxford, UK: Wiley-Blackwell, 1996.

Squazzoni, Flaminio. *Agent-Based Computational Sociology.* New York: John Wiley & Sons, 2012.

Squazzoni, Flaminio, ed. *Epistemological Aspects of Computer Simulation in the Social Sciences.* Berlin: Springer, 2009.

Stankov, Lazar, and Jihyun Lee. "Nastiness, Morality and Religiosity in 33 Nations." *Personality and Individual Differences* 99 (2016): 56–66.

Stavrova, Olga, Daniel Ehlebracht, and Detlef Fetchenhauer. "Belief in Scientific – Technological Progress and Life Satisfaction: The Role of Personal Control." *Personality and Individual Differences* 96 (2016): 227–236.

Steffen, P.R., K.S. Masters, and S. Baldwin. "What Mediates the Relationship Between Religious Service Attendance and Aspects of Well-Being?" *Journal of Religion and Health* 56, no. 1 (2017): 158–170.

Stephens, Mitchell. *Imagine There's No Heaven: How Atheism Helped Create the Modern World.* New York: Macmillan, 2014.

Sterelny, Kim. "Religion Re-Explained." *Religion, Brain & Behavior* 7 (2017): 1–20.

Stewart, Evan M. "The True (Non)Believer? Atheists and the Atheistic in the United States." In *Sociology of Atheism,* 7:137–160. Annual Review of the Sociology of Religion. Leiden: Brill Academic, 2016.

Stinespring, John, and Ryan T. Cragun. "Simple Markov Model for Estimating the Growth of Nonreligion in the United States." *Science, Religion and Culture* 2, no. 3 (2015): 96–103.

Stoet, Gijsbert, and David C. Geary. "Students in Countries with Higher Levels of Religiosity Perform Lower in Science and Mathematics." *Intelligence* 62 (2017): 71–78.

Stolz, Jörg. "Institutional, Alternative, Distanced, and Secular: Four Types of (Un)belief and their Gods." *Nordic Journal of Religion and Society* 30, no. 1 (2017): 4–23.

Streib, Heinz, and Constantin Klein. "Religious Styles Predict Interreligious Prejudice: A Study of German Adolescents with the Religious Schema Scale." *International Journal for the Psychology of Religion* 24, no. 2 (2014): 151–163.

Strozier, Charles B., David M. Terman, and James W. Jones. *The Fundamentalist Mindset: Psychological Perspectives on Religion, Violence, and History*. New York: Oxford University Press, 2010.

Strulik, Holger. "An Economic Theory of Religious Belief." *Journal of Economic Behavior and Organization* 128 (2016): 35–46.

Sumerau, J. Edward. "Some of Us Are Good, God-Fearing Folks." *Journal of Contemporary Ethnography* 46, no. 1 (2017): 3–29.

Sumerau, J. Edward, and Ryan T. Cragun. "'I Think Some People Need Religion': The Social Construction of Nonreligious Moral Identities." *Sociology of Religion* 77, no. 4 (2016): 386–407.

Sumerau, J. Edward, Alexandra C.H. Nowakowski, and Ryan T. Cragun. "An Interactionist Approach to the Social Construction of Deities." *Symbolic Interaction* 39, no. 4 (2016): 577–594.

Sun, Ron. *Cognition and Multi-Agent Interaction: From Cognitive Modeling to Social Simulation*. Cambridge University Press, 2006.

Sun, Ron, and Pierson Fleischer. "A Cognitive Social Simulation of Tribal Survival Strategies: The Importance of Cognitive and Motivational Factors." *Journal of Cognition and Culture* 12, no. 3–4 (2012): 287–321.

Svedholm, Annika M., and Marjaana Lindeman. "The Separate Roles of the Reflective Mind and Involuntary Inhibitory Control in Gatekeeping Paranormal Beliefs and the Underlying Intuitive Confusions." *British Journal of Psychology* 104, no. 3 (2013): 303–319.

Svensson, Isak. "Fighting with Faith." *Journal of Conflict Resolution* 51, no. 6 (2007): 930–949.

Svensson, Jonas. "God's Rage: Muslim Representations of HIV/Aids as a Divine Punishment from the Perspective of the Cognitive Science of Religion." *Numen* 61, no. 5–6 (2014): 569–593.

Swann, William B., Michael D. Buhrmester, Angel Gómez, Jolanda Jetten, Brock Bastian, Alexandra Vázquez, Amarina Ariyanto, et al. "What Makes a Group Worth

Dying for? Identity Fusion Fosters Perception of Familial Ties, Promoting Self-Sacrifice." *Journal of Personality and Social Psychology* 106, no. 6 (2014): 912–926.

Swann, William B., Angel Gómez, John F. Dovidio, Sonia Hart, and Jolanda Jetten. "Dying and Killing for One's Group: Identity Fusion Moderates Responses to Intergroup Versions of the Trolley Problem." *Psychological Science* 21, no. 8 (2010): 1176–1183.

Talbot, Elizabeth, and Colin Arthur Wastell. "Corrected by Reflection: The De-Anthropomorphized Mindset of Atheism." *Journal for the Cognitive Science of Religion* 3, no. 2 (2017): 113–124.

Taves, Ann. "Reverse Engineering Complex Cultural Contexts: Identifying Building Blocks of 'Religion.'" *Journal of Cognition and Culture* 15, no. 1–2 (2015): 191–216.

Teehan, John. "Cognitive Science and the Limits of Theology." In *The Roots of Religion: Exploring the Cognitive Science of Religion*, edited by Roger Trigg and Justin L. Barrett. Surrey, UK: Ashgate, 2014.

Teehan, John. *In the Name of God: The Evolutionary Origins of Religious Ethics and Violence*. Malden, MA: Wiley-Blackwell, 2010.

Teehan, John. "The Cognitive Bases of the Problem of Evil." *The Monist* 96, no. 3 (2013): 325–348.

ten Kate, Josje, Willem de Koster, and Jeroen van der Waal. "The Effect of Religiosity on Life Satisfaction in a Secularized Context: Assessing the Relevance of Believing and Belonging." *Review of Religious Research* 59, no. 2 (2017): 135–155.

Terrizzi, John A., Russ Clay, and Natalie J. Shook. "Does the Behavioral Immune System Prepare Females to Be Religiously Conservative and Collectivistic?" *Personality and Social Psychology Bulletin* 40, no. 2 (2014): 189–202.

Terrizzi, John A., Natalie J. Shook, and W. Larry Ventis. "Religious Conservatism: An Evolutionarily Evoked Disease-Avoidance Strategy." *Religion, Brain & Behavior* 2, no. 2 (2012): 105–120.

Thiessen, Joel, and Sarah Wilkins-Laflamme. "Becoming a Religious None: Irreligious Socialization and Disaffiliation." *Journal for the Scientific Study of Religion* 56, no. 1 (2017): 64–82.

Thomson, Nicholas D. "Priming Social Affiliation Promotes Morality – Regardless of Religion." *Personality and Individual Differences* 75 (2015): 195–200.

Tobia, Kevin Patrick. "Does Religious Belief Infect Philosophical Analysis?" *Religion, Brain & Behavior* 6, no. 1 (2016): 56–66.

Toft, Monica Duffy, Daniel Philpott, and Timothy Samuel Shah. *God's Century: Resurgent Religion and Global Politics*. New York: Norton, 2011.

Tolk, Andreas. ed., *Ontology, Epistemology, and Teleology for Modeling and Simulation*. Berlin: Springer, 2013.

Tremlin, Todd. *Minds and Gods: The Cognitive Foundations of Religion*. Oxford University Press, USA, 2010.

Trippas, Dries, Gordon Pennycook, Michael F. Verde, and Simon J. Handley. "Better but Still Biased: Analytic Cognitive Style and Belief Bias." *Thinking & Reasoning*, (2015): 1–15.

Twenge, Jean M. et al. "Generational and Time Period Differences in American Adolescents' Religious Orientation, 1966–2014." *PLoS ONE* 10, no. 5 (2015): 1–17.

Unterrainer, Human-Friedrich, and Andrew James Lewis. "The Janus Face of Schizotypy: Enhanced Spiritual Connection or Existential Despair?" *Psychiatry Research* 220, no. 1–2 (2014): 233–236.

Upal, M. Afzal. "The Structure of False Social Beliefs," *Proceedings of the 2007 IEEE Symposium on Artificial Life* (2007): 282–286.

Vaden, Victoria Cox, and Jacqueline D. Woolley. "Does God Make It Real? Children's Belief in Religious Stories from the Judeo-Christian Tradition." *Child Development* 82, no. 4 (2011): 1120–1135.

Vail, Kenneth E., Jamie Arndt, and Abdolhossein Abdollahi. "Exploring the Existential Function of Religion and Supernatural Agent Beliefs among Christians, Muslims, Atheists, and Agnostics." *Personality & Social Psychology Bulletin* 38, no. 10 (2012): 1288–1300.

Vail, Kenneth E., Zachary K. Rothschild, Dave R. Weise, Sheldon Solomon, Tom Pyszczynski, and Jeff Greenberg. "A Terror Management Analysis of the Psychological Functions of Religion." *Personality and Social Psychology Review* 14, no. 1 (2010): 84–94.

Vaish, Amrisha, Esther Herrmann, Christiane Markmann, and Michael Tomasello. "Preschoolers Value Those Who Sanction Non-Cooperators." *Cognition* 153 (2016): 43–51.

Valdesolo, Piercarlo, and Jesse Graham. "Awe, Uncertainty, and Agency Detection." *Psychological Science* 25, no. 1 (2014): 170–178.

Valdesolo, Piercarlo, Jun Park, and Sara Gottlieb. "Awe and Scientific Explanation." *Emotion*, 16, no. 7 (2016): 937–940.

van Cappellen, Patty, Vassilis Saroglou, and Maria Toth-Gauthier. "Religiosity and Prosocial Behavior Among Churchgoers: Exploring Underlying Mechanisms." *The International Journal for the Psychology of Religion* 26, no. 1 (2016a): 19–30.

van Cappellen, Patty, Maria Toth-Gauthier, Vassilis Saroglou, and Barbara Fredrickson. "Religion and Well-Being: The Mediating Role of Positive Emotions." *Journal of Happiness Studies* 17, no. 2 (2016b): 485–505.

van Der Tempel, Jan, and James E. Alcock. "Relationships between Conspiracy Mentality, Hyperactive Agency Detection, and Schizotypy: Supernatural Forces at Work?" *Personality and Individual Differences* 82 (2015): 136–141.

van Elk, Michiel. "Perceptual Biases in Relation to Paranormal and Conspiracy Beliefs." *PLoS ONE* 10, no. 6 (2015): 1–15.

van Elk, Michiel. "Paranormal Believers Are More Prone to Illusory Agency Detection than Skeptics." *Consciousness and Cognition* 22, no. 3 (2013): 1041–1046.

van Elk, Michiel, and André Aleman. "Brain Mechanisms in Religion and Spirituality: An Integrative Predictive Processing Framework." *Neuroscience and Biobehavioral Reviews* 73 (2017): 359–378.

van Elk, M., D. Matzke, Q. Gronau, M. Guan, J. Vandekerckhove, and E. Wagenmakers. "Meta-Analyses Are No Substitute for Registered Replications: A Skeptical Perspective on Religious Priming." *Frontiers in Psychology* 6 (2015): 1664–1078.

van Elk, M., B.T. Rutjens, and F. van Harreveld. "Why Are Protestants More Prosocial Than Catholics? A Comparative Study Among Orthodox Dutch Believers." *International Journal for the Psychology of Religion* 27, no. 1 (2017): 65–81.

van Eyghen, Hans. "Two Types of 'Explaining Away' Arguments in the Cognitive Science of Religion." *Zygon* 51, no. 4 (2016): 966–982.

van Leeuwen, Neil. "Religious Credence Is Not Factual Belief." *Cognition* 133, no. 3 (2014): 698–715.

van Leeuwen, Neil and Michiel van Elk. "Seeking the Supernatural: The Interactive Religious Experience Model." *Religion, Brain & Behavior*, in press.

van Slyke, J.A. "Can Sexual Selection Theory Explain the Evolution of Individual and Group-Level Religious Beliefs and Behaviors?" *Religion, Brain & Behavior*, 7 (2017): 1–4.

van Tongeren, Daryl R., Don E. Davis, Joshua N. Hook, and Kathryn A. Johnson. "Security Versus Growth: Existential Tradeoffs of Various Religious Perspectives." *Psychology of Religion and Spirituality*, 8, no. 1 (2015): 77–88.

van Tongeren, Daryl R., Sabrina Hakim, Joshua N. Hook, and Kathryn A. Johnson, Jeffrey D. Green, Timothy L. Hulsey, and Don E. Davis. "Toward an Understanding of Religious Tolerance: Quest Religiousness and Positive Attitudes Toward Religiously Dissimilar Others." *The International Journal for the Psychology of Religion* 26, no. 3 (2016a): 212–224.

van Tongeren, Daryl R., Joshua N. Hook, Don E. Davis, Jamie Aten, and Edward B. Davis. "Ebola as an Existential Threat? Experimentally-Primed Ebola Reminders Intensify National-Security Concerns among Extrinsically Religious Individuals." *Journal of Psychology and Theology* 44, no. 2 (2016): 133–141.

Varese, F., and M. Yaish. "The Importance of Being Asked: The Rescue of Jews in Nazi Europe." *Rationality and Society* 12, no. 3 (2000): 307–334.

Visala, Aku. *Naturalism, Theism and the Cognitive Study of Religion: Religion Explained?* New York: Routledge, 2016.

Voas, David. "The Rise and Fall of Fuzzy Fidelity in Europe." *European Sociological Review* 25, no. 2 (2009): 155–168.

Voas, David, and Mark Chaves. "Is the United States a Counterexample to the Secularization Thesis?" *American Journal of Sociology* 121, no. 5 (2016): 1517–1556.

Vonk, Jennifer, and Jerrica Pitzen. "Religiosity and the Formulation of Causal Attributions." *Thinking & Reasoning* 22, no. 2 (2016): 119–149.

Vüllers, Johannes, Birte Pfeiffer, and Matthias Basedau. "Measuring the Ambivalence of Religion: Introducing the Religion and Conflict in Developing Countries (RCDC) Dataset." *International Interactions* 41, no. 5 (2015): 857–881.

Wallin, David J. *Attachment in Psychotherapy*. Guilford Press, 2007.

Wardekker, J., A. Petersen, and J.P. van Der Sluijs. "Ethics and Public Perception of Climate Change: Exploring the Christian Voices in the US Public Debate." *Global Environmental Change: Human And Policy Dimensions* 19, no. 4 (2009): 512–521.

Warner, R. Stephen. "In Defense of Religion: The 2013 H. Paul Douglass Lecture." *Review of Religious Research* 56, no. 4 (2014): 495–512.

Wathey, John C. *The Illusion of God's Presence: The Biological Origins of Spiritual Longing*. Amherst, NY: Prometheus Books, 2016.

Watson, Peter. *The Age of Atheists: How We Have Sought to Live Since the Death of God*. New York: Simon and Schuster, 2014.

Watts, Joseph, Oliver Sheehan, Quentin D. Atkinson, Joseph Bulbulia, and Russell D. Gray. "Ritual Human Sacrifice Promoted and Sustained the Evolution of Stratified Societies." *Nature* 532, no. 7598 (2016): 228–231.

Weber, Samuel, Kenneth Pargament, Mark Kunik, James Lomax, and Melinda Stanley. "Psychological Distress Among Religious Nonbelievers: A Systematic Review." *Journal of Religion and Health* 51, no. 1 (2012): 72–86.

Webster, Gregory D., and Ryan D. Duffy. "Losing Faith in the Intelligence – Religiosity Link: New Evidence for a Decline Effect, Spatial Dependence, and Mediation by Education and Life Quality." *Intelligence* 55 (2016): 15–27.

Weeks, Matthew, and Alex Gilmore. "The Implicit Associations Between Religious and Nonreligious Supernatural Constructs." *The International Journal for the Psychology of Religion* 27, no. 2 (2017): 89–103.

Wen, Nicole J., Patricia A. Herrmann, and Cristine H. Legare. "Ritual Increases Children's Affiliation with In-Group Members." *Evolution and Human Behavior* 37, no. 1 (2016): 54–60.

White, Claire. "The Cognitive Foundations of Reincarnation." *Method & Theory in the Study of Religion*, 28, no. 3 (2016): 264–286.

Whitehouse, Harvey. *Modes of Religiosity: A Cognitive Theory of Religious Transmission*. Walnut Creek, CA: AltaMira Press, 2004.

Whitehouse, Harvey. "Rethinking Proximate Causation and Development in Religious Evolution." In *Cultural Evolution: Society, Technology, Language and Religion*, edited by Peter J. Richerson and Morten H. Christiansen, 349–363. Cambridge MA: MIT Press, 2013.

Whitehouse, Harvey, and Ian Hodder. "Modes of Religiosity at Çatalhöyük." In *Religion in the Emergence of Civilization: Çatalhöyük as a Case Study*, edited by Ian Hodder, 122–145. Cambridge: Cambridge University Press, 2010.

Whitehouse, Harvey, and Jonathan A. Lanman. "The Ties That Bind Us." *Current Anthropology* 55, no. 6 (2014): 674–695.

Whitehouse, Harvey, and Robert McCauley, eds. *Mind and Religion: Psychological and Cognitive Foundations of Religion*. Walnut Creek, CA: AltaMira Press, 2005.

Whitehouse, Harvey, Ken Kahn, Michael E. Hochberg, and Joanna J. Bryson. "The Role of Simulations in Theory Construction for the Social Sciences : Case Studies Concerning Divergent Modes of Religiosity," *Religion, Brain & Behavior* 2, no. 3 (2012): 182–201.

Wichman, Aaron L. "Uncertainty and Religious Reactivity: Uncertainty Compensation, Repair, and Inoculation." *European Journal of Social Psychology* 40, no. 1 (2010): 35–42.

Wigger, J. Bradley, Katrina Paxson, and Lacey Ryan. "What Do Invisible Friends Know? Imaginary Companions, God, and Theory of Mind." *International Journal for the Psychology of Religion* 23, no. 1 (2013): 2–14.

Wildman, Wesley J. *Religious and Spiritual Experiences*. Cambridge University Press, 2014.

Wildman, Wesley J. *Religious Philosophy as Multidisciplinary Comparative Inquiry: Envisioning a Future for the Philosophy of Religion*. SUNY Press, 2011.

Wildman, Wesley J. *Science and Religious Anthropology: A Spiritually Evocative Naturalist Interpretation of Human Life*. Ashgate, 2009.

Wilkins, John S. "Gods Above: Naturalizing Religion in Terms of Our Shared Ape Social Dominance Behavior." *Sophia* 54, no. 1 (2015): 77–92.

Wilkins, John S., Paul E. Griffiths, G. Dawes, and J. Maclaurin. "Evolutionary Debunking Arguments in Three Domains." In *A New Science of Religion*, edited by Gregory W. Dawes and James Maclaurin, 133–146. New York: Routledge, 2012.

Willard, Aiyana K. "Religion and Prosocial Behavior among the Indo-Fijians." *Religion, Brain & Behavior* 7 (2017): 1–16.

Willard, Aiyana K., and Lubomír Cingl. "Testing Theories of Secularization and Religious Belief in the Czech Republic and Slovakia." *Evolution and Human Behavior* 38, no. 5 (2017): 604–615.

Willard, Aiyana K., and Ara Norenzayan. "Cognitive Biases Explain Religious Belief, Paranormal Belief, and Belief in Life's Purpose." *Cognition* 129, no. 2 (2013): 379–391.

Willard, Aiyana K., and Ara Norenzayan. "'Spiritual but Not Religious': Cognition, Schizotypy, and Conversion in Alternative Beliefs." *Cognition* 165 (2017): 137–146.

Wilson, Edward O. *The Meaning of Human Existence*. New York: Liveright, 2014.

Wilson, D.S., Y. Hartberg, I. MacDonald, J.A. Lanman, and H. Whitehouse. "The Nature of Religious Diversity: A Cultural Ecosystem Approach." *Religion, Brain & Behavior*, 7, no. 2 (2017): 134–153.

Win-Gallup. "Global Index of Religiosity and Atheism." Win-Gallup, 2012.

Winkelman, Michael. "Shamanism as a Biogenetic Structural Paradigm for Humans' Evolved Social Psychology." *Psychology of Religion and Spirituality* 7, no. 4 (2015): 267–277.

Winsberg, Eric. *Science in the Age of Computer Simulation*. University of Chicago Press, 2010.

Wlodarski, Rafael, and Eiluned Pearce. "The God Allusion: Individual Variation in Agency Detection, Mentalizing and Schizotypy and Their Association with Religious Beliefs and Behaviors." *Human Nature* 27, no. 2 (2016): 160–172.

Wolfram, Stephen. *A New Kind of Science*. Champaign, IL: Wolfram media, 2002.

Woody, Erik Z., and Henry Szechtman. "Adaptation to Potential Threat: The Evolution, Neurobiology, and Psychopathology of the Security Motivation System." *Neuroscience and Biobehavioral Reviews* 35, no. 4 (2011): 1019–1033.

Woolley, Jacqueline D., and Victoria Cox. "Development of Beliefs about Storybook Reality." *Developmental Science* 10, no. 5 (2007): 681–693.

Wright, Jennifer, and Ryan Nichols. "The Social Cost of Atheism: How Perceived Religiosity Influences Moral Appraisal." *Journal of Cognition and Culture* 14, no. 1–2 (2014): 93–115.

Xu, Xixiong, Yaoqin Li, Xing Liu, and Weiyu Gan. "Does Religion Matter to Corruption? Evidence from China." *China Economic Review* 42 (2017): 34–49.

Xygalatas, D., S. Kotherová, P. Maňo, R. Kundt, J. Cigán, E.K. Klocová, and M. Lang. "Big Gods in Small Places: The Random Allocation Game in Mauritius." *Religion, Brain & Behavior*, 7 (2017): 1–19.

Xygalatas, Dimitris, Eva Kundtová Klocová, Jakub Cigán, Radek Kundt, Peter Maňo, Silvie Kotherová, Panagiotis Mitkidis, Sebastian Wallot, and Martin Kanovsky. "Location, Location, Location: Effects of Cross-Religious Primes on Prosocial Behavior." *The International Journal for the Psychology of Religion*, 26, no. 4 (2016): 1–16.

Xygalatas, Dimitris, Panagiotis Mitkidis, Ronald Fischer, Paul Reddish, Joshua Skewes, Armin W. Geertz, Andreas Roepstorff, and Joseph Bulbulia. "Extreme Rituals Promote Prosociality." *Psychological Science* 24, no. 8 (2013): 1602–1605.

Yilmaz, Levent, ed. *Concepts and Methodologies for Modeling and Simulation*. New York: Springer, 2015.

Yilmaz, Onurcan, and Hasan G. Bahçekapili. "Supernatural and Secular Monitors Promote Human Cooperation Only If They Remind of Punishment." *Evolution and Human Behavior* 37, no. 1 (2016): 79–84.

Yilmaz, Onurcan, and Hasan G. Bahçekapili. "When Science Replaces Religion: Science as a Secular Authority Bolsters Moral Sensitivity." *PLoS ONE* 10, no. 9 (2015): 1–8.

Yilmaz, Onurcan, Dilay Z. Karadöller, and Gamze Sofuoglu. "Analytic Thinking, Religion, and Prejudice: An Experimental Test of the Dual-Process Model of Mind." *The International Journal for the Psychology of Religion*, 26, no. 4 (2016): 360–369.

Yonker, Julie E., Laird R.O. Edman, James Cresswell, and Justin L. Barrett. "Primed Analytic Thought and Religiosity: The Importance of Individual Characteristics." *Psychology of Religion and Spirituality*, 8, no. 4 (2016): 298–308.

Yoon, Eunju, Christine Chih-Ting Chang, Angela Clawson, Michael Knoll, Fatma Aydin, Laura Barsigian, and Kelly Hughes. "Religiousness, Spirituality, and Eudaimonic and Hedonic Well-Being." *Counselling Psychology Quarterly* 28, no. 2 (2015): 132–149.

Zhong, Wanting, Irene Cristofori, Joseph Bulbulia, Frank Krueger, and Jordan Grafman. "Biological and Cognitive Underpinnings of Religious Fundamentalism." *Neuropsychologia* 100 (2017): 18–25.

Žižek, Slavoj. *The Parallax View*. Cambridge, MA: MIT Press, 2009.

Zuckerman, Miron, Jordan Silberman, and Judith A. Hall. "The Relation Between Intelligence and Religiosity: A Meta-Analysis and Some Proposed Explanations." *Personality and Social Psychology Review* 17, no. 4 (2013): 325–354.

Zuckerman, Phil, ed. *Atheism and Secularity: Issues, Concepts, and Definitions. Vol 1.* New York: Praeger, 2010.

Zuckerman, Phil, ed. *Atheism and Secularity: Global Expressions. Vol 2.* New York: Praeger, 2010.

Zuckerman, Phil. "Atheism: Contemporary Numbers and Patterns.," In *The Cambridge Companion to Atheism*, edited by Michael Martin. Cambridge University Press, 2007.

Zuckerman, Phil. "Atheism, Secularity, and Well-Being: How the Findings of Social Science Counter Negative Stereotypes and Assumptions." *Sociology Compass* 3, no. 6 (2009): 949–971.

Zuckerman, Phil. *Faith No More: Why People Reject Religion*. New York: Oxford University Press, 2011.

Zuckerman, Phil. *Living the Secular Life: New Answers to Old Questions*. New York: Penguin Press, 2014.

Zuckerman, Phil. *Society without God: What the Least Religious Nations Can Tell Us About Contentment*. New York: NYU Press, 2010.

Zuckerman, Phil, Luke W. Galen, and Frank L. Pasquale. *The Nonreligious: Understanding Secular People and Societies*. New York: Oxford University Press, 2016.

Zussman, Asaf. "The Effect of Political Violence on Religiosity." *Journal of Economic Behavior and Organization* 104 (2014): 64–83.

Index

Printed in the United States
By Bookmasters